State and Society

State and Society

British Political and Social History 1870–1992

Martin Pugh
Professor of Modern British History,
University of Newcastle

Edward Arnold
A member of the Hodder Headline Group
LONDON NEW YORK MELBOURNE AUCKLAND

First published in Great Britain 1994

Distributed in the USA by Routledge, Chapman and Hall, Inc.
29 West 35th Street, New York, NY 10001

British Library Cataloguing in Publication Data
Pugh, Martin
State and Society: British Political and Social History, 1870–1992
I. Title
941.08

ISBN 0-340-50710-1 (paperback)
0-340-50711-X (hardback)

Library of Congress Cataloging-in-Publication Data

Pugh, Martin.
State and society : British political and social history,
1870–1992 / Martin Pugh.
p. cm.
Includes bibliographical references (p.) and index.
ISBN 0-340-50711-X : $59.95. --
ISBN 0-340-50710-1 (paper) : $19.95
1. Great Britain--Politics and government--20th century.
2. Great Britain--Politics and government--1837–1901.
3. Great Britain--Social conditions--20th century.
4. Great Britain--Social conditions--19th century. I. Title.
DA566.7.P84 1993
941.08--dc20 93-28691
CIP

Typeset by Hewer Text Composition Services, Edinburgh.
Printed and bound in Great Britain for Edward Arnold, a division of Hodder Headline PLC, Mill Road, Dunton Green, Sevenoaks, Kent TN13 2YA by Biddles Ltd, Guildford & King's Lynn.

Contents

Part V. The Era of Reaction and Decline, 1970–1992

List of Illustrations

List of Tables and Figures

Tables

Figures

Part I

The Loss of Confidence, 1870–1902

1

The Retreat of the Industrial Revolution

It is hardly surprising that many mid-Victorians exuded pride and complacency in being British. In the 1860s Britain appeared to be a singularly favoured nation. She enjoyed a remarkably stable form of government, huge wealth arising from her superiority in commerce and manufacturing, one of the largest empires the world had known, and was largely free from external threats to her security.

Yet there was an element of illusion in this mid-Victorian triumphalism. In geographical terms Britain was the smallest of the Great Powers. With an army that was quite inadequate for the defence of her imperial possessions, she was lucky that the major powers were too preoccupied elsewhere to challenge her during the Victorian era. Britain's population, though growing rapidly, reached only 37.4 million by 1890 – behind that of France, Germany, the United States, Russia, Austria-Hungary and Japan. Italy alone, whose claim to the status of a Great Power was dubious, had fewer people.

What, however, made the crucial difference to Britain's strength was her capacity to mobilize her human and physical resources for sophisticated and profitable economic activities more effectively than her rivals. After embarking on industrialization in the second half of the eighteenth century, she had ridden each wave of innovation – first cotton textiles, then iron and steel, then railways – comfortably ahead of the other European states. Thus, by the middle of the nineteenth century her population had become uniquely industrialized; by 1871 only 11 per cent of her labour force worked in agriculture, and the proportion continued to fall. Her people had also become unusually urbanized – 65 per cent of them in 1870 and 78 per cent by 1901. This was two-and-a-half times the proportion in France and Germany. As a result of her early lead, Britain reached her peak as an industrial power in the 1860s. By that time she produced (to take but a few examples) half the world's coal, over half the iron and steel, and nearly half the cotton goods, and possessed over a third of the world's merchant shipping. Russia, with over three times as many people and a huge land area, was an underdeveloped country by comparison.

Britain made the most of her limited resources in other ways. Unlike all the great imperial states, apart from the United States, she largely avoided the diversion of her wealth into the armed forces, which received only 2–3 per cent of gross national produce in the mid-Victorian period. Such expenditure was widely considered 'unproductive'. Parsimony also reflected the traditional fear of a large standing army as a domestic political threat. Yet for all this, in time of crisis Britain did prove quite capable of expanding her military effort by means of extra taxation and extensive borrowing. This was a purely temporary expedient. In peacetime her island position enabled her to concentrate on the Royal Navy, an economical form of defence, while maintaining an army of 240,000 men which, in view of Britain's huge interests, was extremely small. Victorian free-traders liked to argue that, as more and more nations were drawn into Britain's beneficent trading network, the causes of war would steadily diminish and military aggrandizement become a thing of the past.

Of course, the British Empire represented a major complication in this vision of international peace. Both formal possessions and informal commercial interests had to be defended and even expanded. The non-imperial mid-Victorian period saw the addition of the Punjab, Sind, Burma, and Hong Kong to British territory. Indeed, although the Victorians thought of themselves as presiding over an era of peace, the fact is that the Queen's men were actually fighting somewhere in the world in most years of her reign. But only the Crimean War in the 1850s involved a *European* opponent. On the whole, wars were short and economical engagements by a handful of men with native peoples in Asia and Africa. It is striking how little impression even a great threat like the Indian Mutiny of 1857 made on British policy. After the revolt British troops in India were increased, but from only 40,000 to 65,000, a tiny force for a huge Empire. British soldiers continued to be outnumbered more than two to one by the Indian sepoys on whom the British Raj relied. Indian troops were often used in campaigns in other parts of the Empire to supplement British forces. Moreover, they were paid for out of *Indian* revenues, not by the British taxpayer. The nineteenth-century empire was run on a shoestring.

The Beginnings of Decline?

It was during the 1870s that contemporaries seriously began to consider whether Britain had already passed her peak. After several buoyant years at the start of the decade, the economy suffered the sudden collapse of a speculative boom. By 1874 quantities of cheap wheat had begun to arrive from North America, severely undercutting the prices of British farmers. The slump of the 1870s was followed by another in the mid-1880s and by a third in the first half of the 1890s. Such cyclical fluctuations had, of course, characterized the mid-Victorian era too, but now they were regarded as symptoms of a long-term phenomenon – the 'Great Depression' of 1874 to 1896. This view gained credibility from a series of royal commissions, which gathered evidence from those farmers and manufacturers who were doing badly.

However, one must disentangle the various threads in the pattern of economic change. In one perspective British producers were experiencing pressures felt all over the world in a period of deflation. From the 1870s to the late 1890s, prices fell by about 40 per cent. This squeezed profits, put some out of business, and seemed to mark a basic change from the buoyant mid-Victorian era, when business had been stimulated by mild inflation. But had Britain suffered disproportionately from the vicissitudes of the late nineteenth century? Had her long-term decline as an economic power really begun?

In many ways this seems improbable. Measurements of British gross domestic product show a continuous expansion of a fairly rapid order; by 1890 it was over 50 per cent greater than in 1870, for example. On the other hand, the average annual *rate* of growth was somewhat slower than in the pre-1870 period. Moreover, productivity, i.e., output per head of population, diminished significantly, especially in the period 1900–13. This suggests that contemporary complaints from the 1870s give a misleading impression about the *timing* of decline, but not, perhaps, about the underlying trend.

There is also plentiful evidence to suggest that Britain's economic performance deteriorated in relation to that of other advanced countries. However, this is true with respect to Germany and the United States rather than the other great powers. Between 1870 and 1913 the annual per capita rate of growth in manufacturing ranged from 0.6 per cent to 0.9 per cent in Britain; but in Germany the range was 1.7–3.9 per cent, and in America 2.1–3.2 per cent. The disparity was greatest after 1900 in spite of the Edwardian boom. A good indication of Britain's status is provided by the figures from manufacturing capacity. By 1880 she was still the world's leader, having increased her share in the 1860s and 1870s partly as a result of deteriorating performances of China and India. By 1900 Britain, Germany, and the United States stood out as the top three producers; but America had comfortably overtaken Britain, and Germany was rapidly catching up. Of course, their faster rate of growth largely reflected the fact that they had started from a lower level of production than Britain. The exploitation of their greater resources and the application of British techniques and capital inevitably entailed very rapid growth.

Of course the figures for national output provide us with sweeping generalizations about a complex pattern of change. Britain's performance varied considerably from one sector to another. In some industries any talk of depression and decline seems premature. For example, British shipbuilders continued to enjoy great success up to 1914, both in exporting and in supplying the British merchant fleet and Royal Navy. The cotton textile industry also continued to make profits and to export over half its total output. But as one moves across the spectrum the picture begins to look less rosy. Railways are an example of a major industry whose profitability was declining, albeit only slightly, after 1870. The companies have been criticized for investing too much in dubious branch lines and duplicate facilities, and for having insufficient regard for the rate of return on capital. However, it is fair to say that much of their expenditure could not be avoided, especially on essential improvements in safety and the provision of extra capacity; they were also subject to legal controls on

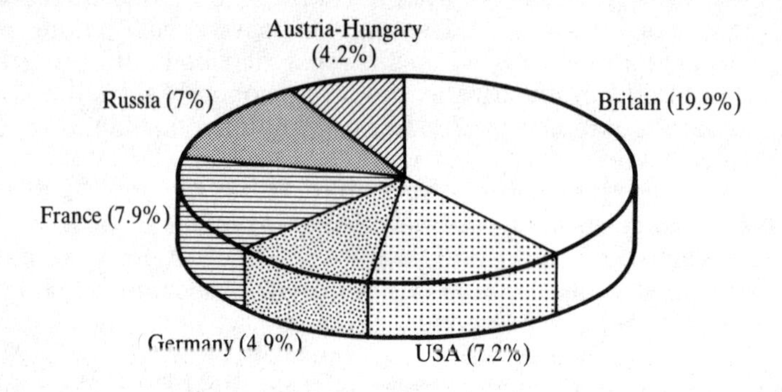

1880

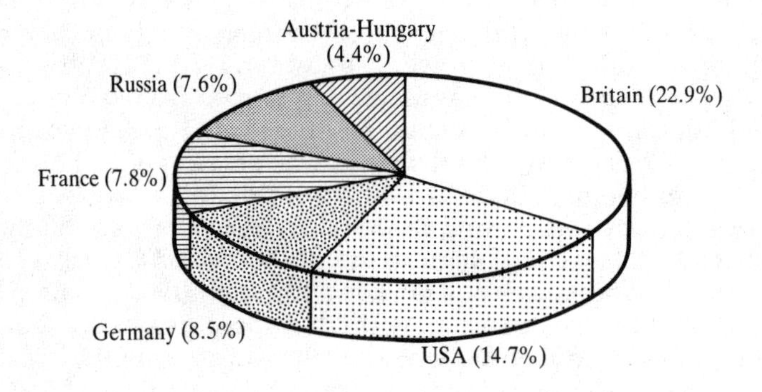

1900

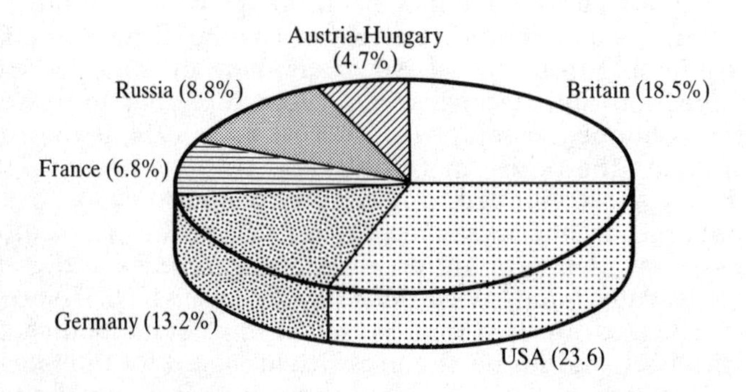

Figure 1.1 Relative Output in Manufactures Amongst the Great Powers*

* share of world output shown as a percentage

charges to their customers, and to pressure to improve wages. Similarly, the coal industry continued to expand output up to 1914, but its productivity declined steadily. This was partly a consequence of the deteriorating physical conditions in some of the older coalfields, but it also reflected a general failure by the many small companies to invest in mechanical methods and thereby improve efficiency. The coal industry's success was precariously founded on the urgent demand on the part of other rapidly expanding countries, but sooner or later its inefficiency would lead to a loss of markets. Even more than the railways, coal suffered from the fragmented pattern of private ownership which by the turn of the century had become inappropriate for such strategic industries.

Market loss was already taking place in agriculture, at least among the wheat-producers of southern and eastern England, whose prices were drastically undercut by large-scale foreign farmers now able to take advantage of cheaper transport by rail and by sea. Here, too, improved technology in the form of combine harvesters was available but little used. However, while many farmers saw profits dwindle and went bankrupt, those who concentrated on dairying, livestock, fruit, and market gardening profited from growing domestic demand and better prices. The worst example of decline was in the iron and steel industry. Manufacturers found that their prices were being severely undercut by German and American producers, largely because they had been too slow to adopt the more efficient technology available to them. As a result, exports diminished greatly.

None the less, taken as a group, the traditional industries on which Britain's industrial revolution had been built – textiles, shipbuilding, railways, iron and steel – did not perform too badly. They continued to make profits even if their hold on world markets weakened. The real flaw in Britain's economy was that it remained too heavily dependent upon this narrow range. Cotton and metals, for example, generated no less than 70 per cent of all British exports in 1870. Yet the world demand for the goods that Britain traditionally supplied was not very buoyant in the late nineteenth century. Thus in this period there was an urgent need to shift resources into the faster-growing sectors that formed the cutting edge of the new industrial revolution – electricity, chemicals, machine tools, and motor cars. However, in none of these did Britain dominate as she had once done in the old staple industries. In particular, British firms failed to repeat what had been crucial to success in the first industrial revolution, namely, the mass production of a limited range of items. Both motor cars and machine tools were characterized by a large number of small companies, each manufacturing a wide range of products in small quantities, and thus at higher prices, than their competitors. British power stations were smaller and less efficient than those abroad, and as a result the British economy did not benefit from cheap electricity as early as others. In chemicals, once a British success story, only 11 per cent of world output originated in Britain by 1914, while Germany and the USA together accounted for nearly 60 per cent.

By the 1890s the consequences of this loss of competitiveness began to make themselves apparent. Whereas formerly Britain had supplied nearly

Table 1.1 Britain's Balance of Payments in 1900 (£ million)

Imports	Domestic exports	Re-exports	Net overseas earnings	Net invisible trade
523	291	63	103	109
Overall Balance				
38				

all her manufactured goods herself, she increasingly imported them; by 1914 25 per cent of all imports were manufactured items. Conversely, manufactured goods diminished as a proportion of British exports from three-quarters to a little over half between 1870 and 1914. In this way, the process that had made Britain such a wealthy country began to falter, for the greatest profit was to be made on sophisticated manufactured items, not on primary products like coal which loomed larger in British trade. Neither of the obvious solutions – improvements in the efficiency of the old industries and heavy investment in the growth sectors – was adopted. Contemporaries usually failed to diagnose the real problem, preferring to fix upon simpler explanations and remedies. Agricultural problems were attributed to unusually bad weather and to outbreaks of disease. Lost markets could be blamed upon rival countries who cheated by imposing tariffs or by dumping their own goods at artificially low prices in Britain. To some extent, however, the underlying deterioration in Britain's status as an exporter of manufactures was obscured. For as she was pushed out of the competitive markets of Europe and North America, she compensated to some extent by selling more in the easier markets of the less developed economies such as Latin and South America or China, as well as in the British Empire, where she enjoyed special advantages. Moreover, Britain's declining competitiveness did not, in this period, lead to any serious balance-of-payments problems (see Table 1.1), because her deficits on visible trade were comfortably offset by the indirect exports earned through shipping, insurance, and the interest and dividends from investments made in foreign countries, all of which increased substantially between the 1870s and 1914.

Britain and Free Trade

The accepted diagnosis for Britain's modest relative decline in the late Victorian era is that she failed to invest her resources in the growth sectors, thereby missing the greatest opportunities for profits and employment. But why? It is tempting to explain the situation in terms of Britain's pioneering role as the first industrialized society. Latecomers, it may be conjectured, have the advantage of being able to invest in the most recent and efficient technology. Yet this view has not commanded much support. As the most developed economy, Britain should have been in an excellent position to supply capital and re-equip herself at regular intervals so as to keep ahead of rivals. In the real world, however, economic decisions rarely grow out of purely rational economic considerations. Social and political factors have a very real bearing on British performance. As the first industrial nation, the British began to reap the benefits in the form of higher living standards earlier than other peoples. This may have been all the more true because of the system of responsible parliamentary government she enjoyed. The late Victorian era witnessed huge increases in imported food, in luxury goods, and in expenditure on entertainment, travel, and holidays. The influence of the consumer should not be regarded merely as an indulgence or as a cost borne by the economy; British consumers had made the initial phase of the industrial revolution possible. Moreover, concessions to consumers to maintain living standards were to serve governments well in times of crisis such as the First World War. British society was to prove itself steadier under strain, and better able to maintain industrial output, than those of Russia, Germany, Italy, or France.

It was during the 1880s that some British businessmen and politicians began to question the policy of free trade on the grounds that it no longer served Britain's best interests. Certainly the imposition of new tariffs by Germany in the 1870s and by France and the USA in the 1890s contributed to the difficulties faced by some British exporters. The feeling that foreigners played unfairly by dumping goods at specially subsidized prices in Britain was directed at the Germans in particular, and culminated in the famous book by E. E. Williams, *Made in Germany*, published in 1896. On the whole, however, protectionism was of fairly marginal importance in Britain's economic decline, for the industries suffering from foreign competition had been undermined primarily by their own inefficient methods.

None the less, by the 1880s the call for 'fair trade', which meant that Britain should force foreigners back to genuine free trade by threatening retaliatory tariffs, was on the face of it plausible and patriotic. Yet Britain remained loyal to free trade up to the First World War. There were a number of reasons for this. Manufacturing industry was never united on the subject of tariffs, and many major interests like cotton textiles, which were dependent on world markets both for raw materials and for sales, strongly opposed protectionism. In addition, all manufacturers had to weigh the likely effects of tariffs in raising food prices and thereby subjecting such firms to severe pressure for higher wages. From the political perspective, the argument that Britain's huge urban population depended on the maintenance of supplies of cheap food from abroad seemed compelling, to say the least. It grew more so during the 1880s, as standards of living for working-class families rose as a result of the falling price and greater range of food available in the shops. Any check to this trend was likely to prove fatal for the government responsible.

Beyond this, it was argued that Britain's unique dependence on the state of the world economy for her prosperity made the maintenance of free trade a national interest. By freely importing food and raw materials, especially from the less developed countries, Britain helped to maintain their capacity to purchase her own manufactured items. The trade with India alone financed over 40 per cent of Britain's deficits. Thus Britain's interest lay in maximizing the level of world economic activity, both by her direct trade and by financial investments. As we have seen, overseas investments, which rose by around six times between 1870 and 1914, made a vital and growing contribution to national solvency, as did earnings from shipping and insurance. The conclusion is inescapable: contemporary opinion was correct in its belief that free trade was economically and politically justified for late Victorian Britain. The most conspicuous cost of free trade was the shrinkage in agriculture; but the shift of resources out of areas of low productivity, and the concentration on the forms of agriculture that were most viable in British conditions, were entirely rational, though the readjustments were inevitably painful for some.

The Role of the State

The debate over tariffs raises the wider question of the significance of the state in British economic development, a theme which has not received the attention it deserves. Of course, there were strong empirical reasons why state intervention remained minimal during the mid-Victorian era. Contemporaries, impressed by the extraordinary rate of growth of British wealth and industry since the late eighteenth century, argued that governments had helped to promote this by abstaining from interference and keeping the burdens on productive industry as light as possible. Many Victorians regarded state action in the light of eighteenth-century and earlier practice: it implied the granting of special concessions and privileges to groups or individuals which tended to promote inefficiency and work to the detriment of the public interest. Thus a body like the East India Company had come under repeated attack; and its monopoly rights had been whittled away when the Company's charter came up for renewal by Parliament. Similarly, one of the arguments used against tariffs was that they almost inevitably led to corrupt deals between government and vested interests, with the result that a few profited at the expense of the country as a whole.

What must be emphasized is how peculiarly British this anti-state ideology was. American and European governments routinely dealt in special concessions, protective tariffs, and subsidies for major works of construction from which the British largely abstained. But there were many other ramifications to the British hostility to the involvement of the state, some of which had damaging economic consequences. For example, Britain was slow to develop a national system of elementary or secondary education, not to mention the higher scientific and technological training of her population, by comparison with Germany and France. By the late nineteenth century the effects of this began to emerge in the form of an inadequate supply of skilled personnel in sectors like chemicals and engineering. It was also conspicuous that, whereas other governments used their diplomatic representatives abroad to promote the interests of their exporters, the British gave little or no help of that kind. Non-interference in business was one thing, but British attitudes often amounted to sheer neglect.

Indeed, at the heart of government there was no minister for industry who might have co-ordinated a positive strategy. The most relevant department, the Board of Trade, was dominated by a belief in free trade and a certain suspicion of manufacturing. However, failings must be attributed to the inertia or ignorance of the political heads of the department as much as to the civil servants. When Lloyd George became President of the Board of Trade in 1905, he swiftly took up several questions affecting the interests of businessmen and surprised them by his willingness to consult. For example, he reformed the law on patents so as to deter foreign companies from taking out patents merely with a view to checking British competition. But Lloyd George was clearly exceptional. Indeed, businessmen had some reason for regarding such legislation as was undertaken as distinctly unhelpful to them in the late Victorian period: the 1870 Tramways Act and the 1882 Electric Lighting Act, for example, inhibited investment, and the railway legislation of 1888 and 1894 restricted the right to increase charges. These, however, were relatively minor examples. Apart from free trade, the most important economic policy affecting British business was the maintenance of the gold standard. Yet here we have a most striking case of the detached stance adopted by government.

The Banks and the Question of Investment

British monetary policy was determined not by the government but by the Bank of England. But the Bank remained a private body run by a governor and a board of self-appointed representatives very few of whom represented manufacturing industry. Reflecting the interests of financiers and merchants, the Bank of England took a rather cosmopolitan, or, to its critics, unpatriotic, view of economic policy. Officially it endeavoured to protect the country's reserves of gold, thereby to maintain the international value of the pound sterling. Unfortunately, the gold reserves were invariably too low for the job, and as a result the periodic loss of gold was regularly met by an increase in the bank rate which was designed to attract overseas funds back into Britain. This worked well for the financiers whom the Bank represented, but not for British manufacturers. They naturally suffered from the frequent fluctuations in bank rate; it actually changed every six weeks on average between 1851 and 1885, and every eight weeks up to 1913. Inevitably the effect was to exacerbate periods of depression by pushing down the level of production and employment. In other European countries, in contrast, the central banks considered it their responsibility to *minimize* sharp fluctuations in the trade cycle; but the interests of manufacturers were simply not a priority for the Bank of England.

No doubt the harmful effects of fluctuations in the cost of capital on British industry should not be exaggerated; although companies paid more for short-term capital then their Continental rivals, long-term loans were usually cheaper. However, the attitude of the Bank throws some light on the wider question of the supply of capital to British industry. Given what is known about the failure to invest in the most efficient technology, which hampered several sectors of the economy, it is tempting to see shortages of capital as an underlying explanation for Britain's decline in the late nineteenth century.

The experience of Britain does seem to have been distinctive in that her central bank showed little interest in provincial manufacturing companies, while the banks in general were reluctant to lend to industry. Both the joint-stock banks and the country banks provided only short-term loans rather than the long-term finance required by industry. All this stood in contrast to the more supportive practice of financial institutions in Germany, for example. Now, during the early stages of industrialization this weakness had not been of great significance. It had proved possible to launch textile industries without much capital, and what was required could be obtained by personal connections rather than institutional sources. But the hand-to-mouth methods of the early entrepreneurs were scarcely adequate later in the nineteenth century, as industry became more sophisticated and the technology required grew more costly. One consequence was that a relatively high proportion of British firms remained small scale by comparison with Germany and failed to reap the advantages of mass production.

Another strand in this question is the paradox that, while much of domestic industry suffered from insufficient investment in the late Victorian era, Britain invested huge and rapidly growing sums abroad. There is thus a *prima facie* case that this approach both diverted resources from home industries and accelerated the pace of technological innovation in foreign countries. Some of the evidence appears to bear this out. Between 1871 and 1913 Britain invested only 5–7 per cent of her gross national product at home, whereas the USA and Germany invested 12 per cent. Meanwhile Britain's net foreign investment rose from around £1,200 million in the 1870s to £4,000 by 1914. Of this, approximately 40 per cent went to the Empire and 60 per cent to foreign countries. By 1914, 44 per cent of foreign capital in the world was British-owned.

But was this necessarily damaging to Britain? One must remember that by 1870 a high proportion of the new foreign investment was simply a matter of recycling the dividends now flowing in as a result of earlier investments abroad. Also, much depends on whether capital obtained a higher rate of return abroad than it would have done at home. If it did, then capital exports would have enlarged the national income and improved the growth rate of the economy. In a well-developed economy such as Britain's there were, after all, some sectors, such as transport, where further investment no longer yielded significant gains; in these cases diversion of resources to, say, Argentine railways could be perfectly rational. However, there is no consistent pattern of higher earnings on foreign as opposed to domestic investments; rather, a good deal of variation from year to year. It can certainly be argued that a really major diversion of capital into certain domestic industries would have yielded significant results in the long term. But it is generally the case that capital was not in short supply. Rather, there was a failure to make the most of the opportunities that existed both on the part of the financial institutions and on the part of businessmen.

The Problem of Entrepreneurship

Because of the intractability of the narrowly economic explanations for British economic problems, increasing attention has been given to *social* factors as an alternative or complementary line of explanation. This rests upon the assumption that there is a strong connection between a society's economic performance on the one hand and its social attitudes and cultural values on the other. In the case of Britain this has generated a number of ideas centred around the hypothesis that, in spite of the early Industrial Revolution, the spirit of entrepreneurship was never particularly strong in Britain, and that by the last third of the nineteenth century the economy suffered from a deterioration in entrepreneurial talent.

From the very beginning the industrial process attracted criticism on account of the ugly and dirty urban development it entailed. By the 1870s this had resulted in a back-to-the-land movement which took the form of attempts to preserve footpaths, attractive countryside, and open spaces in towns, to revitalize agricultural employment, and to impart a rural quality to urban developments. Much interest focuses upon the middle classes and whether they ever really challenged the values of the traditional aristocratic elite effectively. Was their real objective to be assimilated into its ranks? Such a process could not be achieved overnight. First, the middle-class man must make his fortune. Then, so the argument runs, by the second and third generation business families often grew complacent and conservative about their economic interests and lost their entrepreneurial drive. It was tempting to enjoy wealth and leisure, or to use it to enhance one's status or diversify into other occupations – hence the migration by manufacturers out of city centres into salubrious rural enclaves, the purchase of land and country houses. Some sought recognition in the form of knighthoods and peerages, while others entered politics or helped their sons to do so. These forms of behaviour, so the argument runs, were all symptoms of the low status attached to industry in British society; consequently, those who had succeeded as manufacturers merely used their wealth and influence as a vehicle for further advance, but never saw business as the goal. The effect was compounded by a generational shift. The mid-Victorian era saw a major expansion of public schools. Thus middle-class men increasingly sent their sons to be educated alongside young aristocrats, thereby to acquire their values and aspirations. This gentrification of the bourgeoisie could subsequently be consolidated by means of marriages with the daughters of impecunious titled families.

It is certainly possible to produce a good deal of empirical support for the idea of entrepreneurial deterioration; but in several cases the implications drawn from the evidence must be questioned. For example, it seems doubtful whether the appeal of the rural idyll actually explains anything; similar ideals may be discerned in other advanced societies such as that of the USA, but do not appear to have inhibited successful enterpreneurship. Similarly it is undeniable that middle-class men entered Parliament in growing numbers during the late Victorian and Edwardian era. They also began to win a substantial share of national honours; whereas before the 1880s few middle-class men received peerages, by the end of the century they accounted for over 40 per cent of all new creations. However, it is by no means obvious that such political success was necessarily associated with the decline or neglect of business enterprise. Political influence and connections could be used to promote economic interests.

The significance of education is also difficult to ascertain. No doubt the prevailing characteristics of public-school education – the dominance of the classics and neglect of science – was its narrowness. Moreover, the schools reflected the aspirations of the *professional* middle classes in the sense that their object was to prepare boys for prestigious careers in the Indian civil service, the army, politics, the law, the Church, and the home civil service – but not manufacturing. On the other hand, it is difficult to pin down the effects. There is an obvious problem over timing. If entrepreneurial decline is thought to be evident by 1870, this seems too early for public-school expansion to have made a major impact, though it might be relevant to post-1900 weaknesses. It is by no means clear that boys were either handicapped by their education or actually alienated from business. It has often been pointed out that the British aristocracy played a substantial part in the early Industrial Revolution, and its influence need not necessarily have been unhelpful. From the perspective of the middle-class businessman, the wider range of contacts resulting from public-school education could be seen as strengthening the family business, not as undermining it.

There is, however, a more basic problem with the entrepreneurial explanation. Is it possible to measure business skills so as to sustain generalizations about decline? At any point in time one can produce examples of family firms going downhill; but this would be true in any occupation. The significant thing is whether the evidence of a perfectly natural deterioration in some businesses is matched by evidence of new businesses growing from small beginnings, under the leadership of fresh recruits to the ranks of entrepreneurs.

In fact there was an impressive range of new enterprises in late Victorian Britain which is difficult to square with the pessimistic view of British businessmen. Several of the sources of entrepreneurship are easily overlooked. For example, many men moved into the south and east of the country, taking advantage of falling land prices, to set up as dairy farmers and market gardeners supplying the huge London market with fresh produce. Another neglected area of enterprise is that of women. Among the middle classes this typically took the form of careers as novelists; but innumerable working-class women who took lodgers to help make ends meet moved on to become proprietors of guest houses and small shopkeepers. Perhaps the most notable evidence of the latter was the rapid spread of the fish-and-chip shop during the late Victorian period. One must also remember the influx of entrepreneurial talent amongst the foreign immigrants of this period. This included Jews such as Sir Ernest Cassel and Montague Burton, John Brunner (of Swiss origin), Alfred Mond (a German), and several Americans, such as William Waldorf Astor and Gordon Selfridge.

Among the large-scale successes of the late nineteenth century one would have to include the chemical works established in Cheshire by Brunner and Mond which eventually gave rise to ICI. The 1890s particularly saw a major expansion of bicycle production by firms like Raleigh, using the traditional metalwork skills of the towns of the Midlands. William Lever advanced in the classical manner from the small beginnings of his father's shop to become the leading manufacturer of soap and the architect of Port Sunlight. In addition, a number of men made their fortunes through the development of chains of grocery stores, such as Hudson Kearley (the International Stores) and Thomas Lipton, who bought his own tea plantations so as to cut out the middle men and reduce the price of tea to the consumer. Finally, the world of popular newspapers generated a number of successful businessmen, of whom Alfred Harmsworth of the *Daily Mail* is the best known.

It should be noted that Lipton received a knighthood and Kearley, Harmsworth, and Lever peerages; moreover, Mond, Brunner, Lever and Kearley were all involved in politics (as Liberals) and several became MPs. It is not clear that this distracted them unduly from their role as businessmen. But how much positive significance may be placed upon a catalogue of new businesses? At the least it undermines assumption about the late nineteenth century as a period of unrelieved entrepreneurial failure. The successful men were repeating the practice of the early industrial entrepreneurs by marketing a cheap, mass-produced item. However, the examples of success also tell us something about the wider problems of the economy. First, nearly all these new enterprises were supplying *consumer* goods to the domestic market relatively free from external competition. Second, many of them managed to start out and to expand with only modest amounts of capital. In this sense they were not unlike the early cotton textile manufacturers. But their success could not easily be repeated in industry generally, which increasingly required major initial investment, sophisticated technology, and skilled manpower.

Industrialization in Perspective

The varied symptoms of incipient economic decline, albeit in the context of absolute economic strength, do seem to justify some reassessment of the traditional emphasis placed on the British Industrial Revolution. Both because of Britain's role as the first industrial nation and because of assumptions about the *revolutionary* character of industrial change, it has always been natural to see the process as pervading all aspects of nineteenth century British society. However, from the perspective of the late twentieth century, when roughly a hundred years of decline has left Britain no longer a great power, it is much easier to see how exaggerated the traditional view has been. In many ways it was possible for Britain to become an industrialized country without experiencing a thorough disruption of her exciting ideas, institutions, and practices.

To take the most obvious aspect, it is clear that the impact of industrialization on the organization of the economy was less dramatic than conventional accounts suggest. It comes as a surprise to find that even in manufacturing industry the average British workshop employed only twenty-nine men as late as 1898. And so much attention has been attracted to industries like textiles and coal that one can easily overlook the largest occupation in the country – domestic service – which grew in size during the nineteenth century. In many ways, therefore, pre-industrial forms of economic activity survived stubbornly right up to 1900. This point is complemented by recent studies of the wealth generated by British industrialists in the nineteenth century. Obviously many fortunes were made, but they were on a more modest scale than is often appreciated. By the end of the century the predominant sources of wealth amongst the millionaires or half-millionaires in Britain were land, banking, and commerce, not manufacturing industry. Of course, the pattern was in a process of change; during the twentieth century the traditionally wealthy families often lost their pre-eminence. But in 1900, after a century of industrialization, the basic structure of the British social pyramid had not been significantly changed.

As far as the character and quality of the British business elite is concerned, it is not necessary to assume a decline to recognize that in some respects their significance was always overestimated. The success of the original cotton entrepreneurs reflected a concatenation of favourable circumstances. They, and their successors, enjoyed the advantage of being not so much the most efficient as the *only* mass producers at least for many years. In this perspective, British economic supremacy was a rather special and brief phase, enjoyed rather fortuitously by a small country that depended heavily on transporting vital raw materials cheaply over long distances. Moreover, even in her Victorian heyday Britain's strength in the world economy reflected her role in finance and commerce rather than simply manufacturing. On the direct exchange of goods she enjoyed only a modest surplus; it was the indirect earnings that made her overall trade balance look strong.

The fact is that even at her peak Britain often showed herself slow to exploit the economic potential that beckoned. Perhaps we are too accustomed to the critical view of the ruthless Victorian capitalist as a keen-eyed hawk quartering the skies, perpetually ready to swoop on an opportunity. In reality he was often a cautious and slow-moving individual. This is abundantly clear if one looks at the broad record in what might be expected to have been a thoroughly exploited area – British India. The chief source of wealth derived from India in the nineteenth century was the land revenue laboriously gathered in by the civil service. But neither the East India Company before 1858 nor the British government thereafter showed a great concern for economic development. Private individuals worked fitfully at unlocking the country's great resources. Britain's tea continued to be obtained overwhelmingly from China until the 1890s. The jute industry was similarly left undeveloped for many years. It was not until the 1860s, when the Civil War in America interrupted Lancashire's cotton supplies, that India's output of cotton began to be extensively increased. Most striking was the fact that it took the mutiny to make the development of a railway system a priority. Even so, private entrepreneurs were so reluctant that they had to be led by the Indian government, which guaranteed a 4.5 per cent rate of return on capital for the construction of a railway network. Taking overseas investment as a whole, about three-quarters of it went into loans to governments, public utilities, and railways, and very little into manufacturing in the period 1870–1914. If Britain had a talent, it was very much for investment and commerce rather than for industry itself.

This reminds us that British pursuit of economic success was always less single-minded than it must have appeared to outsiders. One traditional reading of nineteenth-century history had Britain falling to a series of challenges from the industrial middle class, as demonstrated by the extension of the vote in 1832 and the repeal of the Corn Laws in 1846. Yet there was never a complete clash of values and ideas between the old ruling groups and the rising classes, partly because the aristocracy itself had long been involved in capitalistic enterprise and continued to shore up its position by this means. In any case, it is now widely recognized that, in spite of isolated victories achieved by the reformers, British government was not taken over by middle-class interests. Parliament and the Cabinet continued to be dominated by the traditional landed and aristocratic families throughout the mid-Victorian era and even later still. Moreover, such increase as occurred in middle-class political representation was as much due to professionals like lawyers as to industrialists. Consequently, it is difficult to show that government policy was moulded by the interests of industrialists.

This may seem a surprising claim. After all, by 1866 no fewer than 215 MPs had connections with railway companies, often as directors. But there is not much evidence that they significantly influenced the legislation dealing with railway interests, especially after 1870. Governments tended to take a wider view and to listen to criticism of the companies. Indeed, some scholars have argued that some industries like railways and electricity were seriously disadvantaged by government intervention. Similarly, when in the 1880s pressure developed among some manufacturers for protection against foreign competition, the party-political interests gave very little support, and the question of 'fair trade' had to wait another twenty years before it rose to the top of the political agenda. Britain was very far from giving priority to her manufacturing interests.

2

Not Quite a Democracy

Victorian Britain enjoyed an unusual, possibly unique, form of government. On the one hand it operated a liberal, parliamentary system based on public debate and freely conducted elections; governments could be criticized and regularly overthrown without recourse to violence. Britons took pride in the independence of their judicial system and in the civilian control over the armed forces. On the other hand, British politics bore the heavy imprint of traditional, autocratic elements. Participation was, in practice, limited by wealth and class. Parliament, the Cabinet, and local government were still dominated by men endowed with landed wealth and aristocratic titles. The majority were unable to vote, and women were entirely excluded from the franchise. Elections were frequently uncontested because of deals between rival parties or interests, and were invariably marred by the extensive corruption and intimidation of voters. Finally, a good deal of power continued to be exercised by the hereditary House of Lords and the monarch.

This ill-assorted system was to some extent the result of the fact that, unlike other Western states, Britain lacked a written constitution. From time to time it was judged that the mixture of institutions and conventions through which government actually worked was unsatisfactory; innovations were made which grafted elements of liberalization upon the traditional structure but never thoroughly overhauled it. The period from the late 1860s to the turn of the century clearly witnessed a definite transition towards democracy, but this occurred without any sense of traumatic upheaval. The extent of change was to some degree obscured by the continuity of personnel. Statesmen such as W. E. Gladstone – an MP since 1832 – Lord Salisbury, and Joseph Chamberlain were engaged in a process of unceasing adaptation to changes in issues and institutions.

Power was divided between three institutions: the House of Commons, the House of Lords, and the monarch, though not in equal proportions. The essential basis of a government's power consisted in its ability to command a majority in the Commons won at a general election. Provided this majority held together, a government might enjoy up to seven years in office, though in practice elections took place more frequently – 1865, 1868, 1874, 1880, 1885, 1886, 1892, 1895, 1900 – in order to suit the interests of the party leaders. Legislative programmes occupied a less central role than under twentieth-century administrations; and Victorian governments commonly brushed off defeat on a bill and continued in office. The likelihood of an upset was greater both because the party whips excercised only limited control over back-benchers and because the 500 peers in the upper chamber felt free to use their powers.

In this period the Lords contained a high proportion of leading politicians. Among the prime ministers, for example, Salisbury and Rosebery sat in the upper house, and Disraeli first in one then in the other, while only Gladstone remained in the Commons throughout. From the 1860s the peers actually became more important in checking legislation from the Commons. This reflected the evolution of two more coherent and disciplined parties. As the propertied interests rallied increasingly behind the Conservatives, so that party obtained an overwhelming hold in the upper house, with the result that major Liberal legislation, notably the bill for Irish Home Rule, was thrown out. However, before 1885 this obstructionism complicated but did not actually kill the Liberal programme. For although the peers liked to claim their right to restrain a government whose policy lacked a popular mandate, they appreciated the dangers of such a strategy. If provoked too far, the Liberals might seek to arouse popular support not simply over the immediate issue but against the House of Lords and its privileges. Such a prospect loomed in 1884, when the peers initially rejected a franchise reform bill; but they backed down before the controversy developed too far.

The monarch still excercised considerable influence in the choice of prime ministers, though subject to the qualification that it was unwise to select one who did not command the loyalty of the party majority in the Commons. Increasing party discipline inevitably curtailed the monarch's room for manœuvre. Contrary to traditional belief, Queen Victoria found it hard to adjust to her dwindling role. This was the result of her excitable temperament, her partisanship, and her limited understanding. Age and experience only made her increasingly truculent towards her prime ministers and the parties. She had less trouble with the Conservatives in the late nineteenth century. In 1885, for example, she chose Salisbury over Sir Stafford Northcote without causing controversy. But she grew increasingly hostile towards the Liberals, and Gladstone in particular. Following their victory at the 1880 election, she tried to offer the premiership to Lord Granville and Lord Hartington. But this was to ignore the reality of the situation; Gladstone had effectively recaptured the party leadership, and the Queen had to back down before the will of the party. However, Queen Victoria continued to be guided by partisanship. She checked the promotion to Cabinet of such men as Sir Charles Dilke and Joseph Chamberlain, both of whom were suspect for their radicalism and republican sympathies. Worse, she undermined Gladstone as Prime Minister by showing his letters and her replies to the leader of the opposition; and she several times consulted Salisbury in order to ascertain whether a dissolution of Parliament would suit Conservative Party interests. Luckily for the Queen, the public criticism of her in this period focused upon the costs of the civil list and her neglect of her duties after the death of Prince Albert, rather than upon her tendency to allow her partisanship to lead her to exceed her constitutional role. It was, thus, the accession of King Edward VII in 1901 that really marked the emergence of the modern monarchy.

Edward acted much more as a figurehead above the controversies of party politics; as long as his elected government retained its majority, he would not obstruct or undermine its work.

Gladstone and Victorian Liberalism

From the 1870s onwards, Victorian politics revolved increasingly around two coherent political parties. The traditional bipartisan character of politics became anachronistic, both because of the polarization between Gladstone and Disraeli and because of the steady evolution of two elaborate party machines equipped with the professional organization, local branches, large membership, annual conferences, and official party policies that characterize modern parties.

Yet up to the 1860s Liberals and Conservatives still had a good deal in common. In Parliament both were dominated by gentlemen of independent means who did not depend on politics for their livelihood. Both retained their character as parliamentary parties and, as yet, devoted little of their time to cultivating support in the country. They had also pursued somewhat similar policies during the mid-Victorian era, and the major divisions often occurred within rather than between the parties.

There were, however, significant differences between them. The Liberals enjoyed a much better claim to be the *national* party, at least at the beginning of our period. Since 1846 they had been almost continuously in office, and were not defeated until the election of 1874. Under the leadership of Lord Palmerston the party became associated with a bold, patriotic foreign policy, free trade, and cheap government – and thus with national prosperity and success. Moreover, the Liberals boasted a more broadly based support in the country, including the commercial middle classes, Whig landowners, small shopkeepers, and skilled artisans. They spanned the divide between Anglicanism and Nonconformity; and their parliamentary majorities were drawn from Wales, Scotland, and Ireland, not merely England.

In spite of its frequent immersion in internal controversies, the Liberal Party also held a coherent overall view of politics. Perhaps the central element in this was free trade. The importance of free trade is often overlooked because after 1846 it largely ceased to be a matter of controversy. In fact, free trade was far from being an exhausted issue, for it still remained to whittle away the remaining duties on food, and the Conservatives, though they generally acquiesced in free-trade policy, remained susceptible to revivals of protectionism such as occurred during the 1880s and after 1903. Moreover, for Liberals free trade was much more than an economic expedient; it was regarded as a great *moral* cause because it helped to promote international peace through the maximization of commercial relations between the countries of the world. The pursuit of peace also complemented the policies of financial retrenchment and low taxation which mid-Victorian Liberals generally espoused.

In addition, most Liberals sympathized with policies designed to extend individual freedoms and remove artificial privileges. In foreign affairs this involved support for those, like the Greeks and Italians, struggling to achieve self-determination and representative government against authoritarian and clerical regimes. At home, a traditional tolerance towards religious variety led Liberals to advocate granting political and legal rights to Catholics, Jews, and Nonconformists. Most of all, the extension of individual freedoms through parliamentary reform occupied a central place in popular Liberalism. Whereas many Conservatives regarded government as a necessary evil, Liberals emphasized its power for improvement. Theirs was fundamentally an optimistic creed. Their chief reservation about government was that unchecked power inevitably led to abuse and corruption – hence the need for scrutiny by Parliament and a free press.

The outstanding leader of Victorian Liberalism was W. E. Gladstone. To the twentieth-century mind he seems a remote figure; his combination of high-mindedness, intellect, and religiosity is almost wholly absent from modern politics. Yet his strength lay in his capacity to reflect the typical concerns and values of Victorian society. Though he sprang from commercial wealth, Gladstone played the role of the traditional aristocrat-statesman with aplomb. He possessed roots in provincial England, an estate in Wales, a constituency in Scotland, and an abiding, if somewhat abstract, fondness for the Irish. As a young man – the 'rising hope of the stern, unbending Tories' – he had opposed the 1832 Reform Bill. But his apprenticeship under Sir Robert Peel set Gladstone on the road to Liberalism via support for the repeal of the Corn Laws. As a staunch free-trader and sympathizer with liberal-nationalist movements in Europe, he gravitated naturally towards the Liberals in the 1850s. As Chancellor of the Exchequer he occupied a key role in government, and came to be seen as the inevitable leader of the party after the death of the elderly Palmerston and Lord John Russell. His obvious intellect and his massive industry made him the outstanding figure among the ranks of the conventional parliamentarians.

At the same time, Gladstone's growing reputation as a reformer and his notable oratorical powers excercised an appeal over the radicals both in Parliament and in the country. This was the result of his work as Chancellor in assisting the popular press by abolishing the duties on paper, and also his public support for an extension of the parliamentary vote to working men. Gladstone had been genuinely impressed by changes in the labour movement since the decline of Chartism in the 1840s. The development of a responsible trade union organization and the readiness of working men to save money in Post Office savings accounts helped to convince him that there was nothing to fear from a judicious increase in the number of voters. This conviction was strengthened by his dismay at the behaviour of the 'upper ten thousand', whose political influence tended, in his view, towards profligate and self-serving policies. Thus Gladstone found himself in tune with the popular Liberal cry of 'Peace, Retrenchment, and Reform'. By the 1860s he had grasped the fact that many working-class, radical leaders shared his own belief in checking government extravagance and keeping taxation low; he calculated that to enfranchise the top level of the working class would not only benefit the Liberal Party as against the Conservatives but also strengthen his hand against the vested interests that dominated Parliament up to the 1860s. In this sense there was a natural alliance between Gladstone and the working-class politicians in the 1860s and 1870s.

Religion constituted another very important pillar in Gladstone's rise. By comparison with our own century, the Victorian era was much absorbed by religious concerns; and this certainly coloured politics throughout the nineteenth century. There were broadly three reasons for this. First, the Churches became alarmed at the extent to which, as the British population migrated to the towns, it escaped the influence of Christianity. The Church of England in particular began to invest a good deal of money in building new churches, organizing Sunday schools, and establishing elementary schools in order to recover the lost congregations. But when the state began to intervene in educational provision, the voluntary schools inevitably became a matter of political controversy.

Second, there had been a sustained rise in support for the Nonconformist Churches which, as measured by church attendance, threatened to overtake the Church of England. This in itself raised the question whether it was still appropriate for Anglicanism to enjoy the status and other benefits of being the established church. Nonconformists also resented the legal discrimination which, for example, excluded them from the ancient universities and denied them burial by their own rites in consecrated ground. They increasingly resorted to political action to obtain redress, which took the form of promoting pressure groups that worked under the aegis of Liberalism.

Finally, the late nineteenth century witnessed fresh controversy over the reassertion of the Catholic Church in Europe generally and in Britain; this took the form of the spread of 'ritualism' within the Church of England, new recruits to Catholicism, and the influx of Catholics from Ireland. The result was that politicians devoted a large proportion of their time to containing the effects of intractable religious controversies.

Gladstone was a High Anglican who took his religion much more seriously than most politicians. He was just as happy to write a religious tract as to draw up a parliamentary bill. There can be no doubt that he had a capacity to articulate late Victorian moral and religious issues and to give a lead that neither his Tory opponents nor Liberal rivals like Joseph Chamberlain possessed. As a result, several of Gladstone's great crusades, such as Irish Disestablishment in 1868–9 or the Bulgarian Atrocities in 1876–80, had strong moral-religious overtones. This proved to be a considerable political strength, and it made him in effect the chief spokesman for what was dubbed the 'Nonconformist conscience' in this period. His opponents, of course, resented what they saw as his sanctimoniousness; it was not, as one complained, that Gladstone always had the ace up his sleeve that irritated, but his assumption that God Almighty had put it there!

The Impact of Parliamentary Reform

Since the Great Reform Act of 1832 there had been no major changes in the electoral system in Britain. Even that measure had left a total electorate of only 1.3 million by the 1860s – about one in five adult men. Consequently the decades since 1832 had seen little change in the personnel of Parliament, which continued to be dominated by traditional landed figures. Many radicals felt that, until the electorate was expanded, Parliament would continue to be unrepresentative and uninterested in their grievances.

For this reason the death in 1865 of Palmerston, who had been seen as an obstacle to reform, aroused great expectations amongst the radicals. Russell and Gladstone attempted to meet this demand by introducing a modest franchise bill in 1866. Although this suffered an ignominious defeat owing to a revolt among the Liberals, Gladstone and Russell signalled their determination to return to the question of reform by resigning. They clearly anticipated returning to office after another general election in which they would be able to win a mandate for their bill. Meanwhile Lord Derby and Disraeli took office. But since they had no majority they eventually decided to try to keep the Liberals divided by proposing a reform bill of their own. The surprising outcome was a much more sweeping reform than that envisaged in 1866. This was the result of Disraeli's dependence upon Liberal MPs to keep his bill afloat, which led him to accept amendments which enfranchised many working-class householders and lodgers in the boroughs. The effect was to increase the number of electors from 1.3 million to 2.4 million by 1868 and to 3.1 million by 1883 – about one adult male in every three.

The 1867 Reform Act proved a major catalyst for further political change throughout the late Victorian era. Some of the changes were direct effects in the shape of consequential legislation; others were indirect and long-term developments arising out of the reactions and responses of the parties to the wider franchise.

Since many of the new voters were working men, they became vulnerable to influence by employers – for voting continued to be a public act. The introduction of the secret ballot in 1872 was intended as a remedy, though its immediate effects were slight except in Ireland and in parts of Wales, where landlord influence had often been intimidatory. Contemporaries also took alarm at the increase in expenditure in the form of bribery of the new electors, not to mention excessive drinking and violence that often followed. In the wake of the 1880 general election, Gladstone therefore enacted a Corrupt and Illegal Practices (Prevention) Act which introduced strict limits on expenditure and imposed severe penalties on politicians found guilty of malpractices. This proved to be a fairly effective measure, although two qualifications have to be made. Fighting parliamentary elections and cultivating a constituency, even by legitimate means, continued to be an expensive business; and this handicapped working-class organizations during the 1880s and 1890s. Also, the parties circumvented the legal restrictions on local expenditure by raising huge central funds, often by means of donations from wealthy businessmen in return for knighthoods and peerages; this money could be spent in the name of the party rather than the individual candidate without breaching the law.

Tackling corrupt practices was clearly seen as a precondition for further extensions of the franchise to working men. Radical Liberals regarded the 1867 Reform Act as an illogical measure, in that a worker who qualified for a vote as a householder if he resided in a borough lost it if he moved to a county constituency. They believed that extension of the new franchises to counties would enable them to tap extra support, and by 1884 Gladstone agreed to meet this demand in order to check the growing disillusionment among radicals with his government. Conversely, the Conservatives, fearful of losing seats in the counties, used their majority in the Lords to delay the bill. However, they gladly accepted a compromise: in return for passing the franchise bill they were given a scheme to redistribute the constituencies in 1885. This turned out to be another very radical change.

The post-1885 system involved a step towards equal-sized constituencies, in that new seats were to be created for units of population of 50,000. Moreover, the traditional two-member constituencies were largely abandoned. Now both counties and boroughs were carved up into single-member seats. The effect was to create constituencies dominated by a single economic interest or social class in many cases, and in the long term this promoted the growth of a class-based pattern of politics.

Less obvious but equally important were the stimulating effects of franchise reform in 1867 and 1884 on the strength and organization of radical Liberalism. When Gladstone won the 1868 election he at once came under pressure to tackle the grievances of a host of radical groups. However, Gladstone, who was in any case far from sympathetic to all the causes, determined to impose his own priorities. He followed up his deliberate campaign on Ireland during the election by introducing bills to disestablish and disendow the Church in Ireland in 1869–70.

After this triumph, however, the government became embroiled in a series of reforms which were often dear to rank-and-file Liberals, but offended many wealthy supporters and vested interests. Several of these initiatives promoted individual opportunities by stripping away established privilege. For example, the University Tests Act opened up teaching fellowships at Oxford and Cambridge to non-Anglicans, the practice which allowed officers to buy their commissions in the infantry was abolished, and trade unions were granted the legal status enjoyed by other organizations.

A greater challenge to traditional vested interests was W. E. Forster's 1870 Education Act, which introduced a state system of elementary education though without abolishing the existing voluntary provision of schools by the Churches. Now elected school boards would be able to levy a special rate to build and run elementary schools where the existing provision was judged to be inadequate. Although such a reform was long overdue, the Church resented the competition from the state which would inevitably raise the standards required in schools. On the other hand, many Nonconformists resented the continuing role of the Anglican Church in schools because their own religious teaching would enjoy a subsidy from public funds. This controversy rumbled on for many years. In the short term it led to a revolt against Forster's Bill by back-bench Liberals, and to the setting up of a new Nonconformist pressure group, the National Education League.

Internal party friction was also generated by the government's attempts to tackle another radical cause – temperance reform. H. A. Bruce's Licensing Bill provided for the introduction of licensed hours, police inspection of public houses, and penalties against the adulteration of beer. None of this was popular with brewers and publicans; and, as consumption of alcohol reached its peak in the early 1870s, we may assume it was unwelcome in the country at large. Yet this attempt at regulation fell far short of the aims of the temperance reformers, who wanted to eliminate drinking by the suppression of public-house licenses; by purging drink of its worst excesses, Gladstone's legislation seemed likely to make this ultimate objective less attainable. Again, this controversy ran on vigorously up to 1914.

Middle-class feminists formed another pressure group whose expectations had been raised by the political reforms of the 1860s. Since the 1850s a group of women including Barbara Leigh Smith, Emily Davies, Harriet Taylor, and Lydia Becker had been co-ordinating campaigns for changes in the marriage law, women's education, and employment. For women, too, the parliamentary vote appeared to be the key to achieving these wider objectives; and feminists had some grounds for believing that, as Liberals justified the vote for men on the basis that they paid rates and taxes, so they would be prepared to enfranchise similarly qualified women. However, the women's suffrage amendments to the major reform bills by John Stuart Mill in 1867 and William Woodall in 1884 were rejected. Gladstone disliked women's suffrage on principle, but also feared that the inclusion of a women's clause in a franchise bill would give the House of Lords a good excuse to delay it. Consequently, the feminists, just like all the other radical campaigners, were obliged to work away at converting Liberal politicians to their cause. They enjoyed much more success than has usually been recognized. By the turn of the century the majority of Liberal MPs were suffragists. Under Liberal governments, women won the local government vote in 1869, the Married Women's Property Acts in 1870 and 1882, the raising of the age of consent in 1885, and the abolition of the Contagious Diseases Acts in 1886. Like the Nonconformists, the women grew steadily closer to the Liberal party organization (though they were less well integrated) through the formation of the Women's Liberal Federation in 1887. But as with the other causes, the failure to achieve women's suffrage generated friction amongst Liberal activists for many years.

After three years of innovation and change, Gladstone's government began to suffer the fate of most reforming administrations; it lost momentum through sheer exhaustion, disappointment among its supporters, and the controversy stirred up among the vested interests affected by reform. Many Whigs and middle-class Liberals were alarmed at what seemed to some an attack upon property by the legislation for Irish land reform and Irish Church disendowment. The suspicions of the wealthy were reflected in growing opposition in the House of Lords, and in a withdrawal of support for the Liberals by some Whig magnates at the 1874 election. On the other hand, many radicals were determined to push Gladstone further towards reforms such as the disestablishment of the Church in Wales and

England. His government's defeat in the Commons in 1873 presaged the loss of the Liberals' majority in the subsequent general election.

Despite these controversies and setbacks, the provincial radicals felt optimistic about their long-term prospects because they calculated that the extension of the electorate was exerting a stimulating effect on Liberal organization. Both traditional parties were, in fact, changing from small parliamentary elites into extraparliamentary movements. By the 1870s the Liberals were building up a substantial paid membership and an organization in every constituency. Popularly referred to as 'caucuses', these associations claimed to represent rank-and-file opinion and thus to have the right to select candidates who were committed to supporting radical causes in Parliament. The greater the dismay over Gladstone's reforms during 1868–73 the stronger the incentive to develop this organization. In effect the new National Liberal Federation, created in 1877, incorporated a range of radical pressure groups, such as the National Education League, the Reform League, the United Kingdom Alliance, and the Church Liberation Society, into a centralized, national party structure.

The chief inspiration behind the NLF was Joseph Chamberlain, a Liberal Lord Mayor of Birmingham and typical of the provincial radicals who were working their way up through the Liberal hierarchy in this period. He anticipated that, by compiling a programme of reforms and priorities for future Liberal governments, the NLF would effectively accelerate the radicalization of the party. Its work in the constituencies and at elections would help to change the composition of the parliamentary party and thereby realize the objectives of those who had agitated for the franchise in the 1860s.

From Chamberlain's perspective, Gladstone's leadership represented an asset to the radical cause – but only up to a point; for Gladstone not only dissented from many parts of the radical programme, he also disliked the emergence of an assertive party organization outside Parliament. However, he exercised such a powerful hold over popular emotions that for some years he managed to fend off the threat posed by Chamberlain. In 1877 Gladstone appeared to accept the NLF, but in effect he attempted to capture it by attracting its activists into his new crusade against the persecution of Christians by the Turkish rulers of Bulgaria (see p.95). His campaign gained further momentum from Disraeli's imperial wars in South Africa and Afghanistan during 1878–9. Gladstone's attack on these policies resulted in his triumphant re-election as MP for Midlothian in 1880. Although neither Chamberlain nor the Whigs approved of Gladstone's campaign over the Bulgarian Atrocities, they were powerless to stop him. The Liberal victory at the 1880 election was attributed, rightly or wrongly, to Gladstone, and he returned to the premiership once again in an apparently strong position.

However, this triumph led only to a frustrating period in which Gladstone became embroiled in Irish and imperial questions and domestic reform was neglected. Radical dissatisfaction steadily mounted, to such an extent that by 1885 Chamberlain was ready to join with Liberals from the Whig side of the party to drive him from the leadership. To this end Chamberlain launched a radical campaign in the country, pitched at the privileged classes 'who toil not, neither do they spin'; in this he advocated Church

BEARDING THE BUCCLEUGH.

"THESE ARE MIDLOTHIAN'S VOTERS NEW,
AND, SOUTHRON, *I'M* THE BOLD BUCCLEUGH!!"

Gladstone facing the faggot votes in the Midlothian Campaign

disestablishment, graduated income tax, taxation of landed wealth, and elected local government. He clearly hoped to capitalize upon the votes of the newly enfranchised labourers in the counties at the 1885 general election to undermine Gladstone's hold on the party. To a considerable extent the party was indeed moving in Chamberlain's direction. During the 1880s Whigs had already begun to leave the party, and a majority of the MPs supported at least part of the NLF programme by 1885. Whereas in 1868 only 64 MPs had been Nonconformists, there were 96 by 1886 and 177 by 1892. It was in order to arrest any slide towards Chamberlain that Gladstone launched a bold initiative early in 1886 by revealing that he had become converted to a policy of Home Rule for Ireland. This precipitated a split which drove many Whigs and a few radicals, led by Chamberlain, out of the Liberal Party as Liberal Unionists; but Gladstone none the less retained his position as Liberal leader until his retirement in 1894.

The Conservatives under Disraeli and Salisbury

Twentieth-century writing on the Conservatives has often given a misleading impression by emphasizing the liberal or progressive elements and personalities in the party at the expense of the traditionalists. While Victorian Conservatives did adapt to change out of sheer necessity, this went against the grain, and their usual role was simply to resist change and to defend privilege. The party of the 1860s was, after all, the party that had unceremoniously ejected Sir Robert Peel from the leadership after his reforming premiership which culminated in the repeal of the Corn Laws in 1846. The effect of the split in that year had been to leave the Conservatives a somewhat narrow, sectarian party based on the English rural counties. Among the backwoodsmen a strong protectionist element remained within the party which made it difficult to recover the confidence of the electorate.

However, 1867 marked the beginnings of a comeback. The tactical triumph of Derby and Disraeli in passing the Reform Bill gave rise to a pervasive myth, popular among those Conservatives who wanted their party to become more progressive, that Disraeli had always *intended* to make a bid for a popular constituency for Conservatism and to democratize his party. In the words of *The Times* obituary, he 'found the Tory workingman as the sculptor finds the angel imprisoned in a block of marble'.

On the contrary, Disraeli had no such ambitions, and most Conservatives felt highly reluctant to indulge in reform. The euphoria over their parliamentary triumph rapidly evaporated in 1868 when the party slid to defeat by a margin of 116 seats. Salisbury, one of the ministers who had resigned over the reform bill, bitterly attacked Disraeli for having betrayed Conservative principles in a foolish attempt to outbid the Liberals. Nor was Disraeli himself quite the bold reformer later claimed by advocates of 'Tory democracy'. Far from having a broad vision about creating a new source of Conservatism, he simply hoped to make some gains for his party by redrawing the constituency boundaries to its advantage and by preventing Gladstone from reintroducing another damaging bill. He certainly shared the apprehensions of the traditionalist leaders about the emergence of an extraparliamentary organization. In 1867 the National Union of Conservative Associations had been established to promote the work of the existing Conservative working men's clubs; but it was confined to propaganda and local organization and had no role in policy-making. Just to make sure, in 1870 Disraeli created the Conservative Central Office, whose officials and funds were effectively controlled by the party leader, to run the organization.

On the other hand, Disraeli undoubtedly had a shrewd eye for a political opening, and he saw his opportunity in the shift of the Gladstonian Liberal Party towards a more radical position after 1867. The more the Liberals turned their backs on middle-of-the-road Palmerstonianism, the more credibly could the Conservatives appeal to the respectable, middle-class electorate. Harbingers of the new strategy appeared in Disraeli's speech at the Free Trade Hall, Manchester, in April 1872, in which he attacked the government for allowing radicalism to get out of control and for threatening the Church, the Crown, and national security. Later that year, at the Crystal Palace, he made some celebrated but very brief references to social reform; but the real significance of the speech lay in his criticism of the external policies of the Liberals, notably the attempt to withdraw troops from New Zealand. This was blown up into an absurd claim that Gladstone intended to dismember the British Empire. Yet Disraeli remained very cautious about his prospects of power. After the defeat of Gladstone in Parliament in 1873 he declined to take office, as he wanted to have a better chance of winning an election. When that election came, in 1874, it did produce a Conservative majority, the first since 1841, though the party still polled a minority of the votes – only 1.09 million compared to 1.28 million for the Liberals. The mid-Victorian mould of politics was not broken yet, though it was clearly cracking.

From 1874 to 1880 Disraeli enjoyed real power for the first time in his long career. Much attention has focused upon the remarkable concentration of social reforms enacted in these years, including the Artisans Dwellings Act, the Health Act, the Sale of Food and Drugs Act, and the Workmen and Employers Act. However, the Conservatives had been elected more through a middle-class revolt than through new working-class support. There is, in fact, little reason for thinking that the reforms attracted working-class recruits to Conservatism. With the exception of the legalization of peaceful picketing, the measures made no tangible impact on working-class life. This is hardly surprising. Most of the reforms grew out of civil servants' proposals to tighten up or codify legislation passed by previous governments; they did not represent any considered or distinctive political strategy on Disraeli's part.

If the social reforms held any general significance for the Conservatives, it consisted – as with the 1867 Reform Act – less in the intrinsic importance of the measures than in the fact that they had been passed at all. The record restored to the party a reputation for competence in government which it had not enjoyed since the days of Peel. This impression was compounded by the careful approach to finance and economics. Unluckily for Disraeli, his premiership coincided with the influx of cheap imported grain. But he firmly resisted the farmers' pressure for tariff protection – a step that would have been calculated to revive fears about the Conservative Party. Moreover, the government adhered to sound Liberal finance. The duty on sugar was lifted; liability for income tax was raised from £100 income to £150; extra grants were given to local authorities to help keep the rates down; and the social reforms were very cheap for central government. All this was shrewdly calculated to reassure both traditional supporters and middle-class recruits alienated by the Liberals' radicalism.

If Disraeli now seems in many ways an ephemeral influence upon Conservatism, he did leave a lasting mark in his approach to foreign and imperial affairs. Seizing upon Gladstone's departure from existing policy, he adopted the flamboyant style of Palmerston and attempted to build for the Conservatives a reputation as the patriotic party. The opportunity to do this arose out of the conflict between Turkey and Russia in the Balkans in 1876. Disraeli took the view that Britain's national interest dictated that she should oppose Russia and support the Turks, because British communications with India would be jeopardized by any extension of Russian influence in the Mediterranean. By 1877, when Turkey was losing her war with Russia, this led Disraeli to dramatic gestures such as summoning troops from India, and brought the country dangerously close to a major war. Luckily for the Prime Minister, several other powers were keen to resolve the issue by means of a conference. This met at Berlin in 1878 and forced Russia to disgorge some of her gains, thus giving the British a diplomatic triumph at little cost.

A succession of other initiatives and controversies kept foreign affairs very much to the fore in the 1870s. Disraeli purchased nearly half the shares in the Suez Canal Company for the British government; he raised Queen Victoria to new heights as Empress of India; and his representatives abroad engaged in two reckless wars of imperial expansion in South Africa and Afghanistan. Although Disraeli had not himself taken the initiative in these latter cases, it looked as though he had. Since every one of his imperial policies attracted criticism from Gladstone, it soon began to appear that a new polarization between the two parties had developed. From this time onwards the Conservatives made it a deliberate strategy to associate themselves with the cause of the Empire, which they had previously disparaged, the monarchy, the armed forces, and hostility to foreigners in general; they took to portraying their opponents as unpatriotic, even traitorous.

In spite of both his new Palmerstonianism and his social reforms, Disraeli went down to a heavy defeat at the general election of 1880, and he died the following year. Although Gladstone's 1880 government was a very troubled one, the Liberals still went on to win the election of 1885, helped by the new county electorate. This casts some doubt on the claim that Tory fortunes had been restored by or under Disraeli. The real turning-point in late Victorian politics came in 1886. It was the three Conservative victories of 1886, 1895, and 1900 that decisively broke the long-standing Liberal hold on power.

How are we to account for this? To a large extent the Conservatives were simply lucky in being able to pick up the windfalls from the divisions on the Liberal side. Above all, Gladstone's dramatic conversion to Home Rule in 1886 precipitated a split that had been long in the making. It resulted in a breakaway party, comprising 78 Liberal Unionist MPs, at the 1886 election. In the country many Whigs and middle-class voters now shifted to the Conservatives, which made the party much stronger in areas like Scotland, the West Midlands, and London. For purposes of propaganda the Conservatives made extensive use of the violence of the supporters of Home Rule in order to discredit the Liberals by association; and the Gladstonian threat to the Union with Ireland served to corroborate the wider claim that only the Conservatives could be relied upon to defend Britain's strategic and imperial interests.

It proved easy to keep this message in the public eye during the 1880s and 1890s because these decades witnessed the emergence of new imperial

Table 2.1 General Election Results (Seats Won), 1865–1900

	1865	1868	1874	1880	1885	1886	1892	1895	1900
Conservative	294	274	356	238	250	316	268	341	334
Liberal Unionist						78	47	70	68
Liberal	364	384	245	353	334	191	273	177	184
Irish Nationalist			51	61	86	85	81	82	82
Labour							3		2

powers and a succession of colonial conflicts. Famous episodes such as the death of General Gordon at Khartoum in 1885, when a relief force sent by Gladstone arrived too late, were perennially cited as proof of Liberal weakness. In fact, the acquisition of new territory in Africa proceeded as freely under Liberal governments as under Conservative, but the latter clearly exploited the trend more successfully. The Liberals, by comparison, were much more divided over the merits of imperial expansion. Not only the Whigs and Chamberlainites who had left the party, but also some of those like Lord Rosebery who stayed, supported 'forward' policies in Africa, in opposition to the so-called 'Little Englanders' who saw this as costly and unnecessary. These disagreements culminated in the Boer War of 1899–1902. Salisbury capitalized on the crisis brought about by that war by holding a general election two years early, in 1900, when the Liberals were divided and he could reap the advantage of patriotic loyalty towards the government.

However, the Conservative revival cannot be explained entirely, or perhaps even primarily, in terms of either luck or the impact of external affairs, helpful as these were. In social terms the key lay in the drift of middle-class support to the party. No doubt issues like Home Rule, Disestablishment, and infringements of property rights contributed. There is also some evidence that, in time, some of the Nonconformist enthusiasm for Liberalism weakened partly because many grievances had been resolved and because Methodists in particular were less concerned about Disestablishment. Wealthy Nonconformists increasingly saw themselves as part of the political and social status quo represented by the Conservatives, rather than as outsiders.

The cautious financial policies adopted by Conservative governments helped to consolidate the new alliance with the middle classes. Until they were blown off course by the Boer War, Salisbury's Chancellors contrived to restrain expenditure and taxation more successfully than the Liberals. This made electoral sense and suited Salisbury's doctrinal dislike of government intervention. The major innovation was a deliberate policy of reducing local taxation by giving local authorities grants-in-aid for specific purposes, financed from the Exchequer. The relief for rates on agricultural land and voluntary schools were blatant attempts to reward Tory supporters on Salisbury's part. In this sense the party clearly forfeited the Disraelian claim to represent the two nations; it chose instead to consolidate its position as the party around which wealth and property could rally.

By itself this might have been a dangerously narrow strategy. Indeed, the modest Liberal revival which gave Gladstone a victory in 1892 underlined the need to do more than rely upon the errors of their opponents. Conservatives had to make the most of their resources. In this they were considerably helped by the redrawing of the constituency boundaries in 1885. As Salisbury anticipated, this preserved some of the influence of landowners from the spread of urban Liberalism into the counties. More importantly, it created dozens of new Conservative strongholds in suburbs and seaside resorts that were largely middle-class in character; for a century to come, these were to be the bedrock of the party's electoral success.

Even so, Conservatives were obliged to work much harder at cultivating and mobilizing their support, especially now that much of the traditional expenditure was illegal. The party trained a body of professional agents both to register Conservative voters and to organize campaigns. It also developed constituency associations affiliated to the National Union throughout the country. However,this official party structure was significantly augmented by a mass membership enrolled under the auspices of the Primrose League, founded in 1883 by Lord Randolph Churchill. By the 1890s the Primrose League had recruited approximately a million members, of whom half were women, and its local habitations covered nearly all parts of the country and each social class. They gave the Conservatives a welcome supply of volunteer activists, thereby cancelling out the advantage the Liberals had traditionally enjoyed in the constituencies. But the Conservatives also capitalized very effectively on the *social* activities of the Primrose League as a means of extending the reach of the party and holding the supporters together in between elections.

None the less, the party had ultimately to sink or swim on the basis of its political message. On the whole, the leaders shrank from any 'Disraelian' appeal in the sense of social reforms. Salisbury's rooted aversion to improving legislation militated against this; instead, the party's message was cast more in negative terms. Conservatives could, for example, attack Liberal interventionism and improvement on such issues as temperance by championing the rights of the drinking classes. Essentially, Conservatives played the role of a conservative force by defending the status quo and exploiting traditional causes now threatened by radicals. This involved maintaining the Church Establishment and religious education, defending property and the British Empire, rallying to the Queen, and attacking foreigners in general and the Irish in particular, on the grounds that immigration threatened British jobs and reduced the supply of cheap housing. This was not an infallible recipe; Conservatives won late Victorian elections more because the Liberal turnout was low than because their own vote was very high. But it clearly held together a wide range of support in all social classes sufficiently to keep the party in power for the best part of twenty years. In view of Salisbury's dire prognostications about the trend towards democracy in the 1860s, it was a surprisingly successful experiment in adaptation to political change.

The Rise of the Labour Movement

Although much of the controversy in the late Victorian period took the form of arguments between the aristocracy and the middle classes, the organized working class formed an increasingly important third element in the debate. Earlier struggles over parliamentary reform before 1832 and the campaign against the Corn Laws had produced some collaboration between middle-class radicals like Richard Cobden and John Bright and working-class political leaders. By the 1860s the latter made their voices heard both through trade unions and through pressure groups like the Reform League. Although Chartism had dwindled since 1848, many of its

Table 2.2 Membership of Trade Unions in 1888

Metals, engineering, and shipbuilding	190,000
Mining and quarrying	150,000
Textiles	120,000
Building	90,000
Transport	60,000
Clothing	40,000
Printing and paper	30,000
Other	70,000
Total	750,000

ideas, and including manhood suffrage, secret ballot and payment of MPs, continued to be advocated by radicals. During the 1860s the demonstrations organized by Ernest Jones and Edmund Beales in support of franchise reform effectively continued the moral-force tradition of Chartism. The backing of John Bright and Gladstone, albeit heavily qualified, gave the them a vital bridge into Parliament and the Liberal Party which their predecessors had not enjoyed. The enfranchisement of manual workers in 1867 was largely attributed to Gladstone rather than to Disraeli. Politically aware working men looked to Gladstone for further instalments of reform. This was quite obvious in the case of the Agricultural Labourers' Union led by Joseph Arch in the 1870s. After its initial success the union quickly lost members, but instead it devoted its efforts to securing the enfranchisement of county labourers as the best way forward.

In spite of Britain's early industrialization, the development of a mass trade union movement proved to be a very slow and difficult business. A large number of workers continued to be employed in quite small workshops or in occupations like domestic service, where union organization was very unwelcome. Many workers were hampered by the availablity of surplus supplies of labour which enabled employers to break strikes or ignore unions altogether. Trade union membership repeatedly collapsed in the wake of unsuccessful strikes. Unions were also handicapped by periods of economic depression; conversely, boom periods such as the early 1870s or the late 1880s often led to an upsurge in union fortunes, as the men enjoyed a temporary advantage when bargaining for better pay and conditions. For most groups of workers, suffering low pay or irregular employment, union membership was hardly a feasible proposition. But by the 1860s a fairly stable movement had emerged, based largely on skilled men in iron and steel, engineering, shipbuilding, and similar trades. The craft unions levied high subscriptions, registered as Friendly Societies, and offered a range of welfare benefits that made membership a real attraction.

But how far could this organized body of men exercise political influence? Clearly the reforms of 1867 and 1884 had enfranchised not only skilled workers but some who were in fact quite poor. Some of the new single-member constituencies were obviously dominated by working-class electors, especially mining seats in South Wales, Yorkshire, and north-east England. Yet it was not easy to translate this potential into direct political influence. The first stage in the assertion of working-class political ambitions was under way in the 1860s when the Trades Union Congress was established. Its immediate concern from 1869 onwards was to improve the legal status of trade unions, which was then the subject of an investigation by a royal commission set up in 1867. The evidence the commission received about the responsible conduct, sound finances, and benefit schemes of the craft unions helped to convince sceptical politicians of the advisability of granting the unions the legal status they sought; this was achieved by Gladstone's 1871 legislation. Subsequently, the right to engage in peaceful picketing in support of a strike was conceded by Disraeli in 1875.

In this way the unions began to function as a pressure group. The next step was to gain a direct footing in Parliament; but many union leaders felt this to be over-ambitious at this stage. One must remember that in a total labour force of some twelve million, union membership in the 1880s was still only three-quarters of a million, and 1.5 million by the early 1890s. Their leaders by no means lacked political awareness and ambitions, but they recognized that their members already felt allegiance to the existing political parties. Cotton textile workers, for example, were notoriously divided between Conservative and Liberal loyalties. In Lancashire, Birmingham, and London's East End, the Conservatives appeared to have won large numbers of working-class votes, but the leaders of the organized working class were for the most part enthusiastic Gladstonian Liberals. Free trade, financial retrenchment, parliamentary reform, and avoidance of costly foreign adventures won their emphatic endorsement. The fact that many of these politically aware working men were also teetotallers, staunch Nonconformists, and advocates of self-help made them natural adherents of Victorian Liberalism. They took their place along with the other radical groups that gathered beneath the umbrella of the Liberal Party. But if they were not socialists, they were not Chamberlainites either; this they demonstrated by their loyalty to Gladstone over Home Rule. Consequently, from Gladstone's perspective there could be no real objection to direct parliamentary representation for such working men. The breakthrough came in 1874, when two trades unionist officials were elected as 'Lib-Labs', that is, as candidates who were unopposed by the local Liberals and took the Liberal Whip in Parliament. By the 1880s a dozen Lib-Labs had been elected, of whom the best known were Thomas Burt, the Northumberland miner, William Abraham, a Welsh miner, Henry Broadhurst, a stonemason, and Joseph Arch, the leader of the agricultural labourers.

However, it proved difficult to extend working-class representation beyond the Lib-Lab bridgeheads. The explanation for this lay partly in Liberal fortunes after the 1886 split with the Liberal Unionists, which left the party much worse off financially. As a result, few local associations were keen to adopt working men as candidates because of the expense involved both in fighting elections and in the annual revision of the electoral register. As yet, only a few trade unions felt able to bear these costs, and, since MPs received no salary until 1911, the working man would himself have to find alternative means of support.

None the less, during the 1880s and 1890s working-class politicians maintained their pressure on the Liberals. Keir Hardie stood for election in Lanarkshire as an independent, but reminded the voters of his credentials as a Gladstonian. It is now clear that the Liberals missed an important opportunity in this period. The future leaders of the Labour Party largely enjoyed a close relationship with Liberalism; they agreed with its basic ideas and often served as agents or secretaries to Liberal politicians. Some such as Thomas Burt and John Burns did in fact become lifelong Liberals; but Hardie, Ramsay MacDonald, Arthur Henderson, and George Lansbury, who nearly became Liberal MPs, were kept waiting too long and opted for an independent party instead.

There was also an ideological dimension to the movement towards independent labour representation. There was no hard-and-fast division between Liberals and working-class politicians over land reform, the eight-hour working day, or even the state ownership of coalmines and railways. Liberals ranged all along a spectrum from individualism to collectivism. Similarly, the early Labour leaders included men like Philip Snowden, who was dedicated to *laissez-faire* liberalism, and George Lansbury, who came increasingly under the influence of socialist thinking; both served as Labour MPs in the Edwardian period. But there was a growing realization of the failings of the capitalist system. The prevalence of poverty and underemployment in the 1880s and 1890s undermined confidence in Gladstonian economics and inevitably loosened the party's hold. 'I came to the conclusion', wrote Lansbury, 'that Liberalism would progress just so far as the great capitalist moneybags would allow it to progress, and so I took the plunge and joined the S[ocial] D[emocratic] F[ederation].'

By the late 1890s, when trade union membership had grown to almost two million, the prospects of establishing a new political party on the resources of the unions had improved greatly. For the expansion of the movement had involved some modification in its composition: it now included more semi-skilled workers as well as a number of leaders, such as Will Thorne of the gasworkers and Ben Tillett of the dockers, whose views were more socialist than Liberal. They stepped up the pressure on the TUC to adopt more radical policies and to break with Liberalism.

However, the goal of separate labour representation continued to be elusive during the 1890s. The trade unions showed themselves to be suspicious of attempts by socialists to tap their funds. Hence they had little to do with the Social Democratic Federation. The SDF had made a considerable impact by its role in the great demonstrations by unemployed workers in Trafalgar Square in 1886 and 1887. These events attracted a good deal of sympathy, both because of the plight of the workers and because of the repressive tactics used by the police to disperse the gatherings. In parliamentary elections the SDF continued to poll very few votes, however, and it never showed signs of developing into a mass movement. Most trade unionists preferred to participate in politics by forming trade councils in the major towns; this was a modest but economical way of bringing pressure to bear on local authorities and sponsoring the election of working men to municipal councils.

The multiplication of elective local authorities after 1870 provided an excellent apprenticeship in politics for many working men. It was an area in which the Independent Labour Party, formed in 1893, enjoyed considerable success. As its name suggests, the ILP was intended to pioneer the return to Parliament of working men free from any pacts or deals with other parties. It attracted most of the outstanding orators of the labour movement, including Keir Hardie, Ramsay MacDonald, and Philip Snowden. Like the ILP, these men had a foot in both Liberal and socialist camps. Their socialism involved relatively little emphasis on economics or state intervention; rather, they stood for free trade, land reform, and graduated taxation. For them socialism was a matter more of morality, a vision of a society based on co-operation and brotherhood. This owed a good deal to the influence of the Christian moral code and to the arts of the lay preacher. Even the left-wing George Lansbury argued that Christ had been a social revolutionary.

The ILP was also a more democratic party than the SDF. It attracted the more independent-minded socialists, and its emphasis on political reform made for continuity with the radical Liberal tradition. Indeed, the ILP claimed to be extending the work that was now being neglected by leading Liberals; Lansbury was one of those who reacted against the failure of the parliamentarians to respond positively to women's demand for the vote. The ILP was also rather less working-class in composition than its rhetoric suggested. A large proportion of the activists in this period were lower-middle-class men, often drawn from the families of manual workers who had acquired the education necessary to become clerks, schoolteachers, journalists, commercial travellers, or insurance collectors.

For a time the ILP appeared to be the coming force in politics. At the 1892 general election, three independent working men had won election to Parliament: Keir Hardie in West Ham, John Burns in Battersea, and J. Havelock Wilson at Middlesbrough. But in 1895 all twenty-eight ILP candidates were defeated. The party remained extremely short of funds, and even in local government it enjoyed only pockets of strength in places like Bradford. But its leaders were realistic enough to recognize that they had not found the formula for a major breakthrough. In particular they had failed to tap the strength of the unions.

As the 1890s wore on and the Conservatives established their grip on politics, the various left-wing groups, the ILP, SDF, and the Fabians, showed an increasing willingness to co-ordinate their efforts. Moreover, the experiences of the skilled workers, now facing unusually high unemployment as a result of the depression, created dissatisfaction amongst the old established unions over the political status quo. The effect was compounded by the Lyons v. Wilkins legal case which resulted in the conviction of a striking worker for picketing, thereby undermining the legislation of 1875. Everything seemed to point to the need for a new initiative to increase the influence of the unions in Parliament. Such a move also appeared feasible now that the unions' membership had reached 2 million and their reserve funds had risen from £1.4 million in 1893 to £3.7 million by 1900. This was the situation when the railway workers proposed the setting up of a new political organization at the 1899 TUC session; it was approved by 546,000 votes to 434,000. Thus in February 1900,129 delegates from trade unions and socialist societies met to establish the

Labour Representation Committee whose object was to sponsor 'a distinct Labour group in Parliament who shall have their own whips, and agree upon their policy'. This was the organization that became the Labour Party in 1906.

At first the LRC felt obliged to tread cautiously. Some of the major elements – miners, cotton textile, and building workers – stood outside the new body. Even among its supporters the LRC was as yet regarded as an improved pressure group rather than as a potential party of government. Nor was there any formal commitment to a socialist ideology, which would only have alienated some of the unions. But in order to keep the door open for sceptical middle-class Fabians, the LRC rejected the suggestion that its candidates should be restricted to working men. For the purpose of recruiting these disparate groups, the LRC's federal structure proved to be distinctly advantageous. It enabled each of the trade unions and socialist societies to choose its own representatives to the executive, and to affiliate on behalf of its members at the low rate of ten shillings per thousand. Since only 350,000 members were affiliated during the first year, the LRC's income remained low. There was time to endorse only fifteen candidates for the general election of 1900, of whom two were elected, Keir Hardie at Merthyr Tydfil and Richard Bell at Derby.

The LRC gained fresh momentum, however, from the Taff Vale case in 1901, when a trade union was made to pay compensation for the costs of a strike. This new blow from the courts stimulated a rapid rise in affiliations and an increase in the level of fees charged. In any case, by 1902 the political tide had turned against the Conservatives, and the LRC benefited from this at three by-elections. The success of Will Crooks at Woolwich, David Shackleton at Clitheroe, and Arthur Henderson at Barnard Castle rapidly raised the profile of the new party.

National Decadence

Although parliamentary politics made the 1890s appear a highly Conservative period, beneath the surface Britain was in the throes of a radical reappraisal of many aspects of her political and social life. The ending of the Boer War in 1902 proved to be the final catalyst for change; the ramifications of that conflict penetrated deeply through the Edwardian period and even into the Great War.

At the time of its outbreak, the war seemed only to exacerbate the problems of the radicals. By the mid-1890s the Liberals had patently lost their momentum and direction. Though effectively a Chamberlainite party, they largely lacked the bold, constructive leadership that Chamberlain would have provided. The entrenched position of NLF radicals at local level resulted in excessive concentration on causes like Disestablishment, Home Rule, and temperance which failed to arouse sufficient popular enthusiasm. Although the 'New Liberals' were devising a fresh agenda on social questions, they did not yet occupy influential positions in the party. Working-class radicals were being diverted into labour politics because the NLF seemed deaf to trade-union demands like the eight-hour day. The party's reluctance to support the payment of MPs or women's suffrage suggested an inability to live up to its liberal and democratic principles. Gladstone's resignation in 1894 symbolized the party's dilemma – stuck in the cul-de-sac of Home Rule. But none of the Grand Old Man's immediate successors, Rosebery, Harcourt, and Campbell-Bannerman, offered an alternative source of inspiration. By 1900, when the Liberals were badly divided over the Boer War, Rosebery at last made a challenge to the party leadership to abandon its Gladstonian tradition, especially Home Rule; but this came to nothing.

By contrast, the Conservatives appeared to have less to worry about. After 1886 they managed an electoral alliance with the Liberal Unionists, and consolidated this in 1895 when the Duke of Devonshire and Chamberlain joined Salisbury's Cabinet. What is surprising is that the relationship between the interventionist Chamberlain and the negative Salisbury lasted as long as it did. The explanation lay in the fact that, as Colonial Secretary, Chamberlain relaxed his pressure for domestic reform which had been the chief theme of his earlier career, though he did advocate innovations like old-age pensions. As the Conservatives remained in power until 1905, they could be forgiven for thinking that domestic change could safely be avoided. But in time they became too closely associated with wealth and privilege, and out of touch with half of the nation. As a result, many of them found it difficult to respond to a new political agenda in the early 1900s.

For some years the situation proved less frustrating for Chamberlain than might have been expected, because he became absorbed in colonial questions, the future of the Empire, and ultimately in the South African War. Yet this set in train a series of political disasters. The initial British defeats at the hands of the Boers may have reflected badly on the government's inadequate preparation for war, but they were turned to advantage at the election in 1900. The Conservatives took the opportunity to paint their opponents as 'pro-Boers'. However, the patriotic euphoria soon deflated as the war dragged on, proving increasingly costly in terms of lives, taxation, and the National Debt. This left the government even more unwilling to contemplate domestic social reform. But Chamberlain himself was inspired to seek a fresh initiative in order to break free from the travails of the war. The immediate means to this end lay in extending the duty imposed on imported wheat as a temporary expedient for raising revenue. Chamberlain aimed to elevate this into a general tariff on all foreign goods, but with exceptions for the products of the British Empire. This would serve the aims of external policy by strengthening imperial trade and would also promote a constructive domestic policy by protecting jobs and generating an extra source of revenue. However, the free-traders were well entrenched in the government, and A. J. Balfour, who succeeded Salisbury as Prime Minister in 1902, chiefly wished to avoid a divisive debate over tariff reform. Chamberlain found himself manoevred into resigning from the Cabinet. This set him free to pursue his campaign in the country from 1903 onwards; in the short term, however, tariffs had the effect of dividing the Conservatives and helping to drag them down to defeat in 1906.

Such were the immediate ramifications of the war in South Africa.

But the experience also crystallized existing fears about the deterioration of late Victorian Britain. Concern over the economic competition from Germany was well established by 1900. On top of this came a series of revelations about the poverty of the urban population and the declining middle-class birth rate. A new external threat arose in the shape of the huge Continental armies of the post-1870 era, whose capacity for rapid mobilization underlined the vulnerability of the British Isles, all the more serious in view of the naval building programmes of France, Russia, and, latterly, Germany.

The Boer War plainly exacerbated these fears. In some towns as many as six out of ten volunteers had been judged unfit for military service. Britain had shown herself practically incapable of organizing a major campaign at short notice. As the war dragged on it became clear how vulnerable other parts of the Empire, such as India, would have been if another power had taken the opportunity to intervene. Not surprisingly the effect of this was to foster a pessimistic, self-doubting mood in British society around the turn of the century. It became fashionable to diagnose the causes of 'national decadence' by examining the decline of other great empires in history. It was tempting to make Darwinian assumptions that only the more adaptable and vigorous species survived. As the new century dawned and the old Queen passed away with the South African War still unresolved, many doubted whether British power would last another century.

The immediate targets for the critics of Britain's national decadence were the politicians and civil servants, now derided as gentlemen-amateurs, too absorbed in the game of party politics and the pursuit of orthodox financial policy. Their favoured remedies involved injecting an element of efficiency into government by introducing more experts and businessmen, cutting down on parliamentary interference, rationalizing elective local government, imposing universal military training, and extending state education and social welfare reforms. Such a programme attracted a wide range of support from Fabian socialists, disillusioned Liberals, and imperially minded Tories. In its declining years Salisbury's lackadaisical, aristocratic government seemed to symbolize much that was wrong with Britain; and as Balfour tried to overcome the disintegrating effects of the war it was evident that a new agenda and a different political configuration had already emerged.

3

The Victorian State and Its People

In the daily struggle for subsistence the majority of Victorians viewed their Westminster government as a remote and uncertain element; local authorities seemed much more relevant and immediate. But the most effective means of avoiding poverty and the workhouse was to rely upon the network of family, friends, neighbours, and trade unions. Although generalization is risky, we can say that in the late Victorian period a working man would count himself reasonably fortunate if he enjoyed a regular weekly income of around 25s. During 1909–13, when a group of Fabian women in London investigated the living standards of typical families (Maud Pember Reeves, *Round About a Pound a Week* (1913)), they studied those who received between 18s. and 26s.; life in this range was a struggle, but more than bare subsistence could be achieved. However, many men earned less than this because of the casual and irregular nature of the work offered to them. There were also marked regional variations in wages; amongst agricultural labourers, one of the lowest-paid occupations, a man received as little as 14s. by 1900 in Dorset and Wiltshire because little alternative work was available, while in Northumberland, Durham, or Lancashire, the effect of wider employment opportunities was to drive the basic agricultural wage up to 22s. As a result, many families simply found it impossible to support themselves: the official returns for paupers showed over twice as many in the south-west as in the northern counties. However, many families were saved from destitution because of the vital additional income of wives and children. Thus Lancashire, where the textile industry made the two-income family common, saw a relatively high standard of living in spite of the low level of wages.

However, the variation in manual workers' wages continued to be very wide at least until 1914. The most highly skilled workers in iron and steel smelting, boilermaking, shipbuilding, and engineering could earn twice or even three times the wage of a casual unskilled labourer. At 30s. a comfortable life was possible, and a fortunate few might reach £2 a week. They could rent houses with more than four rooms and set aside a front parlour for 'best', equipped with such items of luxury as a piano. At the higher levels the so-called 'labour aristocrats' actually overtook the income of many of those in white-collar, lower-middle-class occupations. Clerks, for example, were typically in the £60–100 a year range, though the most senior might earn £150. Male elementary schoolteachers received on average only £127 even by 1914. A salary of £3 a week was, in terms of social status, a critical one, for it meant that one paid income tax and could employ a servant – an indication of middle-class status.

The Rising Standard of Living

These figures for money wages cannot, in themselves, convey a true idea of popular living standards. Even without changes in wage rates, many working men raised their standards by simply moving out of the lowest-paid jobs into more skilled or more regular work. Whereas there had been 1.4 million agricultural labourers in 1850, there were only 0.97 million by 1911. Of course, increases in money wages did occur, as in the early 1870s, when the economy was expanding; however, it was usually the skilled minority that benefited most from this.

However, from the mid-1870s to the late 1890s a rather different pattern of improvement intervened. The difficulties faced by many employers resulted in a stagnation or even a reduction in money wages. Yet the effects were more than cancelled by a major and sustained drop in the price of most basic items of food. The catalyst in this process was the influx of cheap North American wheat, which virtually halved the price of bread in the shops by the 1890s. For most working-class families bread remained the staple item of consumption, though potatoes formed a growing element, especially in the north and Scotland. But many families responded to cheaper bread by diverting some of the surplus expenditure into other items of consumption, either more nutritious or more novel.

In the same period other sorts of food became cheaper, partly because of the long-term increase in supplies from far-flung parts of the world, or because of further cuts in the duties imposed by the government on tea, coffee, cocoa, and sugar. Tea, as the 'cup that cheers but does not inebriate', was a favourite of Chancellors of the Exchequer. The price of sugar fell by 58 per cent between 1874 and 1900, and annual consumption rose to 80 lb. per head. Supplies of meat also improved significantly, partly through extra imports of live cattle from Europe, partly by the canning of beef from the United States and South America, and partly by the freezing of carcasses of lamb from Australia and New Zealand as well as American beef. Consequently the price of fresh meat had nearly halved by 1900, and it became a regular item of diet for most families for the first time. Improvements in transport and refrigeration also facilitated the rapid spread of fried fish-and-chip shops in this period. The combination of cheap flour and sugar promoted the mass production of biscuits by firms like Huntley and Palmer at Reading and Carr's of Carlisle, while the availability of domestic fruit supplies and imported sugar brought the jams of Chivers and Rowntree within reach of many families. Margarine was developed as a cheap substitute for butter, and tins of condensed, skimmed milk became an alternative to the fresh product, which was both expensive and frequently adulterated.

Overall, the effect of these changes was to increase real wages by about one-third between 1875 and 1900; and since the key to the process was the fall in prices, which affected even the poorest families, one is justified in making a generalization about the late Victorian period as one of substantially rising living standards. It was not until the end of the 1890s that prices rose again, and the inflationary effects of the Boer War led to some falling back in real wages at the beginning of the Edwardian period. This, however, could not reverse the steady advance towards a more varied and nutritious diet for the majority of families. Of course, the working-class appetite for novel and relatively expensive forms of food attracted a good deal of misguided criticism from middle-class Victorians at the time. In fact, many of the economical recipes they recommended were not only unappealing but impractical, either involving lengthy cooking or requiring equipment which poor housewives lacked. Instant meals such as fish and chips at one penny a head were, in fact, good value and very nutritious; potatoes supplied a high proportion of the required vitamins and minerals, while fish cooked in batter retained its food value.

A number of other trends and innovations further promoted improved living standards in this period. Housewives were only too keen to use some of their spare case to purchase materials to keep the home and the family clean. Hence the huge success of William Lever in selling one-penny bars of Lifebuoy soap to the working classes. There was, by contrast, a slight reduction in the consumption of alcohol after the peak in the early 1870s. As a proportion of consumer spending, drink fell from 15 to 12 per cent between the 1870s and the 1890s. The pressure to curtail alcohol consumption at the family level was the greater because the price of this item did not fall in line with others. Another helpful, if marginal, trend was the spread of allotments, which gave some families access to cheap, fresh supplies of vegetables and eggs. More important was the intervention of the government in checking the adulteration of food and drink which was rampant during the mid-Victorian years. Not only was food diluted – sugar with sand and tea with hedgerow leaves, for example – but poisonous substances were commonly added to items like bread and beer in order to create a more attractive colour. Under the impact of successive legislation in 1860, 1872, and 1875, local authorities appointed inspectors to test food and to prosecute offenders. Finally, commercial developments also made a contribution to higher standards. For example, the practice of selling tea in sealed packets with a guarantee of purity, which had been pioneered by Horniman, was copied by Lipton, Brooke Bond, and Lyons. Above all it was the Co-operative stores that made cheap but high-quality food available to ordinary families. In the late Victorian period they were followed by new chain stores including Lipton, the International Stores, and Home and Colonial.

On the other hand, a number of qualifications must be made about the improvements in standards of living. Not everyone could take advantage of the pure and economical food at the Co-op. Poor families were obliged to buy from their corner shop because they needed credit; as a result they paid a higher price for their food. There was clearly an increase in unemployment, although in the absence of reliable and comprehensive statistics it is impossible to say how great an impact it made. Income was not always wisely spent, nor was the food distributed fairly within the family. In particular wives notoriously deprived themselves of expensive items like meat in order to feed the male head of household. Also, the cost of other essentials like housing did not fall. The typical working-class rent bill fell between 4 and 8 shillings a week. There was, however, some improvement in the quality of housing in the form of more gas lighting and water closets. In some areas overcrowding was reduced, but the flood of immigrants and the loss of slum properties in the big conurbations made for the survival of atrocious conditions. Rural housing remained very poor and probably deteriorated because of a lack of new investment. In the towns, too, many of the landlords owned only a few properties, had little spare income, and failed to repair and improve unhealthy buildings. The effects of all this were evident in the poor health of many of the men who volunteered for service in the Boer War, and in the high infant mortality rates. During the 1880s, on average 142 out of every 1,000 babies died in the first year; and in the 1890s the average actually rose to 154 – a reminder that not everyone benefited fully from dietary improvements, and that the changes in food consumption took a generation or more to show results.

The Persistence of Mass Poverty

The Victorians employed three strategies for dealing with the poverty in their society: self-help, charity, and the Poor Law. What is significant about the last twenty years of the nineteenth century is that all three were increasingly recognized as unsatisfactory and inadequate. Consequently the problems of the poor moved a little higher up the political agenda; and although policies had not changed radically by 1900, the politicians had gained a clearer idea of the options available to them. The Victorian debate thus prepared the ground for action in the Edwardian and subsequent years; for example, William Beveridge, a university undergraduate much influenced by the 'discovery of poverty' around the turn of the century, was to leave his mark upon national welfare policies in the post-1945 era.

Concern over poverty was stimulated and sustained by a series of revelations and investigations beginning with Andrew Mearns's pamphlet *The Bitter Cry of Outcast London* (1883), Charles Booth's monumental *Life and Labour of the People of London* (1887–1903), and B. S. Rowntree's *Poverty: A Study of Town Life* (1901). They influenced the debate in three ways. First, and simply, the studies suggested that poverty was much more widespread and persistent than had hitherto been thought. Around 28–30 per cent of the population were now described as being in a state of poverty. This stood in stark contrast to the official figures for paupers, which indicated two to three per cent; and it undermined complacent contemporaries who pointed to the diminution in the numbers of people receiving poor relief since mid-century. Since the reform of 1834, the Poor Law system had run on the principle that those who enjoyed assistance from the rates should be required to enter the workhouse rather than receive a dole, and should there experience worse conditions than those faced outside. This, it was assumed, was a necessary means of maintaining the incentive to lead a self-supporting life. In practice the system had never been operated as uniformly as intended. Some guardians gave outdoor relief to the deserving poor, either out of humanity or because this was cheaper per head than indoor relief. However, orthodox critics charged that outdoor relief simply created more poverty, by encouraging people to apply to the guardians rather than help themselves. Yet if Booth and Rowntree were even approximately right, the bulk of the problem was not being touched by the Poor Law authorities because people feared and resented the humiliating treatment in the workhouse: the separation from one's relations, the uniform, the disfranchisement, and the prospect of a pauper's burial. Consequently the Poor-Law system found itself under attack from both the orthodox and the liberal reformers.

The second contribution of Booth and Rowntree was to focus attention on specific causes of poverty in terms of irregular or cyclical employment, low-paid occupations, and old age. Since these were factors clearly outside the individual's control, much of the poverty could not reasonably be blamed upon moral failings. It was a sign of shifting perceptions that during the 1880s the word 'unemployment' came into vogue to describe an involuntary condition.

Finally, the investigations began to point the way towards more rational ways of tackling poverty. After all, the findings themselves were by no means entirely novel. In the 1850s Henry Mayhew had published similar revelations about the extent of destitution in London. In the 1860s the sudden rise in poverty in Lancashire owing to the interruption of raw cotton supplies by the American Civil War had underlined that men could be the victims of circumstances beyond their control. But Rowntree in particular advanced the discussion by his less impressionistic and more statistical approach. He introduced a measurement or poverty line on the basis of the cost of basic food and shelter for a family of five, which he put at just under 22 shillings a week. He also described the poverty cycle commonly experienced: the young married couple were well off when in work, they fell into hardship as the number of dependent children grew, prospered when the children brought home extra income, and then fell unavoidably into poverty again in old age. This helped to direct attention towards certain aspects of the problem like childhood and old age, for which remedies were forthcoming.

It was not only the Poor Law but private charity that lost credibility in the late Victorian period. Although large sums of money continued to be donated, they were increasingly recognized as inadequate in view of the scale of poverty and the difficulty of distributing resources rationally. Sudden slumps or disasters attracted philanthropic attention in the form of distribution of coal and blankets, but this left the long-term causes of daily poverty untouched. In recognition of this the Charity Organization Society had been founded in 1869. It tackled the problem by means of casework with individual families, not the indiscriminate distribution of relief, in this way pioneering modern social work. But it still reflected the traditional idea that the poor were responsible because of their personal failings, and remained largely blind to the underlying social causes. Nor could the COS do any more than nibble at the edges of the problem.

It would clearly be an exaggeration to suggest that attitudes towards the poor altered radically during this period. Even the reformers retained some belief in the idea of personal responsibility. Rowntree, for example, distinguished between primary and secondary poverty; the latter he attributed to those who enjoyed an adequate income but who still experienced poverty as a result of excessive expenditure on alcohol or a very large family – factors held to be within the individual's control. Nor was the confidence in thrift, abstinence, industriousness, and self-help confined to the middle class. Working men like Arthur Henderson and Keir Hardie argued that many a worker could quickly improve his condition of life by giving up drink. Moreover, the new thinking of Booth and Rowntree cannot be said to have penetrated throughout society; rather, it influenced an educated minority within the middle and upper classes. Consequently the impact of such reformers in terms of policy was limited, at least before 1906, and the old Poor Law survived until 1929.

Popular Attitudes towards the State and Self-Help

Although many Victorians believed that their economic success was conditional upon minimizing the activities of government, their society was, in fact, characterized by a steady expansion of state interventionism (see pp.56–9). What concerns us here, however, is the extent to which this development reflected pressure from below. From the perspective of the post-1945 welfare state it is only too easy to take for granted the idea of government responsibility for the level of unemployment and standards of living; many have assumed – as did some mid-Victorian reformers – that once a popular vote had been granted, collectivist social policies designed to raise living standards and to redistribute wealth would become inevitable. Empirically, however, this appears not to have been the case. Gladstone's governments concentrated more on political, legal, and religious reforms than on social; Disraeli's innovations made little impact; and from 1880 onwards the tendency was towards *less* social interventionism in spite of the increase in the electorate. How are we to account for this?

In the first place, some of the existing expressions of intervention in the lives of working-class people aroused a good deal of antagonism, which naturally blunted the appetite for more. Often the supposed beneficiaries felt themselves to be the targets or victims of those who would improve them, regulate their behaviour, or extract taxes from them. The most persistent object of working class hostility was the Poor Law system. But the move towards universal elementary education after the 1870 Foster Act constituted a growing provocation. Parents often felt less than convinced that able-bodied children should be detained in school until twelve years when they might be earning extra wages for the family. Compulsory attendance meant visits from the inspectors to chase up the parents of truants, and resulted in 86,000 prosecutions in 1892. Indeed, much of the state's activity in connection with children – vaccination, medical inspection, school meals, arrangements for taking them into care – was resented by parents as an infringement of their role.

A second explanation for popular scepticism is that, while some collectivist measures seemed acceptable in theory, the practical effects greatly diminished their significance in the eyes of the working classes. For example, the 1875 Artisans Dwellings Act may loom large in the long-term development of public-housing policy, but all it did in the 1880s and 1890s was to facilitate the demolition of slum property by local authorities. Since the slums were frequently not replaced, the effect was to exacerbate overcrowding. Understandably, then, many working men placed less importance on social reforms than on wage rates, conditions of employment, and unemployment, which were largely, though not entirely, regarded as beyond the control of the government. The most welcome benefits were often those obtained by their own initiatives through membership of Co-operative Societies, Friendly Societies, and trade unions. By 1900 approximately five million Friendly Society policies were held, which, in return for small weekly contributions, offered the 'death benefit' and sometimes sickness benefits and treatment by a doctor. Understandably, these forms of self-help, which gave each family some choice as to the most appropriate scheme, seemed preferable to solutions imposed by the state.

Finance was the third reason for negative attitudes towards collectivist policies. The working-class radicals of the 1860s and 1870s largely endorsed Gladstone's belief in retrenchment and minimal taxation. They appreciated that the bulk of the government's revenue was derived from indirect taxes and duties on consumption and relatively little from the direct taxation of income and wealth. This meant that the burden fell disproportionately upon poor people. At the same time, a high proportion of state expenditure benefited comparatively well-off people because it was devoted to the armed forces, the civil service, interest on the National Debt, pensions for former state employees, and the civil list, which grew steadily on account of all the marriages and births amongst Queen Victoria's numerous offspring. John Bright's famous dictum about the British Empire and foreign policy as a gigantic system of outdoor relief for the upper classes rang true with many radical working men; hence their enthusiasm for the Gladstonian strategy of avoiding expensive entanglements abroad.

Thus when Gladstone's Chancellor of the Exchequer, Robert Lowe, cut both expenditure and taxation in 1869 and 1870, he was responding to the pressure of working-class voters. And Gladstone's pledge in 1874 to abolish the income tax altogether should be seen in the same context. For although liability for income tax did not begin until an annual income of £100, increases in money wages at that time were bringing some skilled workers into the net. Subsequent governments actually raised the threshold to £150, and to £160 by 1894, which effectively excluded manual workers. Thus the emphasis on retrenchment by both parties in the later nineteenth century has to be seen not simply as a wish to pander to the middle classes; many working men, too, regarded low taxation, free trade, and non-interference as in their best interests.

None the less, it is important not to exaggerate these negative attitudes towards the state. They commanded much less support in the early 1900s than they had in the 1870s. Even in the late Victorian period *some* social reforms attracted a positive working-class response, notably Booth's idea for a non-contributory old-age pension. By 1898 the scheme was being canvassed by the National Committee of Organized Labour for the Promotion of Old Age Pensions. Also, some workers showed more interest in reform than others; the low-paid frequently found it impossible to afford self-help schemes or to join trade unions, and therefore stood to gain from greater government intervention. Women in particular tended to be neglected under existing arrangements and were, correspondingly, the beneficiaries of state welfare. Moreover, the reliance upon self-help strategies, which had always had shortcomings, looked increasingly unwise. The emphasis on payment of a death benefit by the Friendly Societies was a misdirection of resources which gave a man a respectable funeral but did nothing for his family. A large number of the policies of both Friendly Societies and insurance companies lapsed when members found themselves unable to keep up the payments, and all their contributions were lost. Also, by the 1890s the finances of many societies were in a precarious state. As members lived longer they were making more demands for benefits during old age when they found themselves too ill to work; consequently the societies' incomings failed to match their outgoings, and bankruptcy threatened. They therefore looked with growing interest to the state pension as a means of relieving them of their burdens. By 1900, then, confidence in the traditional remedies had been considerably undermined and the ground prepared for the innovations of the Edwardian period.

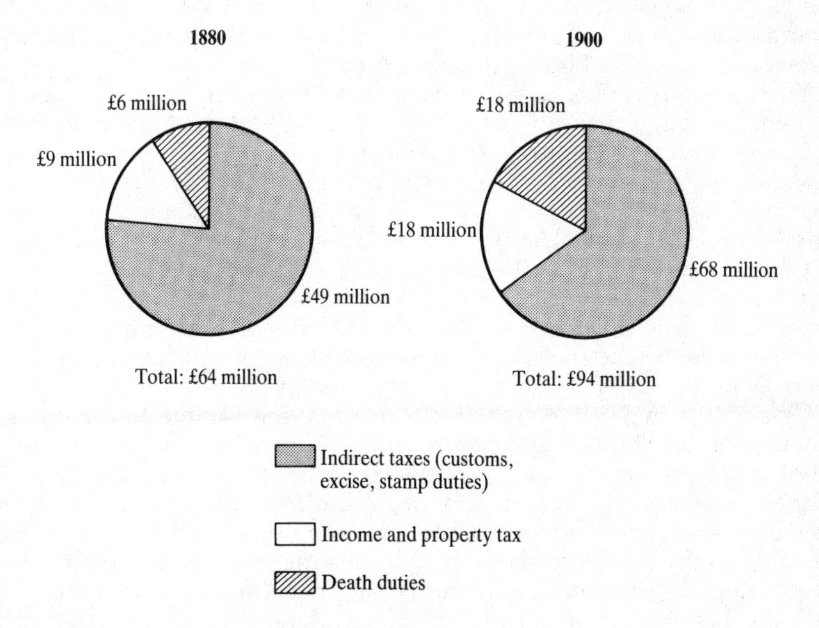

Figure 3.1 Sources of State Revenue (£ million)

The Growth of Local Government

On the whole, Westminster and Whitehall adopted a rather high-minded attitude towards local government in the nineteenth century, regarding it as chaotic and tending towards jobbery and corruption. A typical expression of this view was the New Poor Law of 1834, intended to impose uniform policies on the local boards of guardians; in practice they frequently evaded or defied pressure from the centre. On the other hand, Victorian governments increasingly wanted local bodies to assume responsibilities they themselves shirked, and so, in spite of their misgivings, they significantly extended local democracy after 1870. Up to that point the chief elective authorities were the municipal councils created in some of the large towns, excluding London, in 1835, and the boards of guardians. Annual elections were held in November. After 1870 school boards were also elected and added their own education rate. But county government remained in the grip of unelected landowners serving as Lords Lieutenant and Justices of the Peace. Since the radicals made no secret of their wish to sweep away this bastion of privilege, Salisbury's government decided to forestall a Liberal bill by introducing elected county councils in 1888. However, they attempted to limit the damage and scope for radicalism by restricting the electorate and including a bench of nominated aldermen in the new authorities. The next stage came in 1893, when the Liberals created elected urban district, rural district, and parish councils. Again, Conservative fears about the threat to landed power led the House of Lords to impose severe limits on the legal expenditure of the new councils. Finally, the Conservatives were so upset by Liberal control of the London County Council that in 1899 they tried to weaken the LCC by introducing a lower tier of elected borough councils in London.

All this amounted to a major extension of democratic participation in Britain. Although the local electorate was based on the ratepaying heads of households, which excluded some of the men eligible to vote in parliamentary elections, it did include some unmarried and widowed women who, after 1869, composed 10–15 per cent of the local electorate. There was also some change in the composition of local councillors. Although a number of aristocrats took the field and held on as chairmen of the new county councils, many of them withdrew or suffered defeat as the Liberals won majorities in the early elections. Working men also won places on the new councils, and some agricultural labourers came forward to serve on the parish councils. The abolition of the property qualification for guardians in 1894 brought fresh recruits like George Lansbury onto some poor law boards and, with them, more radical ideas. However, since meetings were largely held during working hours many working men experienced difficulty in serving, and often withdrew after one term in office. The male monopoly also came to a decisive end in this period. Apart from the county councils these local authorities were open to female candidates. By 1900 some 270 women sat on the school boards and 1,147 on Poor Law boards as elected members. Taking all elective authorities together, by 1914 1,007 included some women, while 1,414 did not.

Perhaps not surprisingly, the expenditure of local authorities rose significantly from £36 million in 1880 to £108 million by 1905. The two most costly items of expenditure were highways and public lighting and poor relief, followed by education, the police, public utilities, water, sewage and refuge collection, hospitals, and asylums. This was financed by three means – rates, loans, and grants from the national Exchequer. Although the two latter sources clearly rose absolutely, it is noticeable that rates increased at least as fast, so that by the turn of the century they formed as high a proportion as in 1875 – around 40 per cent. This is why aggrieved local ratepayers looked askance at new legislation that added to the burdens of local authorities, and demanded higher subsidies. Conservatives, who were sensitive to the complaints of the gentry and landowners now suffering from falling agricultural prices, favoured the relief of local taxation by grants-in-aid from national revenue. For example, between 1887 and 1892 they doubled local authority support from £4 million to £8 million; and in the 1890s they granted £1.5 million in relief of agricultural land. Liberal governments were much more reluctant to raise grants-in-aid, however. For them the remedy lay in reforming local finance by widening the sources of direct income. As things stood the burden fell heavily on tenants, while the owners of land and premises, who were often much wealthier people, largely escaped taxation. Many of the radicals adopted the idea of site-value rating as a means of tapping the huge increase in the value of urban land; they felt that landowners made large and unjustified profits from sales of land to municipalities or for commercial development when they had done little or nothing to bring about the rise in value. However, any reform of local finance would have been both complicated and controversial, and so nothing was done; Britain entered the new century with an archaic rating system which was propped up by increasing resort to Exchequer grants.

Not surprisingly, many municipalities resorted to loans from the late 1870s onwards. Birmingham led the way by taking on loans of £1 million in both 1877 and 1878. From a total of £137 million in 1880, local authority loans rose to £448 million in 1903. On the whole central government found this an alarming trend, in spite of the beneficial effects of slum clearance and town improvement schemes. The fact was that municipal policy ran counter to the spirit of Gladstonian finance, which placed the emphasis on reducing the National Debt. Between 1875 and 1900, when the National Debt was being cut, municipal indebtedness doubled.

But the local authorities found themselves trapped between contradictory pressures. New legislation invariably involved additional expenditure; yet the rates fell heavily on persons of modest means like clerks, artisans, and small shopkeepers. A further complication was the Local Government Board, created in 1871. It sometimes exercised its power to refuse to allow local authorities to raise loans for new projects, or imposed repayment terms that made it too expensive. Conservative governments especially disapproved of the larger school boards because they extended their role by providing secondary and adult education. Eventually this was checked by a challenge in the courts – the 'Cockerton' Judgement of 1899. In the 1902 Education Act the Conservatives abolished the school boards and gave their responsibilities to the county councils. It was also open to the Local Government Board to institute official investigations into Poor-Law boards suspected of exceeding their remit, as was done in the case of Poplar in 1906.

Above all, since local government was generally regarded as the most appropriate level at which to handle poverty, it became unavoidably involved in a number of experiments and innovations stimulated by the social investigations of the 1880s and 1890s. Some bodies, like the parish councils which had been expected to promote allotments and smallholdings, simply lacked the resources to do very much. Some school boards initiated schemes to provide meals for needy children. The better-endowed boards of guardians built hospitals, as they had done in the past. But they increasingly resorted to outdoor relief for deserving elderly people, in which they were encouraged by the Report of the Royal Commission on the Aged Poor (1893–5) and by the Local Government Board. This was tantamount to providing an old-age pension. In Bradford the guardians even built houses for the elderly. Some socialist guardians like George Lansbury in Poplar wanted to tackle unemployment by setting up farms to which the surplus urban workforce could be transferred, though this was severely inhibited by lack of resources. In 1886 Joseph Chamberlain, then President of the Local Government Board, urged the municipal councils to organize public-works schemes to relieve unemployment. Although widely attempted throughout the 1890s, this strategy usually provided only unskilled work on road-repairing and snow-clearing which was unattractive to artisans. In any case, since unemployment was concentrated in the poorest boroughs, it was never possible to mobilize enough resources to make a major impact. One important advance was the establishment of a direct labour department by the LCC. As the council undertook to pay the wage rates and to observe the conditions of employment negotiated by the local unions, this met with working-class approval, though the competition attracted criticism from private employers.

The Role of Women in Social Politics

As consumers and housewives, women had a particularly close interest in social policy. They experienced at first hand the fluctuations in prices and housing standards; they often took responsibility for self-help strategies in the family; they had a natural interest in promoting temperance; and they made up a majority of the recipients of poor relief. Although officially treated by the Poor Law as dependants, women were frequently instrumental in keeping their families out of poverty by their additional earnings. Their habit of sacrificing their own diet in order to feed husbands and children, and the debilitating effects of repeated pregnancies, undermined the health of many working-class women; yet they were the least likely to enjoy any form of health insurance or professional medical treatment.

Victorian women were highly vulnerable to poverty because they were so frequently widowed or deserted and left with young, dependent children. From 1846 the guardians could grant outdoor relief to widows, but some chose not to do so because they believed the children would be better off brought up in the workhouse. The Poor-Law authorities were usually reluctant to support a wife if they believed her husband was still alive, though not maintaining his family, because the expectation of relief would diminish fathers' sense of responsibility. If her husband refused to enter the workhouse a woman might be refused help, even if she were destitute; and conversely, while a woman's husband remained inside the workhouse she, too, would be confined. From 1876 wives enjoyed a legal claim to maintenance if legally separated from their husbands, but actually obtaining it was not easy. From the 1870s the local guardians were officially advised to treat single women as if they were able-bodied men, which meant no relief, on the grounds that they had a duty to take work. Between 1871 and 1892 the number of women receiving outdoor relief fell from 166,000 to 53,000. This, however, was not necessarily the result of central direction, rather of local pressure over rising rates. Clearly the treatment meted out to women varied, as it did for men, according to the attitudes and finances of each board. But mothers who did receive support were likely to get 2–3s. weekly plus 1s.–1s. 6d. for each child. In spite of the falling cost of living, this obviously fell far below the poverty line as calculated by Rowntree.

During the late nineteenth century women also began to play a new managerial role in the field of poverty. This grew from their traditional activities in charitable organizations, which led naturally to participation in the extended local government system after 1870. By the 1850s women had already gained an unofficial footing in many Poor-Law unions as workhouse visitors, a role promoted by Louisa Twining's Workhouse Visiting Society. From 1869 middle-class ladies involved themselves in social work under the aegis of the Charity Organization Society. Perhaps the best known of these was Octavia Hill, who carved out a role as the manager of private tenement housing in London; she undertook to collect rents regularly so as to ensure a fair return on capital, but also to repair and improve the properties. As more women attended higher-education courses, they were drawn into the activities of university settlement houses in London and other conurbations, where they endeavoured to improve the moral and educational lives of girls and women. For the more academic like the young Beatrice Webb, there was ample opportunity for scientific investigation of poverty under the auspices of the Fabian Society. Those who relished a more public role participated as 'Hallelula Lasses' in General William Booth's Salvation Army.

However, the major step towards a formal public role for women came with the grant of a local-government vote in 1869. Women proved to be especially successful in winning election to the school boards and Poor-Law boards – perhaps because these were regarded as a natural extension of women's domestic functions – but they also sat on municipal councils, rural district, urban district, and parish councils. Only from the county councils were they effectively barred until 1907. Female representatives made a distinctive contribution because as women they were aware of and able to look into matters of which the men chose to remain ignorant. They encouraged the fostering of children as an alternative to long-term institutionalization. Annie Besant, an early recruit to the London School Board, scored a triumph by persuading her colleagues to initiate the provision of school meals. Although they often stood as independents, many of these women were Liberals and Socialists, and as such they prepared the ground for the adoption of the revisionist thinking in social policy of both New Liberalism and Fabianism at national level after 1906.

Nor was the female contribution exclusively a middle-class one. The formation of the Women's Co-operative Guild under Margaret Llewellyn Davies in 1883 brought the working-class wife's view to bear on social policy. In particular the WCG advocated maternity benefits, home helps, the 'endowment of motherhood' (family allowances), and school meals. Such issues bgan to attract more interest in political circles as a result of growing concern about the decline in the birth rate and the rising infant mortality rates in the 1890s. Although it was tempting simply to blame mothers for neglect and ignorance about child-rearing, a more constructive approach also manifested itself. Some local authorities began to supply hygienic milk and employ health visitors. Civil servants and politicians now became anxious to raise the status of motherhood and to ease the mother's burden; they looked with more sympathy upon the prescriptions of the WCG than the demands of middle-class feminists, and the introduction of the maternity benefit in 1911 and local maternity clinics in 1914 were among the first fruits of WCG pressure.

Socialist Revival and New Liberal Revisionism

Ultimately, all interventionist ideas in the field of poverty and welfare ran up against prevailing assumptions about the operation of the economy. Very broadly, Victorian thinking emphasized the virtues of the free market; competition between individuals tended to maximize production, profit, and employment. The system operated 'as though by an invisible hand' to promote the interests of society as a whole. It followed that a wise government adhered to a policy of *laissez-faire*, for the more that was absorbed by way of taxation the less was available for productive investment in the economy. If the effects of this policy upon those who became the victims of competition seemed unfortunate, the exponents of classical economics drew upon science for their rationale. Herbert Spencer in his *Man Versus the State* (1884) adopted the fashionable Darwinian view that, if the state diverted its resources to protecting the weak, the result would be to drag down the whole of society and check progress.

In practice, the bracing notions of classical economics were less respected than some of the rhetoric of Victorian England suggests. The whole era was characterized by a series of interventionist initiatives by governments; and some argued that around 1870 the tendency to infringe rather than reflect *laissez-faire* grew decisively. Certainly Victorians had become aware of the by-products and shortcomings of free enterprise. The most conspicuous deviation from market forces took the form of 'municipal socialism'. This was largely the work of local authorities in London and provincial cities which adopted a collectivist approach to street lighting and paving, public parks and libraries. But they also ventured into municipal enterprise by buying out privately owned water, gas, electricity, and tram companies.

The motive behind such policies was not necessarily ideological. Water companies, for example, were often felt to be charging a high price for a poor-quality service. It could be argued on moral-political grounds that every person was entitled to a supply of clean water. But it also proved to be more efficient and economical to run these services under municipal control, for, while the initial costs might be high, councils often made a profit which helped to subsidize the rates. Moreover, experience of municipalization strengthened the view that other industries, notably the railways, were really utilities that should properly be subject to some form of regulation and control in the public interest. This is not to say that all provincial councillors were socialists, only that many of the steps towards socialism were, as the Fabians perceived, comparatively small and non-ideological ones.

By the 1880s, therefore, when increasing numbers of middle-class intellectuals began to join the new socialist societies, the notions of Herbert Spencer already looked a little out-of-joint with the times. It was apparent that an excessive reliance on voluntarism or the 'hidden hand' was handicapping Britain in some ways. This was underlined by the awareness of the rapid economic progress being made by Germany. Under Bismarck, Germany conspicuously developed a system of state social welfare in the 1880s, not to mention a superior educational system. The Royal Commission on Technical Instruction commented in 1884: 'our industrial empire is vigorously attacked all over the world. We find that our most favourable assailants are the best educated people.' This emphasis on promoting what came to be called 'national efficiency' was typical of late Victorian socialism, especially as expounded by the Fabians. Their socialism was by no means primarily an expression of humanitarian concern for the poor, nor a vindictive desire to expropriate the wealthy. Its chief thrust was that unfettered private enterprise was a wasteful and inefficient use of resources; guided by short-term profits it resulted in excessive investment in some areas and the neglect of others.

By the 1890s a good deal of collaboration had grown up between middle-class socialists and constructive Liberals, especially under the aegis of the progressive majority on the LCC. But there was also a sustained effort to revise traditional thinking across the whole range of social and economic policy in organizations like the Rainbow Circle (founded in 1893) which involved socialists like Ramsay MacDonald and New Liberals such as Herbert Samuel, Charles Trevelyan, and J. M. Robertson. The common starting-point was that British capitalism had evidently failed to reduce poverty. Liberals were more encumbered by existing, negative Gladstonian ideas; but they argued that state interventionism was consistent with Liberalism where it was designed to promote the greater freedom of the individual. It was insufficient to bestow political, legal, or religious rights on the individual if he remained trapped by his material conditions of life.

From this the authors of the New Liberalism – J. A. Hobson, L. T. Hobhouse, Charles Masterman – urged that the state had a duty to promote the social-economic side of the individual's development. In education and temperance this was already widely accepted. But extending this to social welfare meant a new approach to taxation.

The New Liberals proposed that government should move away from its reliance on indirect taxation towards the direct taxation of income and wealth. It had long been considered by radicals that sources of wealth like land ought to be more effectively taxed, especially as the income derived from it was often *unearned* and thus could properly be used for the benefit of the community. Increasingly they argued that taxes paid should reflect the ability to pay; this was reflected in Sir William Harcourt's death duties of 1894, which were based on a graduated scale which began at 1 per cent on property worth £500 and rose to 8 per cent on property worth £1 million. By extracting a fair contribution in this way, the government would not damage productive investment and would leave private ownership basically intact. Some radicals like Hobson went a stage further by suggesting that the underlying weakness in the economy lay not in investment but in underconsumption, that is, insufficient demand for the goods produced. If extra government revenue were redistributed via social policies to relatively poor people, this would have the effect of increasing the consumption of goods and thus stimulating the economy.

During the 1890s these views were far from being typical of provincial Liberalism. But as Gladstonianism lost impetus and the older generation of radicals were defeated and dropped out, those with more advanced views and different assumptions about the scope of political action gradually rose up through the party. Ultimately the force of circumstances gave more credibility to the novel ideas about poverty and the state's responsibility. Back in the 1850s there had been some grounds for confidence that the continued spread of factories and workshops would generate enough employment for all those willing and able to work. By the 1890s this was evidently not so. With the fading of mid-Victorian optimism, the findings of Booth and Rowntree could not be overlooked as easily as those of Mayhew. Gloomy prognostications about Britain's position in the international economy undermined any claims that poverty would diminish in the future. The national interest clearly provided an added force to the collectivist rationale by 1900. As one Liberal, T. J. Macnamara, observed in connection with social welfare reforms for children: 'All this sounds terribly like rank Socialism. I'm afraid it is; but I am not in the least dismayed, because I know it to be first rate imperialism.'

None the less, in spite of the shocking revelations about poverty and the revisionist views about its causes, it cannot be said that any radical change in the methods of handling the problem had occurred by the turn of the century. Why was this? The immediate explanation lies in local government, which was the scene of most attempts to grapple with the problem. In spite of the experiments, the new elective authorities, and the changes of personnel, reform was handicapped by the constraints of local finance; the burden of rates fell too narrowly. The obvious solution lay in a complete overhaul of local-government finance, but with the Liberals largely out of power this was repeatedly postponed. The chief positive conclusion to emerge was that advanced especially by the Fabians – that the functions of the Poor Law should be separated and redistributed to county councils; and that the intractable problems of unemployment should become a *national* responsibility under a separate ministry, which would provide either employment or relief. Other schemes such as pensions also waited upon a national government initiative, because it was clear that, to be of real value, a pensions scheme would have to be non-contributory; otherwise those most in need, like women, would fail to benefit.

The second part of the explanation lies at the parliamentary level. Though increasingly aware of social problems, politicians were slow to adjust their priorities. Any idea that a working-class electorate would necessitate a radical modification of the agenda proved baseless, as we have seen. An alternative argument suggests that the politicians were moved not by votes but by the threat of physical force, as manifested in 1886 and 1887 by the demonstrations and marches of unemployed workers. However, these were not in fact signs of an anti-parliamentary movement, nor were they sustained. In Westminster and Whitehall priorities proved slow to change. Those departments most closely involved with social questions – the Local Government Board, the Board of Trade, and the Board of Education – enjoyed a low status in the Cabinet hierarchy. As late as 1905 the young Winston Churchill disdainfully rejected the LGB: 'I decline to be shut up in a soup kitchen with Mrs Sidney Webb!'

Under the Conservatives, therefore, legislation in the later years of the century was limited to such matters as grants for allotments (1887), smallholdings (1892), and free elementary education (1891) – a concession to Chamberlain. The Liberals' brief 1892–5 ministry showed signs of radical thinking with the death duties scheme and the grant of compulsory purchase powers to local authorities; but several measures, including an employers' liability for accidents bill, were emasculated by the House of Lords. A number of official inquiries marked the progress of ideas. In 1885 a Royal Commission on Housing made new proposals, but Lord Salisbury refused to move beyond purely permissive legislation for local councils. The Royal Commission on the Aged Poor (1893–5) and the Parliamentary Select Committee on the same subject (1899) strengthened the case for old-age pensions. But, again, Salisbury's governments stuck to the traditional nostrums of classical liberalism: the budget should be balanced, the national debt reduced, revenue confined to indirect taxes needed to maintain basic services like the armed forces, civil service, and police.

The final part of the explanation for the pattern of change lies in personnel and party fortunes. The thinking of Booth, Rowntree, the Webbs, and others affected only an interested minority of middle-class politicians, intellectuals, and social investigators. They were in a position to draw up schemes and to proffer policy options to the politicians, but lacked the power to implement them. But by the Edwardian period young reformers like Charles Masterman, Herbert Samuel, and William Beveridge had climbed up the ladder to become MPs, ministers, and civil servants. The huge change in the composition of Parliament brought about by the 1906 election accelerated this process. It opened the way to a sustained extension of the role of the state between 1906 and 1914.

4

Victorian Values: Myth and Reality

In an interview in 1983 the then Prime Minister, Mrs Margaret Thatcher, declared:

> I was brought up by a Victorian grandmother. You were taught to work jolly hard, you were taught to improve yourself, you were taught self-reliance, you were taught to live within your income, you were taught that cleanliness was next to godliness . . . All of these things are Victorian values.

All politicians plunder the past in the interests of current party propaganda. But the practice should not be confused with the writing of history, which involves careful separation of the empirical wheat from the polemical chaff. The idea of Victorian values as resurrected by Mrs Thatcher is riddled with flaws. Quite the oddest aspect is the rather dubious political connections implied by her use of the term. To the extent that these Victorian values were real they were associated with pious mid-Victorian Nonconformist Liberalism. Victorian Conservatism, on the other hand, drew much of its strength and character from quite different values; it was less pious and more populist, less concerned about improvement than about the pursuit of pleasure in the form of sports, gambling, and a man's right to his drink. More importantly, as evidence accumulates it becomes clearer that the virtues later associated with Victorian society arose to a considerable extent from *prescriptive* literature; they were not an indication of actual behaviour. In particular they represent the aims and notions of some middle-class men, conscious of both a frivolous upper class and a dissolute working class. The unregenerate sections of society shared obvious common interests; the slowness of the political elite to tackle the problem of drunkenness was not unconnected with its own fondness for alcohol. In spite of this, however, it must be conceded that in some ways the trend was towards approved middle-class forms of behaviour; in this chapter we will examine the ceaseless struggle between 'virtue' and 'vice' in the late Victorian period.

The Growth of the State

Victorian Britain resounded to the triumphs of improvement through individual endeavour. Samuel Smiles, author of the best-selling *Self-Help*

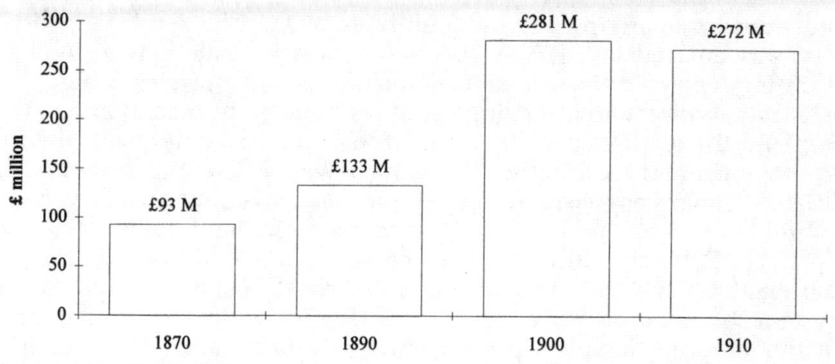

Figure 4.1 Total Government Expenditure (£ million, at current prices)

(1859), and science, in the person of Charles Darwin, underlined the message that the progress of society – and its decline – rested upon the vigour of its individual members. As Chancellor during the 1860s, Gladstone had declared himself in favour of leaving money to fructify in the pockets of the people, and in 1894 when he finally left politics he claimed: 'of one thing I am, and always have been, convinced – it is not by the state that man can be regenerated, and the terrible woes of this darkened world effectually dealt with.'

In spite of all this, however, Victorians increasingly found themselves driven to accept that their society and economy were too complex to be left to run themselves; one of their key achievements was to chart a route that led out of the chaos of competitive individualism towards the sophisticated modern state. A crude measurement of the expanding state is government expenditure. However, for a truer perspective expenditure must be seen in the context of the total output of the British economy. The growth of production, population, and consumption enabled governments to raise more revenue without taking a higher proportion at least up to the 1880s. The twenty years from 1890 to 1910 marked the beginning of a long-term trend towards a bigger state. It will be noted that this did not *precede* the start of Britain's relative industrial decline, and cannot be seen as a cause.

Some indication of the development of government responsibilities may be detected in the establishment of new ministries, including the Local Government Board (1871), the Scottish Office (1885), the Board of Agriculture (1889), and the Board of Education (1899). Expenditure on education from central funds rose from £0.75 million in 1870 to £7 million in 1895. However, the bureaucracy remained surprisingly small; for example, in 1876 the Home Office consisted of a mere 36 permanent officials. In all, the number of civil servants increased from 42,000 in the 1850s to 50,000 in 1881 and 116,000 in 1901. The really major growth in state employment occurred in the Post Office, and in local government in such occupations as teaching. Throughout the mid and late Victorian period, a wide range of interventionist legislation required the employment of additional clerks, accountants, administrators, inspectors, and experts. Moreover, all this interventionism infringed the rights and freedom of individual property-owners and entrepreneurs. The ten-hour working day was enacted in 1847. A succession of acts dealing with food and licensing regulated the sale and production of consumables. Compulsory education brought an unending list of restrictions. In Ireland and parts of Scotland the legislation of 1881 and 1886 deprived landlords of the right to set rents; and local authorities won powers of compulsory purchase. In the 'Congested Districts' of Ireland the government intervened to subsidize a range of economic functions from light railways to seed potatoes. Public Health Acts were passed in 1848, 1866, and 1875, and during the 1860s the state financed the construction of a vast sewage system for London. As early as the 1840s it became clear that in the matter of water supplies private enterprise had failed; as a result, by the 1870s nearly half the local authorities supplied their own water, and by 1900 two out of three water authorities were publicly owned. Cleanliness may have been next to godliness, but it stood rather close to municipal socialism too. Provision of gas followed a similar pattern; and by 1900 two-thirds of the electricity companies were also under municipal ownership. The railways, which had always been regarded as a kind of public utility, found themselves regulated in terms of safety and price controls, and in 1870 the government empowered itself to run the entire system, which it did in 1914. During the late nineteenth century, governments became increasingly concerned about industrial relations; this resulted in the 1896 Conciliation Act, which empowered the Board of Trade to inquire into the causes of strikes and appoint conciliators at the request of the parties involved. This proved to be a modest but important step towards regular government intervention after 1905, usually designed to persuade recalcitrant employers to recognize trade unions.

Of course, this catalogue of interventionist and collectivist measures does not indicate the demise of *laissez-faire* as a general principle; Britain continued to be basically a market economy. But the infringements of *laissez-faire* were so common that one can scarcely depict the late Victorian economy in terms of the pieties of early Gladstonianism.

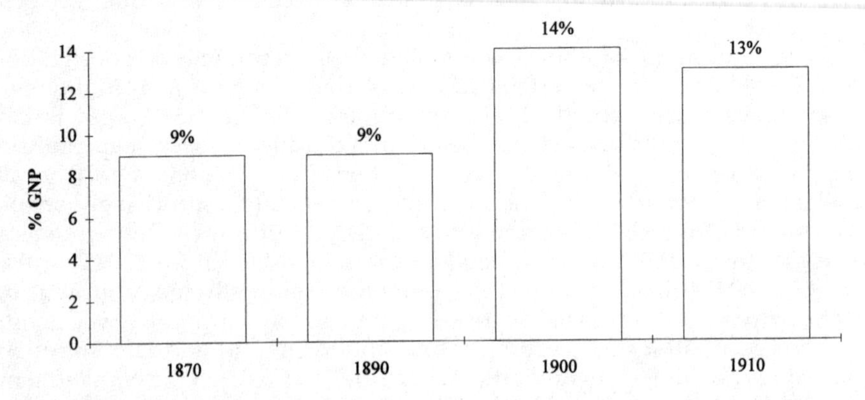

Figure 4.2 Government (Central and Local) Spending as a Percentage of Gross National Product

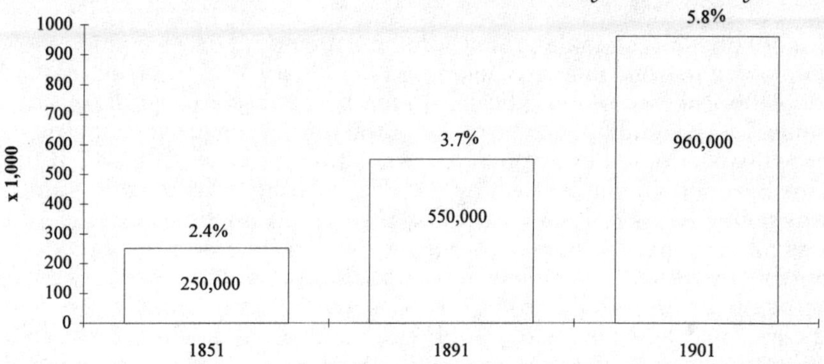

Figure 4.3 Employment in the Civil Service, Local Government, and the Armed Forces (as a percentage of the working population)

The Angel in the House

Nowhere are the perils of the prescriptive literature more apparent than in connection with the role of women. Work was evidently considered a great virtue, but not for women. Coventry Patmore summed up the ideology of the separate spheres in a famous poem entitled 'The Angel in the House'. The ideal woman was not only married and a mother, but was confined to the domestic sphere where she could be decorative, a comfort and support to her husband, and a moral influence on the children; this would be all the better accomplished if she remained free from the distractions of the rough world of employment and public affairs.

All of this amounted to an absurd travesty of the life experience of most Victorian women. For one thing, women constituted nearly 32 per cent of the total British labour force in 1871, though this figure fell slightly to 29 per cent by 1901. Put another way, 55 per cent of single women worked outside the home, as did 30 per cent of widows and 10 per cent of married women. The geographical pattern of employment varied according to the local economy. For example, women made up 60 per cent of the labour force in the Lancashire cotton textile industry; by contrast, regions like the north-east, dominated by coal, iron, and shipbuilding, offered far fewer opportunities for women.

It must be admitted that from the mid-nineteenth century onwards there was a tendency for employment, especially among married women, to diminish. This may reflect the gradual shift of male workers into higher or more regularly paid jobs, which reduced the pressure on wives to work outside the home. Certainly the oral evidence has shown that many women preferred to stay at home provided that their husbands earned enough to support the family. A degree of coercion was also at work. For trade unions increasingly adopted the 'family wage' as their target, and to that end they attempted to exclude women from industrial employment on the grounds that their readiness to accept low wages depressed the general level of pay. To some extent Parliament co-operated in this by legislating to restrict working hours or to bar women and children from certain types of work, ostensibly for humanitarian reasons.

Notwithstanding male pretensions, most working-class women took up employment from their mid-teens until the early years of marriage; but thereafter they were periodically driven by necessity to work outside the home. Moreover, beyond the range of the official census lay a multitude of expedients adopted by women to earn extra money while based at home. This included taking in washing, accommodating lodgers, child-minding, and selling items of food and drink; this latter sometimes developed into a small shop in a woman's own front room. While no firm figure can be put on this activity, it is clear that countless families were saved from hardship by the vital income of housewives. Even among the middle classes, many wives and daughters maintained a charade as purely decorative and leisured ladies. On a lower-middle-class income it was impossible to employ more than one servant to take over the heaviest work of carrying coal, cleaning fires, and washing; but this left a good deal of time to be spent on cooking, cleaning, and needlework by the ladies of the house.

Far and away the most important paid occupation for women was domestic service, which employed 1.5 million by the 1870s: after this came textiles, with around 700,000, and the clothing trades, with nearly as many. The food and drink industries offered a growing number of jobs to women, while in agriculture their role diminished sharply. They benefited most from the expansion of white-collar employment, as shop assistants, typists, civil servants in the Post Office, and elementary schoolteachers. From the 1870s these were all growing sectors. Women teachers increased from 80,000 in 1862 to 171,000 by 1901, and women clerks from 279 in 1861 to 57,000 by 1901. Taking female teachers, nurses, clerks, civil servants, and shop assistants together, the total rose from 184,000 to 562,000. None of these offered high-paid employment, but most represented an improvement in status and conditions for young women whose mothers had been restricted to agriculture, the 'sweated trades', and domestic service. In most cases the key to entry lay in elementary education, which was imposed equally upon both sexes. Women profited more than men from this, for in terms of literacy they had traditionally lagged behind: by 1900 both sexes enjoyed a literacy rate of around 95 per cent.

Women's presence in the supposedly male sphere of employment was not the only sign of the irrelevance of conventional ideas about their role. During the last quarter of the century upper and middle-class women also invaded the public sphere. One route lay through the multitude of charitable and philanthropic activities. Their significance should not be disparaged, for it turned women into organizers, trained in the techniques of conducting meetings, taking minutes, raising funds, keeping accounts, and learning to speak in public. The Women's Co-operative Guild also coached working-class wives in these skills.

From this it was not a very great step to politics. Many Victorians regarded the act of voting as quite outside a lady's sphere, partly because of the traditional association between drink, violence, and elections. However, the very presence of ladies as municipal voters helped to introduce some decorum, and legal restrictions on expenditure undermined the rougher traditions of the poll. In addition, large numbers of women joined party-political organizations from the 1880s onwards, notably the Primrose League and the Women's Liberal Federation. In this capacity the women became more than loyal followers and organizers of social events, though they were both of these. They also canvassed the male electors, conveyed them to the poll, and, perhaps most startling of all, became skilled platform speakers. The assumption that women would not be tough enough and that their voices would never carry on the public platform soon collapsed as the ladies demonstrated that they could often hold an audience rather better than their dull-witted, jaded husbands. By the early 1900s politicians of all kinds freely conceded that without women it would be difficult to run election campaigns. Indeed, a characteristic of this period is the emergence of some very public and equal partnerships including Annie Besant and Charles Bradlaugh, Sidney and Beatrice Webb, Henry and Millicent Fawcett, Ramsay and Margaret MacDonald, John and Catherine Bruce Glasier, and Philip and Ethel Snowden. The male bastion had been decisively breached.

Marriage and the Family

Contemporaries generally held marriage and motherhood to be the overriding object of a woman's life. Yet this ideal remained unattainable because of the imbalance of the sexes. Higher mortality among baby boys, earlier death among adult males, and the emigration of marriageable young men left women in a majority, and many doomed to spinsterhood. The problem was exacerbated, especially in the middle classes, by the preference for marrying late, in one's twenties or thirties. As a result, by 1871, out of every thousand women aged 20–25 years 652 were unmarried, and of those aged 25–35 294 were unmarried. Marriage, quite simply, was a markedly *less* common experience than it was to become in the twentieth century. Moreover, it appears to have been losing ground during the late nineteenth century; by 1901 single women accounted for 726 per thousand in the 20–25 years age group.

Mid-Victorian feminists felt that society had only itself to blame for the reluctance of girls to marry. In law the position of the married woman was abysmal. The male marriage vow – 'with all my worldly goods I thee endow' – was a bad joke, for the wife's property and income became her husband's; even after a separation he could legally claim any money earned by her. Divorce was both difficult and costly even for men, and impossible for women. The separated wife usually lost access to her children. To cap it all, the unmarried female municipal elector lost her vote when she married! From the 1850s onwards some middle-class women, led by Barbara Leigh Smith, who were known as the 'Ladies of Langham Place', inaugurated a series of pressure groups to seek redress from Parliament. By the 1870s such women as Emily Davies, Lydia Becker, Elizabeth Anderson, and Millicent Fawcett were actively pressing for equality for married women, access to professional occupations, improved secondary and higher education, abolition of the double standard in sexual conduct, and the parliamentary vote. Since they placed a good deal of emphasis on the need for single women, who were often trapped as low-paid governesses, to be able to support themselves, the anti-feminists argued that the real purpose was to attract girls away from marriage and motherhood. There is some impressionistic evidence from the autobiographies of able and educated Edwardian women that as teenagers they had seen life in terms of a *choice* between domesticity, which carried low status, and the much more appealing goal of a career and independence. In fact, a number of reforms were achieved by the Victorian campaigners. In 1857 divorce became a much simpler and somewhat less expensive business, though the law continued to discriminate: whereas a man might obtain a divorce for adultery alone, a wife had to prove adultery plus cruelty or desertion. The 1886 Guardianship of Infants Act gave women some rights over their children, subject to the view of the courts. And the Married Women's Property Acts of 1870 and 1882 secured for a wife both the property held before marriage and any subsequent income.

On the other hand, if a few more women opted out, the great majority actively sought marriage, though this may have been because the life of a single woman was so difficult. At all events, the feminist campaigners represented very small middle-class pressure groups. This reflects the fact that in practice the legal inequalities had little bearing on the lives of married women. Among the upper classes it was customary to use the law of equity to settle money on a daughter which her husband could not touch. Moreover, such women enjoyed the wealth to pursue a range of personal, social, and public interests and activities. They had separate bedrooms from their husbands (and relatively small families) and often enjoyed extramarital affairs or platonic relations with other men. It was a fairly liberated lifestyle. At the other end of the spectrum the law was largely irrelevant to working-class wives. Amongst the respectable working class, marriage, a settled home, and a husband who brought home a regular wage were cherished ambitions. For the less respectable, wives frequently found themselves deserted or not supported; but they themselves often abandoned husbands or faced life as widows. In each case questions of custody or property rarely arose, and there was no formal divorce or separation. Moreover, it has emerged that the wife was in many ways the dominant figure in the working-class family; she managed the money and made the decisions while her husband remained rather a marginal figure. It was, thus, in some middle-class households that the wife felt deprived of a satisfying and influential role, rather than among married couples in general.

Finally, the emphasis on home and family life which is so characteristic of Victorian literature gives a misleading impression in many respects. The physical house was much less secure and stable than might be supposed. For example, home-ownership was by no means a Victorian practice even amongst quite well-off people. Around 90 per cent of homes were rented. The wealthier constantly moved house, while in poor urban districts 20-30 per cent of families moved every year because they had fallen behind with the rent. Nor has the notion of the big, happy Victorian family of grandparents, parents, and children survived the scrutiny of modern researchers. Most families comprised just two generations. Moreover, male mortality and desertion resulted in a very high proportion of single-parent families led by women. The working-class home was often a crowded, noisy, and uncomfortable place from which men escaped to the pub. Parents frequently farmed out their children to relations or neighbours, and they dispatched girls of thirteen years to distant towns to work as domestic servants. Of course the middle classes took a dim view of working-class family life. On entry to the workhouse, husbands and wives and parents and children were split up. At the slightest pretext the children of the poor were removed from their families in favour of institutional care, control by employers, or export to the colonies. Higher up the social scale Victorians placed a great value on family life – but only in a tightly controlled and sanitized form. Once childbirth was over, children could be entrusted to nurses, nannies, and governesses. The parents might then enjoy their offspring for half an hour a day between tea and bedtime when children were presented in a clean condition for inspection. They did not invade their parents' space or inhibit their social lives. In due course they were packed off to boarding schools, to endure all manner of ill-treatment and under-nourishment, and were only fully received back into the family when old enough to be useful; daughters in particular were often kept at home to be unpaid secretaries and housekeepers to their mothers.

Population, Sex, and Birth Control

From only 10.5 million people in 1801 the population of Britain grew to 20.8 million by 1851 and to 37 million by 1901. This was one of the highest growth rates in Europe, and it naturally resulted in a distinctly young population; in 1871, for example, one person in every two was under twenty-one. Up to the late 1870s the birth rate remained consistently high at 32–5 per 1,000 population. Death rates changed little in spite of developments in medical science and public health; indeed, by the end of the century life expectancy, at 44 years for men and 48 for women, was only slightly better than that in the early Victorian period.

However, by the beginning of the 1880s population growth had begun to slow down because of a fall in the birth rate which proved to be the start of a long-term trend. From 35 per 1,000 in the 1870s, the birth rate dropped to 29 in the late 1890s. What caused this change is a matter of some argument. At least part of the explanation lies in the tendency to marry later in life. But some couples had deliberately chosen to limit the size of their families. The census suggests that they were concentrated in certain middle-class occupations, and it has been argued that they were reacting to the growing financial burden of children in terms of the wages of nannies, governesses, and domestics, and the fees for private education. This, however, seems more plausible for groups like clerks who aspired to middle-class status on inadequate salaries; for them, family limitation was an obvious means of maintaining their standard of living. The methods adopted at this time are also a matter of speculation. Traditional expedients included abstinence from sexual relations – comparatively easy for partners who enjoyed separate bedrooms – withdrawal, and prolongation of breastfeeding. As yet the more sophisticated methods made little impact. In the 1870s rubber sheaths for men became available, and women used a variety of pessaries, sponges, and douches; but these were costly and not easy to use. In addition the sheath, which was used chiefly to protect men from venereal disease, was damned by association with extramarital sex.

It should not be forgotten that women themselves had the most compelling motive for adopting birth control. In all social classes many women died in childbirth and thousands had their health ruined by repeated pregnancies. The existence of a demand for birth control is underlined by the advertisements in women's magazines and the variety of substances recommended as abortifacients. Birth control also received huge publicity as a result of the trial in 1877 of Annie Besant and Charles Bradlaugh, who were prosecuted for obscenity when they published an old pamphlet of advice on birth control. However, in view of the couple's own unmarried status and their association with such causes as atheism and republicanism, the publicity may have done as much to hinder as to promote birth control amongst respectable couples.

Indeed, at this time many middle-class women, including feminists, regarded artificial methods of avoiding pregnancy as inherently immoral, for their use was calculated to encourage men to indulge their sexual appetites without fear of the consequences. In this they reflected a widely expressed belief that physical desire was chiefly a male characteristic; ladies were not thought to enjoy sex, or if they did this was a sign that they had been dragged down to men's level. It is extremely difficult for the twentieth century, which has such different assumptions, to know what to make of this. Of course, much of the unfavourable comment made by Victorian women on the subject of sex was made in the context of concern about the consequences. During the Edwardian period, evidence accumulated from wives who badly wanted to be able to enjoy physical relations with their husbands but who were inhibited by the fear of pregnancy and poor health. The limited studies we have of the married life of upper- and upper-middle-class couples suggests that they generally married for love, tempered by suitability, and that their relations were sensual and emotional, not cold and distant. Even Queen Victoria is nowadays recognized as having had a very healthy interest in the opposite sex, in spite of her complaints about the burdens of motherhood.

For the majority of people family limitation was not yet on the agenda. It ran up against the prevailing assumption that no man should be denied his conjugal rights. Even in cases where a husband had contracted venereal disease, doctors usually refrained from advising the wife about the need to avoid sex. Edwardian feminists were to use this peril as a major argument against marriage. One victory was won in 1891 when a Mrs Jackson, who had been confined by a husband with whom she did not wish to live, was set free by the Court of Appeal. Public opinion, however, seemed to show itself sympathetic to Mr Jackson.

However much evangelicals urged the merits of self-control and abstinence, most Victorians seem to have accepted the absolute necessity of sexual outlets for men. Gladstone claimed that, of the eleven prime ministers he had known, seven were adulterous. The postponement of marriage among middle-class men until they were able to support a wife and family led many to resort to prostitutes. The most striking recognition of male sexuality was provided by the Contagious Diseases Acts of the 1860s, which reflected the army's view that the men in the lower ranks could not possibly control themselves, whether married or not. The Acts in effect licensed prostitutes in and around army barracks, and allowed the authorities to confine to hospital those women found to have venereal disease. This antagonized feminists, the Churches, and many respectable working-class families; and in any case the Acts were not effective in limiting the disease. A huge and protracted campaign led by Josephine Butler embarrassed politicians and resulted in abolition in 1886. Another success was won by W. T. Stead, who exposed in the *Pall Mall Gazette* the existence of a traffic in young girls; as a result the age of consent, which had been raised from twelve to thirteen in 1871, was put at sixteen in 1885. These, however, were only minor dents in the double standards of sexuality which characterized Victorian Britain.

Alcohol and Other Drugs

By comparison with the twentieth century, Victorian society was given over to excessive drinking. In London in the 1860s there were 17,000 arrests for drunkenness each year – the largest category of offences – and one house in every seventy-seven was a public house or beershop; small country towns had literally dozens of drinking places. It was possible to drink practically twenty-four hours a day. The problem had been exacerbated by government policies designed to free the trade by lowering the duty on spirits and making it easy to obtain a licence under the Beer Act of 1830. Consumption rose through the late 1860s to reach a peak in 1876, fuelled by higher money wages. Men commonly spent the extra money at the pub rather than passing it on to their wives. Thereafter, the check to money wages and the wider availability of other consumables led to a small drop in consumption to 1881 and a fairly stable level of drinking to 1900. Even so, observers like Booth and Rowntree believed that working-class families commonly spent between a sixth and a third of their income on alcohol. To this was added the development of mass gambling on horses and dogs during the 1880s, promoted by the national press and the telegraph system. Expenditure on tobacco also rose sharply from £3.5 million in 1870 to £42 million by 1914.

Not surprisingly, the evils of drink had stimulated the growth of a temperance movement from the early nineteenth century onwards. The leading pressure group was the United Kingdom Alliance, founded in 1853, whose object was to enact a bill to give the 'Local Option' or 'Popular Veto', that is, the power to prohibit sales of alcohol by a referendum of local ratepayers. Many of the less extreme reformers like the Church of England Temperance Society aimed at moral persuasion and pledges by individuals to abstain rather than legislative remedy. Yet although persistent and well-funded, these organizations seem to have failed. Why?

In the first place the upper and middle classes were in no position to influence behaviour by example, and they hesitated to use their political influence to initiate a process that might be difficult to stop. Indeed, the temperance movement was handicapped by the feeling that interfering reformers were trying to deny the working man pleasures which their own class freely enjoyed. Many workers did, of course, take the pledge to abstain, though some lapsed quickly and little progress was made in the long run.

Secondly, drinking was sustained by a variety of social pressures. The public house offered men a welcome escape from the home, a valuable indoor meeting-place for clubs and societies, and was sometimes the place chosen for the payment of weekly wages. In time the building of temperance halls and hotels, the changes in the music hall, and the arrival of the cinema were to undermine this social function.

Third was the question of health. Manual workers considered that the consumption of a pint of beer first thing in the morning and at intervals during the day was necessary in order to keep their strength up. Doctors often advised their patients to take fortified wines and spirits, either as nourishment or to deaden pain. In addition to alcohol, many Victorians relied on drugs for medical-cum-social purposes. This often began in babyhood, when opium in the form of laudanum was administered to keep them quiet. A popular mixture for this purpose was Godfrey's Cordial, a blend of opium, treacle, water, and spices. Opium became so cheap and so freely available that the majority of working-class families are thought to have used it for a wide range of ailments. The middle classes commonly resorted to morphia to relieve pain. Further, from the 1880s cocaine began to be imported by the ton. Regarded as fairly harmless, it became popular among middle-class men and among the many women who enjoyed ill-health. Invalids kept supplies available for routine use when they felt the need of a pick-me-up. 'It works like magic,' said Mrs Humphry Ward, the highly productive novelist.

While this kind of drug addiction was scarcely recognized as a problem, the prevalence of drunkenness was taken seriously by politicians. But they regularly disappointed the temperance fanatics by drawing the line between drunkenness and drinking. Gladstone, for example, was prepared to check adulteration of beer, impose licensed hours, and have premises inspected. But by making drinking a little more respectable he only antagonized the temperance reformers, who saw that removal of the worst excesses would only make total prohibition more difficult. The temperance groups often intervened in by-elections in order to secure the defeat of government candidates, but the huge expansion of the electorate after 1867 made this a futile strategy. Instead they concentrated on influencing the Liberal Party by working within the National Liberal Federation to commit the party to the suppression of licensed premises by means of a local veto bill. Although Gladstone refused to accept this policy, he could not prevent the Conservatives making political capital out of the link between Liberalism and temperance. Not surprisingly, many of the commercial interests in the form of brewers and publicans threw their support behind the Conservatives. Thus, although the UKA greatly increased its support among Liberal MPs it was never possible to pass a local option bill. When Sir William Harcourt made the policy a key plank in the Liberal programme in the 1895 election it proved to be a dead weight, dragging him and his party down to defeat. Under Conservative governments the most that could be achieved was to reduce the number of licensed premises by offering financial compensation for loss of business. But none of this made much impact on the drinking classes before the First World War.

Work or Leisure?

As we have already seen in our discussion of women, the Victorian belief in the virtues of hard work was shot through with inconsistencies. Work must also be seen from the perspective of the various social classes. When writing about the landowning and aristocratic class, historians often dwell upon their political and economic role; but for many of these men the possession of a substantial income from rents, royalties, and investments simply enabled them to enjoy a life of leisure and indulgence. This could take the form of riding to hounds, horse racing, shooting game, cards, cultivating gardens and grounds, or collecting art and furniture, according to taste, pocket, and season. Most of the year could be devoted to a busy social round, beginning with London dinners and receptions from Easter, taking in such highlights as Henley and Ascot, followed by shooting in August and prolonged house parties in various parts of the country. Improvements in travel by sea and by train encouraged the rich to take the waters in the German and Austrian spa towns, holiday on the French Riviera, in Italy, or Greece, and, for the more adventurous, to visit Africa, India, and the United States.

But a life given over to leisure at a time of dwindling agricultural rents resulted in considerable indebtedness in the late nineteenth century. None the less, living within one's means was not the thing for a gentleman, and only the most financially embarrassed were prepared to give up their lifestyle altogether. Traditionally the younger sons of aristocratic and gentry families were denied more than a bare annual allowance; they fell into debt, married into wealth, and took a career in the army, the Church, the law, or politics, though these often proved expensive options. It was taken for granted that at his public school and college a youth would pile up large debts in the hope that his father would eventually settle his gambling losses and the bills of irate tradesmen. Nor did youthful irresponsibility end with money. Life at public schools was characterized by an obsession with sport, fighting, fagging, and brutal punishments, rather than with the development of good working habits or a love of scholarship. Oxford colleges were often essentially good rowing clubs, and students commonly left without a degree.

It is not difficult to see why the more evangelical of the middle classes took such a dim view of the upper class, and why they believed that their own industry, thrift, and self-control were the foundation of British success in the Victorian period. On the other hand, the immensely long hours worked by most middle-class men were made bearable by the supporting army of servants and assistants so freely and cheaply available. The period from the 1870s to 1914 represented a golden age for the professional and business classes; their salaries bought the services of nurses, nannies, governesses, cooks, maids, gardeners, and chauffeurs. Tradesmen were anxious to bring provisions to one's door, and one enjoyed the services provided by railways, the telephone, and postal deliveries three times a day. In addition, large homes were available at modest prices, country houses could be taken for holidays in the Lake District or Scotland, while substantial villas in Italy were for hire for a few pounds.

Faced with growing opportunities to enjoy life, the middle classes often preferred to speak of rational or improving recreation rather than leisure. This might involve activities in the church or philanthropic bodies, reading, singing, the piano, poetry, sketching, needlework, and collecting things. However, there are many signs that after the 1850s the influence of evangelicanism diminished, and that more middle-class people determined to enjoy themselves, if not indulge themselves, in the manner of the rich. The pretext of spending more time with the family helped justify such light-hearted pursuits as charades, summer holidays, and elaborate celebrations at Christmas. But adults increasingly indulged themselves with bridge, tennis, golf, and attendances at the theatre and concert hall to hear Gilbert and Sullivan, Oscar Wilde, or G. B. Shaw. The later Victorian period saw a huge demand for fiction which was satisfied by such prodigious novelists as Mrs Humphry Ward. With her best-selling books *Robert Elsmere* and *Hellbeck of Bannisdale*, Mrs Ward broke the stranglehold of the circulating libraries and opened the way to the rapid publication of cheap editions of new novels for a popular market. The middle classes liked to think that seaside holidays were good for children because of the constructive interests to be pursued on the seashore. But, with the aid of Thomas Cook and his son John, they increasingly ventured further afield, to Paris, the Rhine, Switzerland, Italy, and, by the 1880s, to Egypt, though they continued to patronize Bournemouth, Eastbourne, Torquay, and Southport too. In many respects the middle classes were now following the pattern of their social superiors; a higher standard of living almost inevitably involved a fuller life of leisure and entertainment, notwithstanding religious and political scruples.

What did concern many Victorians was the role of leisure in *working-class* society. In theory there was not much time for relaxation. The Ten Hour Day Act of 1847 did not apply to all workers by any means; servants, for example, would commonly have to get up at five o'clock and be on call most of the day. Also, many working men spent several hours on the journey to and from work. On the other hand, in many occupations the employment was by the day not the whole week, or was subject to seasonal and cyclical fluctuations which inevitably created periods of enforce leisure. Thousands who resented regular work six days a week in factories or mines traditionally coped by taking Monday off, and many declined to turn up for work while they still had money in their pockets from the previous week. The mid- and late Victorian periods saw a trade-off between the employers, who wished to replace irregular and informal holidays with official ones, and the workers' own desire to lighten the wearing round of toil. By a mixture of private commercial policy and government intervention, the opportunities for leisure were widened. From the 1850s employers in Lancashire textiles began to treat Saturday as a half-day, and to organize excursions by train for their men. As this was gradually copied it undermined the resort to 'Saint Monday'. In 1871 and 1875 Bank Holidays were introduced and spread rapidly to most workers. By the 1880s some northern mills and factories had begun to close for an entire week at a time, thereby promoting the modern summer holiday, though workers were not yet paid for this time off and it was decades before the practice became general. None the less, seaside holidays clearly grew popular in this period; Blackpool, for example, attracted 850,000 visitors in the early 1870s and nearly 2 million by 1890.

But to what end was all this extra leisure for working men? The Bishop of Winchester expressed a typical opinion when he declared that it was 'that they may in their leisure hours raise their own physical force . . . then their family, then their intellectual, and above all their spiritual being'. Yet whether middle-class notions about rational recreation influenced popular behaviour seems doubtful. At all events, church-going continued its decline. Several factors also conspired to prevent improvement. In the first place, working men had the capacity to organize their own leisure activities, whether improving or not. Some of these were, of course, approved by their social superiors, including cycling, choral singing, allotments, leek clubs, horticultural shows, pigeons, and whippets. Secondly, it proved impossible to control the commercial element in leisure. The increase in real wages gave many workers the option to choose to take advantage of professional entertainment. One example is the rapid spread of the music-hall from the 1860s onwards. Though not in any obvious way improving or rational, the music-hall did become more respectable. The introduction of fixed rows of seats led to much less drinking and more attendance by women from the 1880s onwards. As a result, working-class audiences were usually reported as perfectly well-behaved by the end of the century.

A third and major problem for middle-class improvers was the capacity of the workers for taking over activities and changing their character to suit themselves. The simplest example of this was the railway excursion, which reformers naively regarded as a route to respectable leisure; but such trips could end in drink and dissipation, however soberly they had begun. Then there was the case of the working men's clubs. Back in 1862 the Workingmen's Club and Institute Union had been founded as a middle-class initiative for improving purposes. Yet by the 1880s the rank and file had asserted itself by insisting on having alcohol on the premises. This was accepted; clubs could make a 30 per cent profit on beer sales, and their membership collapsed without it. Then there was team sport, much favoured by churchmen on the grounds that it improved health, fostered an acceptance of rules and discipline, and brought the different classes together for a common purpose. To this end the rough, traditional game of football was taken in hand by middle-class amateur enthusiasts who set up the Football Association in 1863. They introduced a common set of rules to facilitate regular competition between clubs from all parts of the country and in 1871 the FA Cup was established. However, the effects of these good intentions left much to be desired. Football soon became so popular that it turned into a mass spectator sport. It was the debilitating effects on whole generations, who merely watched, that provoked Kipling's famous outburst in the aftermath of the Boer War:

> Then ye went back to your trinkets,
> Then ye contented your souls
> With the flannelled fools at the wickets
> And the muddied oafs at the goals.

In addition, it rapidly transpired that the working-class footballers of the northern town clubs could not afford the travel and time off work unless they were subsidized. This undermined the amateur status of the game; and during the 1880s the FA failed in its attempt to ban professional players. The working-class players were too good to be cut out. This lesson was underlined by the famous victory of Bolton Olympic over the Old Etonian team of amateurs and gentlemen. From this time the upper classes retreated, to leave football a proletarian sport. Not only did football fail to unite the classes, but the working-class professionals routinely declined to 'play the game'; kicking the man as much as the ball and disputing the referee's decisions became an integral part of football. On the other hand, cricket did succeed in crossing the class divide both among players and spectators, and the tight grip of the higher classes on the organization of the game ensured respect for rules and less emphasis on professionalism.

Respectability and Improvement

The more historians have studied Victorian patterns of leisure the clearer it has become that most people declined to be guided into improving activities. This is not to deny that recreation lost much of its rough and rowdy behaviour. But the chief trend was simply towards a fuller and more elaborate pattern of leisure; the middle classes clearly became more relaxed and self-indulgent, while the working classes took advantage of higher real wages to enjoy life outside the factory and mill. But social control by the higher classes looks improbable. Drink, as we have seen, proved fairly reistant to pressure for change. Religion remorselessly declined. As early as 1851 the census found only 40 per cent attending a Christian service, and as little as a quarter of the population in the big northern cities. Gradually, in spite of the spread of Sunday schools, the churches were edged out of their role in education and became a marginal factor in the lives of most people.

One characteristic Victorian strategy for social improvement was the development of a system of police forces. There are some grounds for thinking that law and order prevailed more generally by the 1890s than in, say, the 1830s. After a long-term rise, crimes against property and persons began to fall in the second half of the century. However, it is not clear how far the police were responsible. Arrests for offences such as drunken and disorderly behaviour or for prostitution were easy to make, and all the statistics tell us is that the more police were employed the more crime was reported. Many crimes were the outcome of poverty, and it seems likely that, even without the new police forces, the improved wages would have had a beneficial impact. In some areas, such as music-hall and elections, the evident reduction in rowdy behaviour owed a good deal to the growing role of the respectable working class and to the wider entry of women into the public arena.

In some ways the most striking manifestation of improvement is to be found in education. In an earlier era education had been suspect in the eyes of the wealthy because of the association between literacy and political radicalism. But by 1870 the radicals had won their battle for a free and cheap press, and the emphasis of public policy shifted to educating the new voters. As a result, compulsory school attendance after 1880 pushed up literacy rates to around 95 per cent for both sexes. The most conspicuous indication of the increase in writing was the expanding work of the Post Office, which delivered 704 million letters in 1870 and 2,827 million in 1913. Commercial lending libraries catered to the middle class and the number of public libraries increased from 60 in 1875 to 438 in 1918 in England alone. The public taste for light romantic and adventurous fiction was eventually satisfied by the speedy production of cheap editions of new novels by the 1890s. By then reading was much easier during the evening, owing to the spread of gaslight to the majority of homes. Most people contented themselves with one of the mass-circulation Sunday newspapers like the *News of the World* or *Lloyd's News*, and the family papers, including *Titbits* (1881), *Answers* (1888), and *Pearson's Weekly* (1890). There also occurred a major expansion in women's magazines, with 48 new titles between 1880 and 1900 including *Housewife* (1886), *The Mother's Companion* (1887), *Forget-Me-Not* (1891), and *Home Chat* (1895). These papers would not have enjoyed as great a success had they not served as a means for advertisers to reach a more prosperous lower-middle- and working-class market. The *Daily Mail*, which reached a sale of half a million by 1900, after its foundation in 1896, was the foremost example of the advantages of advertising revenue; it soon led to imitations like the *Daily Express*. In a way the *Daily Mail* encapsulates the whole phenomenon of popular culture in the late Victorian era. The upper classes did not exactly warm to it – a paper, as Lord Salisbury reputedly said, that was written for office-boys by office-boys. But at the same time, even a population reared on the *Daily Mail* represented a considerable improvement over the earlier years of the nineteenth century.

5

The British Nation: Unity and Division

In many ways Britain in 1870 seemed a remarkably cohesive nation. For one thing, she enjoyed a compact territory and an admirable system of internal communications. By 1870 the rail branch lines extended as far as Cornwall and the west and north of Scotland; 300 million rail journeys were undertaken every year. There was also a strong sense of pride in Britain's economic achievements, her Empire, and her parliamentary system. At the same time, the growing sense of foreign competition probably heightened popular patriotism. On the other hand, there were deep tensions over questions of religion and race in this period. In the Celtic fringe considerable resentment had developed towards the political dominance exercised by London. And the combination of a declining aristocracy, an ambitious middle class, and an increasingly organized and cohesive working class made for a certain amount of social and political conflict. It cannot be assumed that the forces for cohesion and integration were automatically stronger than the elements of disunity.

Scotland: Nationalism and Imperialism

By the middle of the nineteenth century the Union of 1707 between Scotland and England appeared to meet with general acceptance north of the border. As *The Times* rather complacently put it; 'the separate nationalism of Scotland is happily in these days an anachronism'. One reason for this was that the Union had left intact some of the institutions in which Scotsmen took pride, notably the legal and educational systems. Scotland boasted four distinguished universities, and outstanding achievements by her middle class in medicine, science, engineering, literature, economics, and philosophy. The country enjoyed a higher literacy rate than England, and the separate Education Act of 1872 inaugurated a much more sweeping reform than that of 1870 in England. In Scotland the school boards were not hindered by religious disputes; they moved rapidly towards compulsory education up to thirteen, and were permitted to provide secondary education. In contrast to the decadent and elitist ancient universities of England, those north of the border were more academically orientated and relatively open to talent. Five times as many Scots as English attended universities. Nor had the Union involved the imposition of an alien Church. Scotland's controversies here were purely internal; in 1843 the Presbyterian Church of Scotland had suffered a breakaway by the Free Church, and after 1847 was also challenged by the United Presbyterian Church.

Another important reason for the survival of the Union was economic. Scotland had become a major element in the Victorian success story by taking advantage of the wider market for her coal, iron, ships, textiles, and chemicals. Scots entrepreneurs and workers also moved south to take up new opportunities, where they became key participants in the wider expansion of the British Empire. This affected several levels of society. At one end lordly figures like Dalhousie, Elgin, and Minto occupied grand roles as Indian viceroys and colonial governors, while at the other, poor Scots were recruited into the Highland regiments created by the Westminster government as a means of turning the martial qualities of the people to useful purposes. Moreover, thousands of Scotsmen emigrated to become farmers in Canada, tea-planters in India, missionaries in Africa, as well as doctors, merchants, shippers, and bankers. Of the latter, the greatest was the Jardine Matheson firm, whose interests stretched to India, China, and Japan. Scotland was, in fact, so bound up with empire that great imperial crises like the Boer War generated powerful political support for the British cause north of the border.

This, however, is not to say that a distinctive Scottishness no longer existed. It seemed to express itself more in cultural than in political forms. One manifestation consisted in the romantic revivalism associated with the Highland games, clan tartan, and the wearing of the kilt. This was stimulated by the writings of Sir Walter Scott, dignified by Queen Victoria's decision to purchase Balmoral in 1847, and sustained by the tourist trade. However, it remained very much the culture of the Anglicized Scottish aristocracy and gentry, and left ordinary people, especially the majority living in the urban Lowlands, unmoved.

Indeed, Scotland in the nineteenth century was a society deeply divided by social class; and the workers of the central industrial belt found in sport a different medium for expressing a sense of national consciousness. This was something much stronger than the friendly competitiveness in rugby between middle-class, public-school Scots and their English counterparts. Among working men, professional football aroused a fanatical loyalty which crystallized into anti-Englishness in the late Victorian era. The continued drain of top players to English clubs underlined the growing self-image of Scotland as a poor, semi-colonial territory, and made victory over the English on the football field a vital means of bolstering national pride. However, the nationalist implications of Scottish sport were limited. Football reinforced divisions within Scottish society as much as it united them against the English. The influx of Irish Catholics and Ulster Orangemen into the Lowland cities led to the emergence of teams backed by Protestants (Rangers and Hearts) and teams backed by Catholics (Celtic and Hibernians). Further, the popular obsession with football does not seem to have been closely linked to political nationalism, as it was in Ireland. Although some Scottish Labour leaders like Keir Hardie sympathized with Home Rule, it never became an important objective for them. Ultimately, sport was more of a safety valve than a destabilizing element in Anglo-Scottish relations in this period.

Indeed, in many ways Victorian Scotland appears to have been very well integrated into British politics, especially if measured by the role of the leading figures. Gladstone, with his Scots parentage and his Midlothian connection, became a revered figure. Scotland also supplied a number of other prime ministers, both Liberal (Rosebery, Campbell-Bannerman) and Conservative (Balfour and Bonar Law). Rosebery was typical in combining a romantic Scots nationalism with staunch British imperialism. The incorporation of such men into Westminster politics probably blunted separatist sentiments, though the habit of planting English carpetbaggers in safe seats north of the border (H. H. Asquith, Richard Haldane, John Morley) was probably counter-productive in the long run.

From the 1860s right through to the late twentieth century, Scottish politics became tightly woven into English through the allegiance first to Liberalism and later to Labour. By 1884 reform had boosted the traditionally low number of voters to 560,000, thus opening the way for a popular Liberalism based on land reform, free trade, and a general hostility to the feudal class. In 1878 half of Scotland was owned by 68 men including great magnates like the Dukes of Buccleuch, Atholl, and Sutherland. But even in the Highlands these figures ceased to excercise a dominating influence, and by 1885 the Liberals had captured 62 of the country's 70 constituencies. Of course, Scottish Liberalism was as subject to division as other Scots institutions. Five rebel crofter candidates won Highland seats in the 1880s against the conventional Whig-landlord figures. Here the accumulated grievances since the time of the clearances were exacerbated by poor crops and evictions. This produced a serious breakdown of law and order. After the famous 'Battle of the Braes' on Skye, which was an attempt to stop evictions, the government granted the crofters the right of rent reviews, compensation, and security of tenure previously given to Irish tenants. At the same time, Whig magnates like the Duke of Argyll were already leaving the Liberal Party on account of its radicalism. The 1886 split had major consequences in Scotland because of the Irish Protestant community. This resulted in the election of 17 Liberal Unionist MPs. Finally, there began to develop a gulf between Scottish Liberals and the emerging labour movement. That this was deeper than the division in England is indicated by the failure of the 1903 Liberal-Labour pact in Scotland.

Although Scotland participated in mainstream British politics, from the 1860s onwards there were growing signs of dissatisfaction with the Union. Both Gladstone and Rosebery appreciated that the Westminster administration seemed remote and overbearing, and accepted the case for giving the country a separate administration rather than treating it as a subdivision of the Home Office. In fact, it was Salisbury who appointed the first Scottish Secretary in 1885. But the Irish example inevitably had a stimulating effect on Scots nationalism, and 1886 saw the establishment of a Scottish Home Rule Association. From that time onwards, Liberals on both sides of the border increasingly supported the idea of a Scottish parliament;

even Gladstone saw the advantage in a scheme of devolution-all-round which would free Westminster from a number of troublesome domestic issues. However, although a bill was introduced in 1892 it aroused very little interest. After 1906 the question of Scottish Home Rule revived again, to some extent, as a result of the ripples of the Irish controversy. In 1907 the Liberals set up a standing Grand Committee so that Scots legislation might receive proper attention. Bills to introduce a Scottish parliament were introduced on seven occasions between 1906 and 1914, and in 1913 the Commons gave a majority on second reading. But, while the pressure to resolve the Irish question reached new heights, the Scottish case dwindled once again in the years during and immediately after the Great War.

Wales: Nationalism and Liberalism

By comparison with Scotland, Victorian Wales was culturally cohesive, but hardly any national institutions survived. For 600 years the country had been administered from London like any region of England. Yet Wales was much more than a geographical expression. According to the census of 1891, the Welsh-speaking population composed 54 per cent, and was still 50 per cent in 1901. In the north-western counties of Anglesey, Caernarfon, Merioneth, and Cardigan, some 48–51 per cent of the people spoke *only* Welsh. The other side of the coin was that in the whole of the country three-quarters could speak English, and for many years able and ambitious Welshmen had happily used the English language to obtain better employment. By 1891 228,000 natives of Wales resided in England, notably in Liverpool, Manchester, and London. The Victorian railway system helped to integrate Wales with England, for the main routes ran east–west, linking the north with Manchester, mid-Wales with Shrewsbury and Birmingham, and the south with Bristol and London. But the migration was not all one way. In economic terms Wales comprised two regions, a rather depressed, agricultural north and a south made buoyant by coal, steel, and shipping. By the 1880s eight of the thirteen counties were losing population, while Glamorgan, which contained a third of the total population of 1.5 million, grew by two and a half times between 1861 and 1891. Its immigrants came initially from rural Wales, but also from the west of England; in fact, by 1870 four out of ten were English people attracted by the employment available in the ports and mines of Monmouth and Glamorgan.

Although this English immigration diminished the role of the Welsh language, the country none the less retained its distinctive culture. This was partly a reflection of the dominance of the Nonconformist churches, which claimed over 80 per cent of all worshippers by the 1850s. In contrast to the Scots, the Welsh could boast a culture that crossed the class divide. It found expression particularly in the music and poetry that were the glories of the national eisteddfod. This culture permeated the political groups as well as the different social classes. Thus a leading Liberal like Tom Ellis MP could participate in the eisteddfod alongside the union leader William Abraham from the Rhondda. The sport of rugby also had a unifying effect.

Whereas in England rugby was exclusive, in Wales it became a genuinely popular sport which helped to draw the non-Welsh-speaking immigrants into national life.

In some ways Welsh politics echoed that of Scotland. It revolved around the grievances of small tenant farmers against landowners and the hated tithes; but to this was added a campaign against the Anglican Establishment and a demand for improved education. The 1867 Reform Act undermined the influence of the Conservative gentry, and the increase in the county electorate from 75,000 to 200,000 in 1885 completed the process. By 1880 the Liberals already held 29 of the 33 seats. Wales became the most consistently Liberal region of the British Isles; even in a poor year like 1900 the party retained 28 of 34 constituencies, and in the local elections of 1904 639 Liberals were elected against only 157 Conservatives.

With its strong basis in popular culture, Welsh radicalism had the potential to create a serious rift in British politics. The demand for Disestablishment developed into a nationalist cause, which made Gladstone's prevarication over the issue rather dangerous. Welsh radicals also felt rather tempted to emulate the Irish Land League. Indeed, by 1886 a Welsh Land League had appeared, under the leadership of Thomas Gee, committed to securing fair rents, security of tenure, abolition of tithes, and graduated taxation. Yet the Welsh radicals remained notably loyal to the Liberal leader even in 1886; Liberal Unionism enjoyed nothing like the kind of support it won in Scotland. The explanation is that well before the split in the party the Welsh Liberal MPs had settled down under the leadership of Stuart Rendell, who refused to allow Welsh grievances to be used as a vehicle for separatism. This loyalty to British Liberalism cannot be ascribed to patronage. Indeed, in the late nineteenth century Tom Ellis – a junior Whip – was the only MP to obtain promotion from the back-benches; no one reached the Cabinet until Lloyd George in 1905.

However, the Welsh radicals clearly felt that it was worth staying under the aegis of the Liberal Party. In 1881 they achieved the first distinctively Welsh item of legislation – the Welsh Sunday Closing Act. Other Liberal reforms such as the Church Burials Act were of considerable importance to Welsh Nonconformists. In the long run they also capitalized on Gladstone's sympathy with the culture and education of Wales. The new college at Aberystwyth, formed in 1872, received a state subsidy, and it subsequently joined with the university colleges at Cardiff and Bangor to form the University of Wales in 1893. Moreover, as a result of their victories in the post-1870 school board elections and in the post-1888 county council elections, Welsh Nonconformists managed to seize control of elementary and secondary education. In 1905 the National Library of Wales was established at Aberystwyth, while two years later the Liberals created a Welsh Department at the Board of Education with the object of promoting Welsh language-teaching in schools.

On the land question there was nothing comparable to the special legislation granted to Scottish crofters, though in 1891 the Conservatives made tithes payable by landowners, not the occupants, which helped to defuse the issue. One cause that continued to generate friction was Disestablishment. From 1891 onwards the National Liberal Federation included

Welsh Disestablishment in its programme, and even Gladstone abandoned his opposition. None the less, his 1892 government failed to pass a bill into law, and it was left to the Edwardian Liberal government to try again. Although the House of Commons enacted the Welsh Disestablishment Bill three times during 1912–14, thereby overcoming the veto of the House of Lords, the outbreak of war caused the measure to be suspended so that it did not finally become law until 1920.

It was during the 1890s, when frustration over Disestablishment reached its height, that Welsh separatism looked dangerous. The emergence of a new organization, the Cymru Fydd, founded in London in 1886, provided a vehicle for some of the younger and more impatient Welsh radicals. Under the leadership of Lloyd George, Cymru Fydd took over the North Wales Liberal Federation in the cause of Home Rule, and threatened to create a break from British Liberalism. But there was to be no Welsh Parnell. The initiative broke down on the opposition of the South Wales Liberal Federation, and the nationalist leaders realized that there was not enough general enthusiasm for Home Rule. By 1896 Cymru Fydd was fading and Lloyd George had settled his course in the mainsteam of British politics. The only other challenge to the political status quo came from a new generation of labour leaders based in the vallies of South Wales. Keir Hardie led the way with his victory at Merthyr Tydfil in 1900, and in his wake came James Griffiths, Vernon Hartshorn, Arthur Cook, and Frank Hodges, all leaders of the coalmining community. Yet although they posed a serious threat to Liberal hegemony after 1906, their rise had a similar effect to Liberalism in integrating Wales into the broader pattern of British politics; as a result, Home Rule was relegated to the bottom of the agenda for several decades to come.

Ireland: The Threat to the Union

Ireland in the 1860s was a country of 4 million people governed by a Viceroy in Dublin and a British Cabinet minister, the Chief Secretary, who divided his time between London and Dublin. The Irish were also represented in Parliament by 100 MPs. Over two-thirds of the population were Roman Catholics who dominated the provinces of Leinster, Munster, and Connaught, though in Ulster about 60 per cent were Protestant. The chief links with England, other than political institutions, were the industry of Ulster, which benefited from access to mainland markets, the Established Church, which was Anglican, and the Anglo-Irish landowning class. Magnates such as the Marquis of Lansdowne, the Duke of Devonshire, the Duke of Abercorn, Lord Fitzwilliam, and the Marquis of Bath occupied influential positions in society on both sides of the Irish Sea. Over the decades these landowners had let land to small peasant farmers; but the combination of subdivision of plots and a wet, mild climate dictated a heavy reliance on the cultivation of potatoes.

The chief catalyst of change in the Victorian period was the famine of the 1840s, which resulted in the death of a million people, the emigration of a million, and the loss of a further million by emigration between 1851 and

1861. Much of the country continued to be depressed; the 1870s and 1880s saw poor harvests, falling agricultural prices, and consequent pressure for reductions in rents. The new element in the situation was the Irish emigrant community in the USA which had founded the Fenian Brotherhood. By the 1860s they were capable of promoting violent anti-British activities both in Ireland and on the mainland of Britain, where attempts were made to free Fenian prisoners from jail.

British politicians showed themselves slow to grasp the gulf that was beginning to separate them from the Irish people, and to appreciate the seriousness of the challenge to the Union. As Prime Minister, Gladstone was reluctant even to accept that he required constant police protection from the 1860s onwards. He did, however, at least adopt a constructive approach towards the Irish Question. Though rather ignorant of the people and of the country, which he rarely even visited, Gladstone diagnosed two major problems – religion and land – which he attempted to settle by disestablishing the Church in 1869 and by means of his first Land Act in 1870.

The inadequacy of these measures became fully apparent at the election of 1874, when the recently formed Home Government Association returned 59 MPs committed to the restoration of an Irish Parliament in Dublin. These Home Rulers, acting under the leadership of Charles Stuart Parnell from 1878, became a fixed and sometimes crucial element in British politics down to the 1920s. By eliminating Liberal representation in Ireland, they immediately began to alter the configuration of party politics at Westminster. The Irish Home Rulers benefited greatly from the franchise reform of 1884, which increased the electorate in the counties where their support was strong. But at the same time Gladstone deliberately excluded Ireland from the redistribution of constituencies in order to avoid obstructionism by the Irish members. This left Ireland markedly overrepresented with 103 MPs. From 1885 onwards, between 81 and 85 of these were Home Rulers; they held the balance of power in the Commons in 1885, 1892, and 1910.

British governments found the Irish Question difficult to handle because the nationalist campaign operated on two levels – as a respectable political party at Westminster and as a violent mass movement in the Irish countryside. They therefore oscillated between conciliation and coercion. In 1879 the Irish Land League was founded under the leadership of Michael Davitt to protect tenants from high rents and evictions. Its campaigns resulted in widespread violence, including the murder of the Chief Secretary for Ireland, Lord Frederick Cavendish, in 1882. Meanwhile, Gladstone embarked on a radical reform in the shape of a second Irish Land Act in 1881, which established tribunals to revise the rents of the tenant farmers. This represented an unprecedented interference with the rights of property-owners, and was to have profound consequences. In the short term, however, Gladstone obtained no relief from the dispiriting round of conflict and repression; by 1886 he had determined to break the deadlock by conceding a parliament in Dublin and new legislation to buy out the Irish landlords.

Gladstone's proposals have been subjected to much unfair criticism. In fact they went a long way to meeting the tangled needs of the situation.

He built some safeguards for minorities into the scheme. He granted the Dublin parliament extensive internal powers over the whole country but retained links with Westminster; Ireland would contribute one-fifteenth of the imperial revenue and London would retain control over defence, foreign policy, trade, and currency. By removing Irish representation at Westminster, the scheme pleased those who resented the disruption of the parliamentary timetable. Above all, Parnell accepted Gladstone's proposals. Thus, 1886 offered a chance of both preserving the Union, satisfying nationalist demands, and averting decades of violence and misery in Anglo-Irish relations.

However, the first Home Rule Bill was defeated by thirty votes when the Conservatives combined with 93 rebel Liberals who subsequently withdrew to form the Liberal Unionist Party. In fact Home Rule did not cause the flight of the Whigs, which was already under way, but it certainly accelerated the process. This left the Gladstonian Liberals a smaller and less wealthy party, but a more cohesive and radical one. The split precipitated a general election in 1886, which put the Conservatives in office for most of the next twenty years. In his brief ministry from 1892 to 1894, Gladstone managed to push a second Home Rule Bill through the Commons, only to see it rejected in the Lords by 400 votes to 40.

For the Irish this was a sobering statistic. The new Prime Minister, Salisbury, had little liking for democracy even in England, and had declared the Irish to be as fit for self-government as the Hottentots – a newly 'discovered' tribe of Southern Africa. Several considerations were behind the Conservatives' resistance to Home Rule. Initially they reacted as the party of landed property to a threat to their self-interest. However, this is not a sufficient explanation, for the party subsequently extinguished Anglo-Irish landownership but continued to uphold the Union. They feared the eventual result would be complete separation of the two countries which would weaken Britain's strategic position, especially in an era of naval rebuilding and invasion scares. There were other possible ramifications. Concessions to the Irish might stimulate the disintegration of the Empire; 1885 was, after all, the foundation year of the Indian National Congress, whose active London office enjoyed the support of some of the Irish members. More immediately, Home Rule provided a useful supply of material for Tory Party propaganda designed to damn the Liberals as weak and unpatriotic.

The flaw in all this was that the Irish people had emphatically voted for Home Rule, continued to do so consistently, and were becoming ungovernable. Moreover, in the three southern and western provinces Protestant supporters of the Union constituted only 10 per cent of the population, and in time they reconciled themselves to the inevitability of Home Rule. In 1886 the Protestant majority in Ulster belatedly organized itself in support of the Union, stimulated by Gladstone's bill and the notorious visit of Lord Randolph Churchill who proclaimed that 'Ulster will fight and Ulster will be right', thus initiating a tradition of Conservative support for violent resistance to the will of Parliament. However, by securing the election of 17 Unionists from Ulster's 33 constituencies, they gave the British Conservatives a firmer basis for their resistance to Home Rule.

Under Salisbury's premiership Arthur Balfour made his political reputation as Chief Secretary by restoring a degree of calm to the Irish countryside. This involved repressing the violence, which was dwindling anyway, and offering a number of expensive concessions. As the Conservatives left Gladstone's 1881 Land Act intact, the tenant farmers began to benefit from substantial reductions in rents. In addition they passed a series of measures in 1885, 1888, 1891, and 1903 which allocated Exchequer revenue to buy up Irish estates, which were then split up and sold on generous terms to small farmers. The irony in all this was that the Conservatives were implementing the policy of Davitt's Land League! The long-term effect was to leave Ireland as a country of small Catholic farmers – a recipe for a conservative society. Only outstanding incompetence on the part of British governments could make such a society the home of terrorism. For although Balfour and Salisbury congratulated themselves in the 1890s on having subdued the subversives, they actually made no impact on the strength of Irish nationalism. The Home Rulers maintained their grip on the Irish constituencies and waited for their opportunity to come round again. Meanwhile the danger in keeping the parliamentary nationalists waiting was that they would eventually be outflanked by more radical leaders. By 1900, then, the Union with Ireland had become an artificial imposition which dangerously divided the British state and could not be expected to last.

Religion and Immigration

Most of the religious and racial groups in late Victorian society were regarded as 'minorities'; Nonconformists resented this description, however, as they felt they had already drawn level with the Church of England by mid-century. By the 1870s they were beginning to move from the periphery to the centre of British politics by integrating themselves into the structure of the Liberal Party (see p.24–27). A number of essentially symbolic reforms, beginning with the abolition of religious tests at Oxford and Cambridge in 1871, helped to remove the body of discriminatory legislation. This process continued up, to 1898, when the Marriages Act removed the requirement for a civil registrar to be present at weddings in Nonconformist and Catholic churches. But as early as 1880, when the Church Burials Act was passed, the momentum behind political Nonconformity began to wane. This may seem surprising because the great issue of Disestablishment still remained to be tackled, and, according to the Liberation Society, the number of MPs who supported the cause had risen to 171 by 1885. But in fact, the resolution of most of the immediate Nonconformist grievances and the steady absorption of Nonconformist politicians into the political establishment blunted the sense of alienation among the free Churches generally. Outside Wales the campaign for Disestablishment had lost momentum by the (late 1880s, as is indicated by the dwindling funds and membership of the Liberation Society. Some of the more affluent Wesleyans had already migrated to the Conservatives. Thus, although the number of Nonconformist MPs was still to reach its peak – 210 in 1906 – Nonconformist issues were not quite as central as they had once been. In 1908 H. H. Asquith became Britain's first Nonconformist Prime Minister, and David Lloyd George was to be the second; but both regarded Welsh Disestablishment as essentially a necessary concession to the party faithful, and did not even place English Disestablishment on the agenda.

A similar but less conspicuous progress from the periphery towards the establishment was achieved by the small Jewish community. Back in 1847 the Baron de Rothschild had been elected an MP, but had been prevented from taking his seat because the parliamentary oath taken by all members referred specifically to the Christian religion. The House of Commons abandoned this rule in 1858. It was not until the Parliamentary Oaths Bill of 1885 that confessed atheists were enabled to take their seats, and it is notable that in 1892 some 40 members chose to affirm rather than take the traditional oath.

For politically ambitious Jews the real obstacle lay in the reluctance of the parties to nominate them. Their obvious strength lay in the East End of London where Jews were concentrated, though only a minority of them were actually voters. The Liberals, who were the more tolerant party, returned the influential spokesman for the Jewish community, Samuel Morley, for Whitechapel. Several Jews also sat as Conservatives in such seats as Stepney and Limehouse, though they felt obliged to advocate restrictions on immigration – which meant Jewish immigration. Perhaps the most telling sign of the integration of Jews into British politics was the success of men such as Herbert Samuel, Edwin Montagu, and Rufus Isaacs in winning elections in constituencies where the Jewish vote was negligible, and in holding high office in the Edwardian Liberal governments.

In spite of this, however, anti-Semitism remained widespread throughout society and in all political parties; it was simply not strong enough to undermine loyalty to party at elections. At the top of society Jews were often described as clever but disloyal and cowardly; and the ministerial careers of the three Liberals suffered from the prejudice of colleagues and opponents alike. In popular debate Jews endured crude, stereotyped attacks. On the one hand they were portrayed as exploitative capitalists and blamed for dragging Britain into the South African War in 1899; the Liberals, however, usually defended them as models of enterprise and industry. On the other hand Jews in the East End were depicted as a burden on the community because of the destitute condition in which immigrants arrived; politically, their role as opponents of Tsarist Russia gave them a reputation as dangerous socialists or even anarchists.

A similar pattern of prejudice and political assimilation may be discerned in the Catholic community. English Catholics tended to be concentrated in the counties of Lancashire, Cheshire, Durham, Northumberland, Warwick, and Staffordshire, largely because of the survival of old Catholic gentry families there. However, in politics Catholics were forced into a very marginal role for much of the Victorian period because they were the target of vitriolic 'No Popery' campaigns at elections. Although the emancipation legislation of 1829 had allowed Catholics to sit in both Houses of Parliament, there were only 36 MPs by 1868. However, both Gladstone and Disraeli showed a willingness to incorporate Catholics into the system, and in 1871 the Liberal leader lifted the penalties imposed upon Catholic bishops by Lord John Russell in 1851.

As the Liberals moved increasingly towards radicalism there was a temptation for Catholics, a naturally conservative community, to gravitate towards the Conservatives; however, the party's association with the most rabid Protestantism deterred them for many years. Eventually doctrinal prejudice was overcome by a combination of electoral expediency and social class. A number of titled and gentry families, led by the Duke of Norfolk, found an alternative avenue into Conservative politics in the Primrose League. The Duke himself was awarded the Garter, became Chancellor of the League, and joined the Cabinet as Postmaster-General in 1895. In the context of their Irish policy it was advantageous for the Conservatives to be able to avoid looking like an exclusively Protestant organization. Moreover, there were votes to be won, especially in school board elections, by defending the voluntary Catholic schools. At the 1885 election Cardinal Manning urged Catholics to oppose the advocates of secular education. No doubt the Catholics were still a marginal factor except in some of the northern constituencies, but the community had clearly moved in from the periphery by the 1890s.

Britain enjoyed a long and honourable tradition as a haven for immigrants fleeing from religious and political persecution. During the nineteenth century this tradition was strengthened by Liberal and free-trade sentiment; not until 1905 were any controls imposed on entry into Britain. The novel element in Victorian immigration was economic hardship, which was, of course, the chief reason for the large-scale movement of Irish people. By 1871 the Irish in England and Wales numbered 566,000 and in Scotland 207,000; however, by 1900 the numbers had fallen by a fifth as many immigrants moved on to the United States. Irish immigrants were concentrated in Lancashire, Glasgow, and London, but steadily spread to most industrial towns in search of employment.

How far did the Irish become integrated into English and Scottish society? Many were always more attracted by the material opportunities available in America, and viewed England as a purely temporary refuge. Traditionally many Irish people also found seasonal jobs in agriculture on the mainland, but the permanent immigrants largely undertook semi- or unskilled work as dockers, porters, or building labourers. Like the Highland Scots they were recruited into the British Army, where they accounted for 14 per cent by the 1890s. It seems unlikely that they had a very positive impact on the economy; there was already a surplus of unskilled labour which flowed into the industrial centres. Consequently the Irish presence attracted a good deal of criticism, on the grounds that they depressed wages, caused overcrowding of working-class housing, imposed severe burdens on the poor rate, and added to the drunkenness and criminality of urban society. Not surprisingly, they rapidly became segregated by residence for mutual support, and relied upon the assistance of the existing Catholic network, which was much stimulated by the arrival of so many co-religionists. Some resentment was caused when employers resorted to Irish labour for purposes of breaking strikes; but Irish participation in the great dock strike of 1889 helped to integrate them into the labour movement. In time the Irish community produced major union leaders, including James Connolly and James Larkin.

Finally, the Irish were inevitably drawn into the political struggle. The Conservatives exploited them by a barrage of propaganda designed to arouse disgust over the atrocities committed in Ireland and to tap the Protestant backlash in areas like Merseyside. For their part, the Nationalist leaders, T. P. O.'Connor – the MP for Liverpool Scotland – and John Denvir, set up the Irish National League of Great Britain and the Home Rule League of Great Britain with a view to mobilizing the Irish vote. Though not large, it was frequently crucial in shoring up Liberal majorities in urban constituencies threatened by the Tory revival. Only in 1885 did Parnell advise the Irish to vote against the Liberals so as to curtail their majority and thus force the pace on Home Rule.

Immigration brought many other nationalities to Britain in this period. There were small but longstanding groups of Chinese, Indians, and Africans, most of whom arrived as seamen, merchants, or students. Two of the wealthier Indians served as MPs: Dadhabhai Naoroji (Liberal) in 1892 and M. M. Bhownagree (Conservative) in 1895. But the most widely dispersed nationality were the Germans, who numbered 33,000 by 1871. Many of them worked as waiters, musicians, and pork butchers. But some were employed as clerks, attracted by the experience and better wages available in England. Though seen as efficient and well-qualified, they caused much resentment among English clerks struggling in an oversubscribed occupation. Several German Jews who followed careers in finance or the professions achieved a considerable reputation and left a lasting impression on British life: Charles Hallé, Edgar Speyer, Sir Ernest Cassell, and Sir Ernest Oppenheimer are examples. The entrepreneurs were originally attracted by the relatively tolerant climate of Britain as well as by the greater economic opportunities believed to exist.

During the 1890s the Germans were overtaken by Poles, Lithuanians, and others driven from the western districts of Russia by a combination of persecution and inadequate economic opportunities. The Lithuanians gathered in Scotland, while the Poles, who numbered 82,000 by 1901, concentrated in Stepney, Manchester, and Leeds. Many of them were Jewish, and their arrival helped to swell the total number of Jewish immigrants into Britain between 1870 and 1914 to 120,000. In addition, the Italian community increased from 5,000 in 1871 to 20,000 by 1911. They had originally migrated from rural Italy to the towns of the north, but failed to find work there. In Britain Italians often worked as cooks, waiters, hairdressers, and musicians, or set up businesses in confectionery, ice-cream, and catering.

British attitudes towards these immigrant communities varied widely. At one end of the spectrum, Italians evoked sympathy and respect partly because of their recent struggle for freedom and because they did not compete as directly for employment as some other groups. Many of the wealthier Indians met with tolerance and friendliness in Britain, in marked contrast to the racial prejudice they encountered in India and South Africa. Attitudes towards Germans, on the other hand, deteriorated markedly during the 1890s and 1900s as a result of the growing commercial rivalry, the naval race, and German support for the Boers.

It was the Jews, however, who attracted the most sustained hostility, largely on the grounds that they introduced sweated labour into the clothing trade and cabinet-making, and thus undermined existing businesses. In fact, their enterprises did no more than exacerbate problems of low wages and overcrowding that already existed; but with their growing numbers and their shops and synagogues the Jews were a conspicuous target. As a result the period from the 1880s onwards saw sporadic campaigns against alien immigration led by Tory MPs such as Sir Howard Vincent and Major W. E. Evans-Gordon. Under pressure the government eventually set up a Royal Commission on Immigration in 1903; the Conservatives had, so far, benefited from an anti-immigrant reaction without doing anything about the problem. The result was the 1905 Alien Immigration Act, which attempted to exclude undesirable and destitute immigrants. It required immigrant ships to dock at certain specified ports only, and empowered the authorities to expel unwanted aliens, though appeals could be lodged by those fleeing political or religious persecution. After 1905 it fell to the Liberals, who had opposed the Act, to administer or to repeal it; they chose to operate it humanely by giving immigrants the benefit of the doubt. The Liberals' success in the East End constituencies in 1910 suggests that there was no longer much pressure behind the campaigns against the Jews and others. In time Jewish immigrants became involved in radical and socialist politics and were active trade unionists; the rise of Emmanuel Shinwell as a union organizer and politician is a notable example of this progression.

Monarchy and Patriotism

As the Crown gradually lost its political power during the nineteenth century, its function as a personalized symbol of the British nation became much more important. For the ordinary citizen the King or Queen offered a tangible link with the state and a channel for the expression of loyalty through a series of visits, tours, pageants, and celebrations. Queen Victoria has sometimes been thought to have played this role particularly successfully partly because, as a wife and mother, she embodied the Victorian ideal of family life, and because her Jubilees crystallized the popular feeling for empire towards the end of the century, when external threats encouraged the instinct to rally round the flag. However, we have no real way of measuring the popularity or otherwise of the monarchy until the late 1930s. In any case, many of the features of the monarchy during the last two decades of the Queen's reign actually pre-date her. Under George III, who also enjoyed a long reign, the monarchy came to personify the national cause, and provided a good deal of pageantry and spectacle. It may be that in late Victorian times the public display was more splendid and more widely appreciated because of improvements in communications, but it was not fundamentally different from the earlier period.

In fact the standing of the monarchy seemed to follow a cyclical pattern reflecting the life of the incumbent. After 1837 Victoria enjoyed an easy popularity as a young wife and mother. But her withdrawal from public life following the death of Prince Albert in 1861, which was aggravated by stories about her extraordinary reliance upon her Scots servant, John Brown, reduced her standing to a low point for about a decade. An organized republican movement sprang up, and in Parliament criticisms were voiced by as many as 64 MPs, including Sir Charles Dilke, Joseph Cowen, John Morley, and Joseph Chamberlain. Their attacks focused largely on the expansion of the civil list, which amounted to £358,000 by the 1870s.

However, even at its height the republican movement, with around 6,000 members, seems to have been much weaker than at earlier periods. Possibly because she was a woman, the critics underestimated the resilience of the Queen; certainly when she became more visible the attacks on the civil list faded quickly. Even in the 1860s, royal appearances in the provinces attracted enthusiastic crowds. In the eyes of many middle-class Victorians, the Queen's withdrawal doubtless indicated a proper sense of the importance of mourning. In 1871 when the Prince of Wales nearly died from typhoid, a good deal of sympathy was quickly generated, and this is usually considered a turning-point in the recovery of popularity by the royal family. While the Prince's personal life was disapproved of by some, his patronage of popular sports and entertainments probably endeared him to many sections of the population.

One important prop to the popularity of the late Victorian monarchy was the press. By contrast with earlier periods, the radical republican press had become marginal, outflanked by the mass-circulation newspapers which relied heavily on the royal family for copy. Both the national dailies and the expanding women's magazines promoted popular interest in the monarchy with the extra advantage of photographic illustrations.

However, it is easy to exaggerate these developments. The Queen herself proved to be unco-operative in some ways. She stuck to her widow's weeds for forty years and clung to the seclusion of Windsor and Balmoral. She obstinately refused Gladstone's well-meant requests to give the Prince of Wales a major public role, for example as Irish Viceroy, a position which, because of the dignity and the danger attached, would have done much to assist his national standing. The other limitation was political: the Queen's partisanship became blatant during the 1870s and 1880s, as she abandoned her early Liberalism to side with her Conservative Prime Ministers. In this she was, of course, reflecting the migration of many of her middle-class subjects, so she was not entirely out of step with the country. But she was never credible as a constitutional monarch above politics until the end of her reign.

By this time Victoria enjoyed the great advantage of sheer longevity. Now she was admired for bearing the burdens of a wife, mother, and head of state for six decades. The more Britain seemed to be challenged by foreign powers, the more people felt inclined to rally around a figurehead. This sentiment reached a climax with the Silver Jubilee of 1887 and the Diamond Jubilee of 1897. Clearly the celebrations throughout the country owed a good deal to middle-class organization. But each section of the population participated in its own way. Whereas the middle-class committees often attempted to turn the Jubilees to some lasting and serious purpose, working men appreciated having a day off work and a party. Women as a group were probably most enthused by the mystique of monarchy and by a vicarious pride in the long reign of a woman. But there seems to have been something like a consensus about the desirability of a celebration among all sections of society. Ordinary citizens by no means forgot their grievances and divisions, but by publicly demonstrating their pride in British achievements under Queen Victoria they marked their acceptance of the British system of government.

Aristocracy: Devaluation and Decline

Nineteenth-century observers of Britain's social-class system ranged all the way from Karl Marx, who thought he saw a steady polarization of two antagonistic classes, to Lord Palmerston, who believed that wealth and rank were open to everyone. What was not in doubt was the existence of class. By the 1830s the use of class terminology had become common. It assumed three main groupings. Victorians referred to the 'landed interest' which they equated with an upper or ruling class, the 'middling or industrious classes', and the 'working classes', though they distinguished the latter from the 'residuum' at the bottom of the pile. From time to time sections of these classes were clearly in conflict with one another, though for much of the century the harmony between them was more conspicuous. Persistent references to middle or working 'classes' rather than 'class' suggests that contemporaries did not regard the groups as particularly cohesive or unified. There was clearly a good deal of social mobility across the upper/middle and middle/working class frontiers, as individuals rose and fell in the social scale.

At the tip of the social pyramid stood the landed aristocracy, who were probably at the height of their political and economic success between the 1850s and the 1870s. In this period they comprised between 500 and 600 peers, of whom 430 sat in the House of Lords. To maintain the dignity of his position a peer was thought to require a minimum of around 3,000 acres of land, though most owned far more – the Duke of Devonshire had 200,000 acres and the Duke of Northumberland 166,000, for example. The landed class also included the 4,250 gentry families listed by Burke's *Landed Gentry*, many of whom were knights or baronets. Again, there was a convention that a country gentleman or squire should possess at least 1,000 acres. Traditionally these families enjoyed a near-monopoly of power in the House of Lords, the Cabinet, and in the Commons, and at local level as JPs and Lords-Leiutenant. In 1865 around three-quarters of the MPs came from aristocratic and gentry families. Their position was clearly entrenched by the remarkable concentration of land-ownership in Britain; 80 per cent of the entire British Isles was in the possession of 7,000 families.

Moreover, in some ways this was a rather rigid and exclusive elite. Recruitment into the British peerage was much more restricted than in other European countries. Although many peers enjoyed income from commerce, mining royalties, and urban property, mere wealth was not the key qualification; a stake in the land mattered most of all. Queen Victoria rejected Gladstone's proposal to ennoble Nathan de Rothschild with the words:

> she cannot think that one who owes his great wealth to contracts with foreign governments, or to successful speculations on the stock exchange, can fairly claim a British peerage . . . This seems to her not the less of a species of gambling because it is on a gigantic scale.

On the other hand, economic vicissitudes certainly caused both upward and downward mobility in the landed class. Successful farmers could elevate themselves by accumulating landholdings, withdrawing from labour, building fine houses, and marrying their daughters into established county families. Conversely, the younger sons of peers, who were denied a share of their fathers' landed property, usually found themselves obliged to establish professional careers in the army, the law, the Church, or banking, thereby sustaining regular traffic between the landed and the middle classes.

The 1880s marked a real turning-point in the fortunes of Britain's landed class. From this decade onwards their status, power, and wealth began to be undermined, though not as extensively as contemporary comment suggested. Some squires and aristocrats suffered from sharply declining rents as a result of the agricultural depression, though this was not characteristic of all regions. Henry Chaplin lost heavily in arable Lincolnshire, as did the Duke of Atholl on his huge but unproductive Perthshire moors. This resulted in the sale of millions of acres up to 1914 and at very low prices. The upper classes also took alarm from the radical campaigns for the imposition of taxation on landed wealth – although as yet these had only a limited impact. Irish estates were dispersed, the Crofters' Act impinged upon Highland owners, and Harcourt's 1894 death duties began a more effective system of taxation. It was doubtless this last that prompted Oscar Wilde's Lady Bracknell to lament: 'land has ceased to be either a pleasure or a profit; it gives one position and prevents one from keeping it up.' But the real threat to landed wealth was to come with Lloyd George's budget of 1909 and the First World War.

By the 1880s the political role of the peerage began to suffer serious attack. Undermined by the expansion of the county electorate in 1884, which resulted in the defeat of many county families, a number of the magnates retreated to safer seats in suburbia. Some were content to abandon what had only ever been a dilettante interest in national politics in order to devote themselves to other pursuits or to local government. Even there, some suffered the indignity of defeat in county council elections, though many peers continued to play a major role as county council chairmen. To some extent their personal loss of influence was cushioned by the dominance of the Conservatives and by Lord Salisbury's premiership down to 1902. It was not then obvious that he would be the last peer to serve as Prime Minister. Here, too, the real crisis was to come in the Edwardian period, when the House of Lords lost its veto and was held up to derision by the Liberals.

In some ways the most important reason for an uneasy feeling that the peerage was in decline sprang from insidious internal changes rather than external attack. Its members were conscious of the onward march of some extremely rich men who owed their position to industry rather than land. This fear was expressed in the Tory attacks on the 'radical plutocrats' who supported the social and political reforms of the Edwardian Liberals with generous donations. But well before that time the status of the peerage had been undermined by the entry of non-landed figures. Symptoms of the change could be observed in the Prince of Wales's friendship with Jewish financiers, and in the growing number of marriages between impoverished aristocrats and American heiresses; the families of the Duke of Marlborough and Lord Curzon each made several such American alliances in this period. But most alarming was the dilution of the peerage from the 1880s onwards, as result of a much faster rate of ennoblement and the inclusion of many non-landed men. Prominent examples were Lords Armstrong (armaments), Joicey (coal), Inverclyde (shipping), Glenesk (newspapers), Leverhulme (soap), and Ashton – known as 'Lord Linoleum'. Altogether, of the 200 new peers created between 1886 and 1914, one-third were plutocrats and one-third professional middle-class men, while only a quarter came from traditional landed families. The peerage was being devalued by the political parties, who used ennoblements to attract large donations to their funds and to manipulate Members of Parliament. At the same time, hundreds of extra knighthoods were created and new honours were awarded, including the Order of the British Empire, the Royal Victorian Order, the Order of Merit, and the Imperial Service Order. From 1888 onwards, the New Year as well as the Queen's Birthday and her Jubilees produced lavish distributions of these distinctions. A generation before Lloyd George took a hand, the honours system had already been diluted and partly discredited.

The Pivotal Middle Classes

Although they made up only about 15 per cent of the population, the Victorian middle classes have usually been regarded as the key element in nineteenth-century society; they were at the heart of Britain's economic success and vigorously propagated their values and ideals – enterprise, self-help, domesticity, religiosity – among those above and below them. However, there are two obvious flaws in this. The first is that much of the *prescriptive* literature which carried the middle-class message is a very misleading guide to the actual behaviour of Victorians. They were, for example, on the retreat from religion. Neither upper-class nor working-class women were confined to the domestic life that some middle-class males thought proper. Secondly, we cannot speak of a cohesive middle class at all; for the term covers everyone from merchants to small shopkeepers and farmers, the professions, fiananciers, and manufacturers. The incomes enjoyed by these different groups varied greatly. While the secure centre consisted in those families with an annual income of £300–£1,000, this was a very long way from the position of the millionaire banker at one extreme or the insecure clerk on £100 or less at the other. There was a major distinction between businessmen on the one hand and those in the middle classes who depended upon the state for their employment. Even the business world was divided between free-traders and protectionists. Politically, too, they were split down the middle. Whereas the Home Counties professional man was likely to be a Conservative, his counterpart in Wales or the north-east was typically a Liberal. Even the lower middle class comprised two distinct elements. On the one hand were the pre-industrial small entrepreneurs and shopkeepers, and on the other the low-paid white-collar employees such as clerks, civil servants, elementary schoolteachers, journalists, insurance collectors, and travelling salesmen. Many of these people found their incomes scarcely adequate to support conspicuous forms of middle-class behaviour like servant-keeping.

In short, the middle classes comprised a series of social layers, some of which overlapped with the classes above and below them; there were elements of conflict and of collaboration in these relationships. During the early and mid-nineteenth century, political controversy often took the form of a struggle between members of the middle classes and the aristocracy. 'We are a servile, aristocracy-loving, lord-ridden people', complained Richard Cobden who was a foremost exponent of the view that the landed class was merely parasitic upon the industrious classes – by which he meant middle and working. Naturally, such radicals forged alliances with working men in order to challenge the political power of the idle rich. Even in the late nineteenth century there remained a good deal of life in this pattern, as is evident from the vigorous revival of the question of land taxation down to 1914.

On the other hand, there is plenty of evidence to suggest that by the 1880s this type of conflict had diminished. This was an inevitable consequence of the success of the middle classes in joining the political establishment. They were voters, local councillors, MPs, and even Cabinet ministers. From the 1860s such men as John Bright, Henry Fawcett, and John Morley on the Liberal side, and W. H. Smith and Richard Cross on the Conservative side, were achieving ministerial status. From 1908 Britain's Prime Ministers were invariably drawn from a middle-class background: H. H. Asquith, Lloyd George, Bonar Law, and Stanley Baldwin.

However, the most important sign of the *rapprochement* between middle and landed classes was the concentration of propertied interests in the Conservative camp especially after 1886. This shift at parliamentary level was reinforced by other, economic pressures upon the middle ranks of society; small shopkeepers felt antagonized by the rise of the Co-operatives, and many employers reacted against the steady growth and militancy of trades unions. In short, middle-class behaviour was increasingly defensive and conservative in character. They were no longer – if they ever had been – boldly imposing their own entrepreneurial values upon the rest of society. In some ways they appear to have been subject to the downward spread of aristocratic habits and behaviour. In a period of falling land values it was easy for a businessman to acquire a country house and estate. This might not make a manufacturer an accepted figure in landed county society, but in the next generation it would be difficult to distinguish the two quite so clearly. In this connection the rapidly expanding public schools played an important role, for they provided a common education for sons of middle- and upper-class men, though it seems that the professionals were more attracted than, say, provincial manufacturers. Of course, some of the vehicles for interraction between the two social levels were of quite long standing. For example, middle-class families employed in the public service in India had long enjoyed an aristocratic style of life, and found it comparatively easy to fraternize with those immediately above them. The women of the two classes enjoyed extensive opportunities for mingling by means of the huge network of charitable activities and organizations. Among middle-class girls it became increasingly common to ape their superiors by indulging in coming-out parties, presentations at court, and marrying into gentry or titled families.

The Working Class: Conflict and Respectability

Approximately eight out of ten Victorian people were working-class in the sense that they had manual occupations. Marx and Engels confidently expected that, as Britain was the most developed industrial society, the workers would grow as a proportion of the population, would become increasingly immiserated, and would thus become more class-conscious. Indeed, class struggle culminating in the appropriation of private property was widely feared. Victorian Britain was, after all, characterized by huge inequalities of wealth and income; moreover, up to the 1860s the gap appeared to be widening. In 1867 the top 10 per cent of income-earners enjoyed 52 per cent of all income, while the bottom 40 per cent had only 15 per cent of income. Chartism had come and gone, but during the last thirty years of the century the trade union movement appeared to be in a long-term cycle of growth; its membership rose to 2 million by 1900 and reached over 4 million by 1914. The incorporation of semi-skilled workers was gradually making the movement a less exclusive organization. In spite of the economic depression, these unions became involved in between 500 and 900 separate strikes every year during the 1890s, and anything from 3 million to 30 million working days were lost. Gradually the small workshops which inhibited the development of class consciousness were giving way to large-scale industrial organizations in which employer and employees found themselves separated into two conflicting camps.

However, these developments hardly justified the mid-Victorian Marxist analysis. As Britain's capitalist economy matured, it led to a growth of 'white-collar' employment in the service sector and to the relative decline of the proletariat. Far from being immiserated, from the mid-Victorian period onwards sections of the working class enjoyed greater prosperity; in the short run this contributed to the divisions within the working-class community, though in the long run the majority were to gain access to higher material standards. Rather than a simple consolidation around a common proletarian objective, there was something of a scramble for sectional advantage; moreover, the changing occupational structure permitted a certain amount of mobility between working-class families and the social groups immediately above them.

Traditionally many skilled artisans aspired to become small employers. However, by the later nineteenth century this was less and less feasible, at least in the industrial sector. Many men, and their wives, still pursued this objective by becoming shopkeepers, builders, plumbers, joiners, or painters and decorators, though such enterprises frequently failed and the proprietors remained in the ranks of the working class. A much more important form of mobility lay in the expanding opportunities in such occupations as teaching or in banks and local government where clerks were required. Many of the recruits to such jobs were drawn from working-class families. The key lay in improved elementary education, which equipped the late-Victorian generation with the necessary grasp of English and mathematics for this employment. As in other social classes, women often played a key role in upward mobility. Their work as domestic servants, for example, made them more familiar than men with middle-class habits, and sometimes led them to aspire to a different style of life; men, largely because of their different experience of work, were more inclined to want to maintain class boundaries. Their experience and attitudes sometimes led mothers to encourage their sons to use education as a vehicle for upward mobility. But for many, marriage offered a direct route; in some areas it has been found that up to 30 per cent of the daughters of skilled workers married into the lower middle class, thereby blurring the lines of demarcation between them.

Even within the ranks of the working class there remained some major divisions. Contemporary investigators like Booth and Rowntree identified six subclasses within the working-class community in the towns. But Victorian Marxists focused on one major division, that between the majority and what they called a 'labour aristocracy' at the top. This section composed around 10 per cent of the working class – largely the better-paid or skilled men. It was tempting to regard the labour aristocracy as an obstacle to the development of class consciousness, because its members merely aped the middle classes and tried to maintain differences between themselves and other manual workers.

However, it is far from clear that the labour aristocracy really corresponded to the distinctions understood within the community at the time. The income earned by a single male head of household now seems a very inadequate criterion for assessing status within the working class. One problem is that a man's wages frequently fluctuated during the course of his working life according to age, fitness, or changing skills; therefore the labour aristocracy would not have represented a fixed body of men but a category whose composition changed over time. In the long term the migration of men into the higher- and more regularly paid occupations steadily widened the section at the top of the scale. In any case, however, the position of a working-class family depended upon a much wider range of factors. Even its income was frequently the joint result of the wages of wives and children, not just of husbands. Above all, the way in which the housewife spent the income would determine whether a family was seen as part of the respectable working class. The type of house one rented, the use made of the pawnbroker and credit, the regularity of school attendance by the children, drinking habits, the possession of suits of Sunday clothes

– these were all criteria by which a family's status might be measured. An important sign of respectability was the adoption of self-help strategies such as friendly societies and the Co-operative stores; clearly this was by no means confined to a 10 per cent minority but, on the contrary, had become fairly typical by the end of the century.

Respectability should not be interpreted as a sign that workers had simply become conservative, or unduly keen to emulate the habits of the middle classes. On the contrary, the better-off, literate, and socially mobile working men often took the lead in organizing the workers and joining socialist societies. In contrast, the depressed residuum were too ground down by the daily struggle for survival to find the confidence and optimism necessary to be able to challenge the political status quo. Thus, while the emerging movement of unions, the ILP, and other societies was not the vanguard of a revolutionary proletariat, it undoubtedly represented a powerful sense of pride and independence within the working-class community. Conventional politicians appreciated that they were no longer threatened by the kind of class conflict characteristic of the 1830s and 1840s; rather, the workers posed a challenge broadly within the limits of the political and economic system. To that extent, the Victorians had achieved a relatively stable class society.

6

Isolation and Expansion

In common with other nationalities, the British drew much of their sense of unity from essentially negative and external factors – a mixture of superiority over foreigners and fear about the threat they posed. In the three decades after 1870 the latter emotion began to predominate in Britain, though xenophobia did not reach its peak until the Edwardian years.

The Defence of the Realm

In spite of her resources, Britain remained essentially a peripheral power in European terms for most of the nineeenth century. She lacked a substantial regular army, and her governments, regardless of political complexion, showed themselves reluctant to increase expenditure. Thus in the 1870s the British had 0.25 million soldiers compared to 3.5 million in France and Germany and 4 million in Russia.

This was not an indication that the British were a particularly peaceable people; rather, they continued to channel their aggression and expansion into colonial skirmishes, where the costs were usually modest. Instead, Britain concentrated on maintaining a superior navy. Its purpose was to protect existing commercial interests, to pressurize backward states to engage in trade, to transport soldiers to the colonies in an emergency, and to blockade the enemy's ports in a major war. Victorians sometimes boasted that the essential British policy was one of isolation. Of course, in the ordinary sense of the word Britain was the least isolated country in the world, for her myriad commercial interests involved her in the politics of innumerable other states. But Britain could claim to be isolated in a special sense: she avoided any binding treaty obligations to give military support to the other European powers. Her geographical position made this a relatively easy option. It required the maintenance of the two-power standard for the Royal Navy. With control of the Channel and the North Sea, Britain had little to fear from the Continent and little reason to become entangled in standing alliances. Both Liberals and Conservatives favoured the so-called 'Blue Water' school of national defence, in the belief that it was cheaper and less provocative than the alternatives. This was underpinned by a good deal of expert writing in the late-Victorian period which held that the key to victory lay in command of the sea attained by decisive battles between capital ships.

However, Britain's relations with the European powers were less simple than these principles suggest. For one thing, Britain did undertake certain obligations to foreign states. For example, a series of treaties with Portugal went back several hundred years, but were useful only as a pretext for warning other powers away from territory in Africa claimed by the Portugese. Britain had also given a guarantee of the territorial integrity of Belgium going back to 1839 and reaffirmed in 1870. This reflected a material British interest in preventing the occupation of the Low Countries by a major military power. But Britain was not the sole guarantor and could, if she wished, decline to act except in conjunction with others. In addition, Britain acquired a formal involvement with Turkey in the form of the Cyprus Convention of 1878, which bound her to maintain Turkey's Asiatic territory but did not involve any obligation to fight a European war in her support. In any case this was rejected by Gladstone and the Liberals – which underlined a widely held view that Britain's parliamentary system rendered participation in standing alliances impossible; commitments entered into by one government were likely to be repudiated by its successor. In fact this was an exaggerated view. As in most countries, foreign policy showed a good deal of continuity regardless of changes of regime. All of these commitments were held rather lightly by British governments. 'Our treaty obligations', as Salisbury put it, 'will follow our national inclinations and will not precede them.'

Another and more important qualification to British isolationism is that after the 1870s the acceleration of colonial expansion brought Britain increasingly into dispute, and sometimes collaboration, with European states, thereby exposing the limitations of her diplomatic position. Eventually this led to a formal breakdown of isolation. The other disruptive factor was the Franco-Prussian war, which made the British much more concerned about the security of the British Isles. Since 1815 the absorption of the European powers in their internal problems and the isolationism of the USA had left Britain remarkably free to pursue her own interests. But after 1870 this situation changed fundamentally. The triumph of the Prussian armies in three successive wars marked the emergence of the kind of Continental power not seen since Napoleonic times. It was not just the size of the new European armies but their capacity to mobilize quickly that alarmed the British. Although the sea made a direct threat unlikely, the build-up of the French and Russian navies and the introduction of ships built of steel and powered by steam undermined confidence in naval protection. Though still the largest single navy, the Royal Navy suffered from poor gunnery, training, and tactics, and it was lucky not to have been challenged by a serious rival in this period. Moreover, Britain's growing dependence upon imported food rendered her highly vulnerable to enemy fleets.

For much of the century the underlying weakness of Britain's position had been obscured by the absence of opponents in Europe, the pre-eminence of the navy, and the patriotic bombast associated with Lord Palmerston. But when a serious challenge arose – as with Russia in the

Crimean War in the 1850s – Britain managed to deal with it only by means of an *ad hoc* alliance with other powers; victory in the Crimea was ruinous and the gains dubious, but it would have been impossible without French troops, which made up the majority of the allied force. Even this expedient could be adopted only exceptionally. During the 1860s, for example, when the Prussians engaged in three Continental wars Britain stood on the sidelines. She strongly disapproved when Bismarck grabbed the duchies of Schleswig and Holstein in 1864–5, but in the absence of an ally even Palmerston swallowed his pride. A similar but more serious dilemma arose over Prussia's defeat of France and the annexation of Alsace-Lorraine by the new German Empire in 1871. Gladstone was appalled by this, but since Austria and Russia were both unable or unwilling to co-operate, no practical measures could be taken. Foreign policy now began to damage Gladstone domestically, because the Russians took advantage of the crisis caused by the Franco-Prussian War to renounce the Black Sea Clauses which had been part of the settlement at the end of the Crimean War. These forbade Russia to keep warships in the Black Sea. This area was of strategic importance to Britain because the Black Sea offered Russia a warm-water route via the Bosphorus and the Dardanelles into the eastern Mediterranean, through which much of Britain's trade with India and the Far East passed. Although the government managed to convene a conference in London in 1871, it failed to win the backing of France or Germany; as a result, Gladstone was blamed for allowing the Russians to reverse the verdict of the Crimean War, although the only alternative would have been a declaration of war by Britain alone, a foolish and impractical step.

The Eastern Question became important in polarizing opinion between the two political parties. As Premier after 1874, Disraeli quickly grasped the opportunity to play the patriotic card as Palmerston had traditionally done. The chance arose when Russia went to war with Turkey in 1877, following the massacres of Christian Slavs in the Balkans. Since the Turkish Empire straddled the European and Asiatic shores of the Bosphorus, it provided a convenient buffer to Russian expansion. Disraeli was not alone in believing that the Russians exploited Slav nationalism in order to undermine Turkey and justify intervention against her. However, his concern led Disraeli dangerously close to war with Russia, much to the dismay of his own colleagues; his Foreign Secretary, Lord Derby, resigned. What saved Disraeli was the unease of the other powers about Russia's territorial aggrandisement; as a result they convened the Congress of Berlin in 1878, which modified Russia's gains and left Britain in possession of the island of Cyprus. This triumph for Disraelian diplomacy proved to be an empty one. Cyprus, which lacked an adequate harbour, was of little use in protecting British trade routes. And the establishment of Bulgaria, in whatever form, marked another step in the disintegration of Turkey. Although Gladstone's opposition was expressed in moral-humanitarian terms, it was also much more *realistic* than Disraeli's policy. It may have been wrong to prop up Turkey, but it was also futile; in the long run the creation of nation states in the Balkans offered a better prospect of checking Russia's advance. Moreover, it was Gladstone who subsequently did most to secure Britain's strategic interests in the Mediterranean by the occupation of Egypt in 1882, which enabled her to control the Suez Canal. In fact, by the 1890s the Conservative leader, Salisbury, came to the conclusion that Gladstone had been right and that Britain had backed the wrong horse in Turkey.

The Egyptian question proved to be only one of a series of military and colonial worries during the 1880s. By 1882 there was alarm over reports that the French were digging a tunnel under the English Channel. This gave way to better-founded evidence of a major French naval building programme. Gladstone felt obliged to devote an extra £5.5 million to the navy. But this went against the grain for both parties; they believed that Britain was already reaching the limits of her resources, so that further state spending would damage her capacity to create new wealth. In the event, Britain did fail to maintain her position: whereas in 1883 she had 38 capital ships to 40 for France, Russia, Germany, Italy, the USA, and Japan combined, by 1897 the ratio was 62 to 96.

In the absence of adequate physical force, the British government's preferred alternative was negotiation, but this was only likely to be effective in concert with another power. In fact, the annexation of Egypt had been planned as a joint intervention with France, who had dropped out at the last minute, leaving Britain exposed to charges of imperialism. It was more usual in the 1880s for Britain to seek co-operation with Austria, Italy, and Germany, simply because they had less reason than France or Russia to dispute her colonial interests. However, by the 1890s Bismarck, who had often worked with Britain, was no longer in power; and the strain of perennial colonial disputes began to tell. In this decade Britain experienced problems with Russia on the northern frontiers of India, opposition from France and Russia to expansion in Nigeria, hostile collaboration by France, Russia, and Germany in China, where Britain had a valuable trade at risk, as well as a verbal clash with the American President over British Guiana. The year 1896 brought the notorious Jameson Raid in the Transvaal, which was condemned by the German Kaiser; 1898 saw France and Britain close to blows at Fashoda near the headwaters of the River Nile; and after 1899 the outbreak of war in South Africa drew hostile reactions to Britain from all the Great Powers. Britain, in Joseph Chamberlain's words, was like 'a weary Titan staggering under the too-vast orb of his own great weight'.

In these circumstances Britain's diplomatic isolation looked increasingly like weakness; she had neither the military nor the naval strength to defend her far-flung interests. She had remained aloof from Europe at a time when the other powers had gradually committed themselves to two alliance systems. The Dual Alliance of France and Russia opposed the Triple Alliance of Germany, Austria, and Italy. Now, neither of these had been formulated with Britain in mind; they reflected Franco-German enmity over Alsace-Lorraine and Austro-Russian antagonism over Balkan nationalism. Yet in practice the alliance system operated to Britain's disadvantage. For whenever a new problem arose Britain had no regular ally to back her up, and she could not afford to use physical force to defend all her interests.

As a result, the feeling grew in statesmen like Joseph Chamberlain that Britain ought to seek an ally. The crisis in the Far East during 1897–8, when China became the victim of pressure, especially from Russia, brought matters to a head. On balance Germany appeared to be the best prospect, partly because Britain had co-operated with her in the past, and because she could bring pressure to bear on Britain's most acute problem – Russia. While Britain alone lacked the strength to warn Russia off when she tried to expand in the Far East, the combined influence of Britain and Germany would be a more serious deterrent. But the drawback for Germany was obvious. By antagonizing the Russians she would encourage the French to invoke their alliance and force Germany at some stage into a war in the east and the west simultaneously. Could Britain offer Germany anything to make the risk worthwhile? France remained Germany's most immediate concern. Britain's navy, her proximity to French coasts, and her ability to threaten French colonies would amount to a powerful deterrent if she allied with Germany. Yet the government hesitated. Though a great irritation in the colonial sphere, France was not a problem in the European context. Increasingly, Germany seemed to be the dominant Continental force. The British instinct had always been to promote coalitions against the biggest Continental power; an alliance with Germany entailed a risk of encouraging a future German government to provoke the French into another war and an inevitable defeat, which would further exacerbate the imbalance of forces.

By 1901 the enterprise of an Anglo-German alliance had broken down. The failure was rendered more decisive because negotiations had coincided with the new German naval laws of 1898 and 1900, by which Germany announced her intention of building a navy capable of challenging Britain in the North Sea. But the problem that had stimulated ideas for a deal with Germany remained. Her failure made Britain all the quicker to acquire an alternative Far Eastern ally – Japan. The treaty signed in 1902 bound each country to remain neutral if the other were at war with a single power, but to give active support if the other were engaged with two opponents. Of course this treaty was designed to meet a local problem, and its immediate effect was to allow Britain to withdraw some of her ships from the Far East. But the firm pledge did represent a break with the tradition of isolation. Moreover, the alliance proved to have profound effects upon the European situation in the years before 1914.

The New Imperialism

By the 1860s Britain's imperial possessions comprised three distinct elements: the white dominions, the huge Indian Empire, and a scattering of protectorates and colonies in Africa, the Far East, the Pacific, and South America. Though already large, this Empire was to grow at a remarkable rate in the 1880s and 1890s. It is not altogether clear why Britain never had a long-term plan of expansion. To complicate matters, there is even evidence of anti-imperial sentiment among politicians in the mid-Victorian period. Governments of all shades shared the irritation voiced by Disraeli when he referred to colonies as 'millstones around our neck'. All too often, British settlers got into difficulties with native peoples or other Europeans and had to be bailed out at great cost by the home government. But few went as far as free-trade radicals like Richard Cobden, who argued that the only useful part of empire was trade, and that formal political-military control was an unnecessary cost.

In practice British policy, if it can be so described, lurched between caution and recklessness. As long as they could get away with it, governments avoided imposing formal control, but where British commercial interests were threatened they invariably stepped in, however reluctantly. Very often the Westminster government found itself unable to control events; rather, it was led by the men on the spot. In India a succession of Governor-Generals had ignored instructions and grabbed new territories, though by the 1870s the improvement in communications with London and the tendency to make party political appointments reduced the scope for freelance imperialism. Even the supposedly anti-imperial mid-Victorian era witnessed considerable expansion – Sind and the Punjab in the 1840s, Oudh, Burma, and Hong Kong in the 1850s. The pattern of expansion clearly reflected local pressures and strategic objectives; a new acquisition could be seen as buffer to protect an existing territory or as a key point on a trade route. Indeed, Britain's commerce was so extensive that strategic arguments could be invoked to justify almost any annexation.

During the late 1860s party politics increasingly entered into imperial questions, when the Conservatives attacked Gladstone for seeking to dismember the Empire. Despite such charges, colonial policy continued to be largely a bipartisan affair. But there was some basis for controversy. After 1868 Gladstone's ministers attempted to make economies, partly by withdrawing some of the troops scattered around the world. In particular, by pulling them out of New Zealand in 1869 they intended to serve notice on the settlers to cease provoking the Maoris by encroaching on their land. When in office the Conservatives supported this policy. In the long term Gladstone envisaged that the Empire would evolve into something like the institution that in the twentieth century was called the British Commonwealth. By 1870 it was already accepted that the provinces of Canada and Australia were entitled to internal self-government, preferably under a strong federal system. The example of the American colonies served as a warning against attempting to retain British control for too long. But some Liberals looked further than the white dominions to India. Although obviously autocratic and alien, British rule in India was not an unmixed autocracy. The British people there took it upon themselves to criticize official policy in quite extreme terms in the 1880s, thus offering Indians an excellent example of British politics at work. Moreover, for much of the nineteenth century Indians also enjoyed rights like freedom of the press which in Europe were seen as a key aspect of a liberal system. By establishing an English educational system, Britain had helped to produce a generation of Indians well-versed in the political principles of Western liberalism who were anxious to obtain for India the benefits of British parliamentary government. Indian independence was not on the agenda; but a Liberal Viceroy like Lord Ripon (1880–4) accepted that it was both right and necessary to give educated Indians greater participation in their government. The formation of the Indian National Congress in 1885 was an indication that Indians appreciated that concessions could be won from the politicians in London, who often showed more sympathy than the British residents in India.

After 1874 Disraeli's ministry seemed to mark a reversion to a forward policy in imperial affairs which culminated in wars in South Africa and Afghanistan. In his Midlothian campaigns, Gladstone condemned this as reckless aggrandizement which cost money and achieved no good. Of the British invasion of Afghanistan he urged an audience:

> Remember this, that the lives of those poor savages, as we may call them, in their humble homes amid the winter snow, are as precious in the eyes of Almighty God as are your own.

Historians, like contemporaries, have sometimes forgotten that this attitude was not mere sentimentality, but was consistent with British interests. Since a Russian invasion of Afghanistan was scarcely practical at this time, there was no point in Britain going forward to meet her in central Asia; Afghanistan was best left as a natural buffer zone.

However, the difference between the policies of the parties was somewhat exaggerated. In South Africa the conflict with the Zulus was a typical product of the pressure exerted by the white settlers for land, and the initiative came from the High Commissioner, Sir Bartle Frere, acting contrary to his instructions from London. In India Disraeli had deliberately appointed the reactionary Lord Lytton as a party-political Viceroy, and Lytton adopted policies, such as arbitrary control over the Indian press, which were subsequently reversed by the Liberals. But Lytton exceeded his brief in plunging into an invasion of Afghanistan; Disraeli and Salisbury wanted more influence over Afghan foreign policy with a view to excluding Russian influence, but a reckless war was another thing.

Qualifications must also be made about other forward moves under Disraeli, including the establishment of protectorates over the Gold Coast, the Malay States, and Fiji. He in fact took no interest in these, leaving matters to his Colonial Secretary. Moreover, in the Gold Coast and the Malay States the policy was simply a continuation of initiatives begun under Gladstone. It is true that after his return to office in 1880 Gladstone checked Tory policies in South Africa and India. But, as we have seen, he soon felt obliged to intervene militarily to safeguard British interests in Egypt. As so often, he found himself subsequently caught up in the strategic-political consequences. Egypt's own security was considered to depend upon control of the River Nile, which stretched all the way to Sudan and East Africa. By 1885, when these areas were experiencing Muslim nationalist rebellions, Gladstone had had enough of this, and he ordered a withdrawal from the Sudan. In a classic case of imperial insubordination, General Charles Gordon prevaricated over the withdrawal, only to be massacred by the rebels at Khartoum. As usual, the home government felt obliged to send another force to rescue their representative – it simply arrived too late.

The Egyptian invasion of 1882 was widely held to have inaugurated a new phase of imperialism which *The Times* memorably described as the 'Scramble for Africa'. At first sight, all that was novel about the expansionism was the involvement of new imperial powers (Germany and Italy), the concentration on Africa, and the remarkable speed at which territory was acquired. Britain herself absorbed Egypt in 1882, Somaliland, Bechuanaland, and Nigeria in 1884–5, the Rhodesias in 1889, Kenya, Zanzibar, Uganda, and Nyasaland in 1890, and the Transvaal and Orange Free State after the war of 1899–1902.

Whether this frenetic activity really amounted to a new policy on Britain's part is, however, not obvious. Expansion traditionally reflected economic motivation; the search for cheap raw materials and markets for British goods formed a thread running through all her colonial ventures. It can, however, be argued that after the 1870s the consciousness of the loss of markets, the reintroduction of tariffs, and the lack of employment at home underlined the urgency of finding fresh economic opportunities, and thus accelerated the process of colonization. But much depends upon who was actually responsible for imperial expansion: private entrepreneurs or governments? Several leading politicians did identify an economic rationale behind empire. Lord Rosebery supported British expansion in Uganda in the 1890s, and claimed that Britain was engaged in 'pegging out claims for the future'. Joseph Chamberlain also showed himself aware of the material gains: for example, he saw British control of South Africa in terms of the region's mineral wealth and the opportunities for emigration by British workers. Such men shared a broader vision of 'imperial federation' which implied drawing the British possessions together into a single economic community protected by tariffs, so as to give Britain a large and secure 'domestic' market for her products. The successful developments of great continental-scale economies in the USA and Germany seemed to suggest that large units held the key to future prosperity.

On the other hand, nothing substantial came of the idea of imperial federation; it remained a minority cult in Britain and aroused little interest abroad. Moreover, the statesmen who usually determined British policy – Salisbury and Gladstone – did not think primarily in terms of economics. Britain notoriously failed to use her network of diplomats abroad to promote British commercial interests. Where Britain did enjoy a large trade, as in China, the instinct of governments was to try to preserve free trade, not to impose formal control. In Africa, Salisbury and Gladstone cheerfully ceded huge territories to the other European powers, which suggests that they often had considerations other than economic in mind. There was, of course, good ground for scepticism about the contemporary claims made about the material value and opportunities of the new territory. In Africa only the Royal Niger Company, dealing in the cocoa and palm oil of the western states, proved to be really profitable. The British East Africa Company never paid a dividend to its shareholders, and, in spite of the fanfare attending its launch, Cecil Rhodes's South Africa Company only occasionally paid one. In 1902 J. A. Hobson argued, in *Inperialism: A Study*, that the real purpose of the new colonies was to provide a field for investing the surplus capital generated by European industry. In fact, over two-thirds of foreign investment went either to old-established colonies or to countries outside the Empire, notably the USA and Argentina. Beyond the Gold Coast and Nigeria the African territories were not, in this period, especially valuable. Vast areas like the Sudan, Kenya, Bechuanaland, and the Rhodesias were thinly populated and often semi-desert, lacking the communications necessary to exploit whatever resources they might have had. They were consequently unattractive to Western capitalism.

Whatever the reason, British ministers remained doubtful about the value of new territory. For example, Salisbury attracted criticism from patriotic Conservatives when he allowed the tract of territory between Lakes Victoria and Nyasa to become German East Africa. Enthusiastic imperialists noted that its incorporation into British territory would have given Britain a continuous empire stretching from Cairo to the Cape. But Salisbury was not impressed: the land at issue had little intrinsic worth. Clearly the Prime Minister thought in different terms when partitioning the African continent. Sometimes, as in Egypt, strategic motives came to the fore. But increasingly in the 1880s and 1890s Salisbury simply regarded colonies as useful weapons in the more important game of European diplomacy. It was, for example, desirable for Britain to retain the goodwill of Bismarck, and if this meant helping Germany to her African empire, so be it. To some extent all the great powers used the partition of Africa as a safety valve to reduce tensions in Europe. The disputes that arose over colonies were bearable if they helped avert a really serious conflict in Europe.

The Popularity of the Empire

During the later 1880s and the 1890s, the pace of expansion was so rapid and the eruption of imperial issues so frequent that it seemed as though the Empire had come to occupy a much more central part in British life than hitherto. Controversy subsided for a time, as both parties accepted that expansion was a desirable thing. Liberals like Rosebery and Liberal Unionists like Chamberlain found common ground with Conservatives, while propagandists like Rudyard Kipling articulated the moral justification for Empire in terms of extending the benefits of Western rule to backward peoples (*The White Man's Burden*). Yet it is less clear how far these sentiments spread beyond the political elite. How popular was the Empire in British society?

A wide variety of evidence can be adduced to suggest a considerable and growing enthusiasm for almost anything connected with empire. Since the 1850s the popular imagination had been stirred by the heroic explorations of David Livingstone, H. M. Stanley, and Richard Burton. In their wake came narratives of adventurers in exotic locations, ranging from fiction like Rider Haggard's *King Solomon's Mines* to first-hand reports such as Winston Churchill's *The Malakand Field Force* and *The River War*. Schoolchildren learned a patriotic version of history and geography based on British achievements; and the success of the *Boy's Own Paper* (1879), which attracted sales of a million copies, underlined the growing market for patriotic adventure. From 1896 the *Daily Mail* tapped similar sentiments among adults. This period also saw a number of very popular Imperial Exhibitions and a great expansion of music-hall, of which patriotic-imperial songs formed a staple element. In addition, the development of advertising by the suppliers of cheap consumables often served to emphasize the close connection between better living standards and supplies of Empire products.

A key element in all this was the improvement in technological methods of representing the Empire in tangible and visual forms to millions of people – the novel aspect of imperialism. In addition to the press, books, and entertainment, the magic lantern was employed by large organizations like the Primrose League to portray the cause of Empire in crude but colourful images. Among the examination questions set for juvenile members of the Primrose League were: 'What would be the advantages of European supremacy in China?' and 'Give a short sketch of the war in South Africa, its cause and progress. What good is expected to result from it?' Clearly, many late Victorian children grew up in a climate of opinion that was unambiguously imperial.

However, there must be some doubt about how much these manifestations tell us about the popularity of imperialism. Some can be seen as propaganda by commercial interests or by middle and upper-class leaders, rather than as genuine reflections of the attitudes of ordinary people. It has been argued that the higher social classes were consistently more enthusiastic about empire than the working classes. Also, many of the expressions of imperialism were so vague and generalized that it is difficult to know what they meant; on occasions of national triumph or disaster imperialist sentiment merged into patriotism, xenophobia, racism, and monarchism. Like many popular emotions imperialist fervour was apt to fluctuate, especially when expressed in terms of politics. The classic case is the Boer War, which undoubtedly engendered widespread support for a year or two. But as the war dragged on and the sense of crisis passed, a reaction set in. By 1902 it had become clear that a costly war had left a legacy of high taxation and inflation but no obvious gains. The defeated Boers were as strongly entrenched in South Africa as ever. The opportunities for employment in South Africa dwindled when Sir Alfred Milner, the High Commissioner, imported Chinese indentured labourers to work in the mines. Within a short time enthusiasm had been dissipated, and indeed, popular backing for empire never really recovered from the setback.

On the other hand, between the early 1880s and 1900 the imperial message penetrated deeply into all levels of British society; the critics of expansionism, though voluble, were marginalized for a time. Moreover, the late Victorian generation did enjoy a special and tangible link with empire that is probably of greater significance than the mere propaganda. Between 1870 and 1914 approximately 10 million people emigrated from the British Isles. Many of these were Irish and Englishmen on their way to the USA. But many working men looked to Canada, Australia, and New Zealand for better economic opportunities than those enjoyed at home. Amongst groups like agricultural labourers and dock workers, where surplus labour kept wages low, emigration seemed highly desirable. A number of trade unions promoted emigration schemes in order to strengthen the bargaining power of those who remained at home. In addition there were extensive programmes by charitable bodies like Dr Barnardo's to send children from poor families to Canada. Some Victorian feminists were keen to promote female emigration because of inadequate employment opportunities at home, while anti-feminists urged the same policy in order to redress the sexual balance of the population.

For working men emigration, regardless of destination, constituted an important safety valve, for by the late nineeenth century Britain's domestic economy was patently unable to generate enough jobs to absorb its growing labour force. The traffic in people also had political implications. In rural districts, for example, those who left home were invariably the more discontented, able, and ambitious. This tended to deprive such areas of their natural leadership, and deepened the passivity which so frustrated the radical politicians. One of the ironies of emigration was the role of Scots crofters and others driven out by pressure and poverty. Though highly critical of the British establishment in the home context, in Canada the Scots came to occupy a central place in government, and emerged as a highly loyalist, pro-British element against the pressures exerted by both French and Americans.

The most obvious effect of all the movements of population was to leave most British families with one or more close relatives working in the Empire, while others were involved in its defence through the army or the navy. This made for pride and a sense of common interest which was demonstrated by the response to crisis in 1899 and 1914. But empire was not all adventure and war. A multitude of other interests linked the home population with those overseas. For some the colonies offered a field for moral endeavour, notably in missionary work. Another interest lay in gathering the horticultural riches of the tropics – the rhododendrons of South Asia, the buddleias of China, and the eucalyptus of Australia – to mention but a few that graced English gardens. Another characteristically British interest was sexual opportunity, which the tropics offered in abundant variety. To a large extent, indeed, the British Empire can be seen as the work of homosexuals, active and passive, who found life more congenial there than at home, particularly after the tightening up of the law in 1886.

But perhaps the most important form of interaction lay in sport. The advantages of organized sport – fitness, discipline, teamwork, and loyalty – were regarded as vital in the imperial context. Rugby, cricket, and athletics served to foster local pride, but also underlined links with the mother country. Australians, for example, derived pleasure from taking on upper-class English cricket teams, but the development of institutions like the Ashes in the 1880s symbolized the mixture of competition and friendliness. White settlers often wished to foster and celebrate their British cultural roots, and this became especially important in regions where a non-British element existed, such as the French in Canada or the Dutch in South Africa. Cricket also offered a bridge between the British and native peoples in the West Indies and India. Prince Ranjitsinhji actually played in the English Test team and captained Surrey. 'The Game of Cricket', declared Lord Harris in 1880, 'has done more to draw the mother country and the colonies together than years of beneficial legislation could have done.' In post-Mutiny India, where the British were anxious to cement the loyalty of the princes to the Raj, the involvement of the males of the two races was regarded as serving an important political purpose.

Many of those involved in the higher levels of imperial affairs obviously did not emigrate but became semi-permanent residents of the colonies, often moving from one to another. They comprised the administrators, policemen, solicitors, and merchants. As such groups were closer to the politics of empire, they expressed their imperialism in terms of public policy and principle. But it would be misleading to think that they did not share with the lower classes a material and self-interested approach to Empire. Indeed, without the opportunities it offered there would have been an employment crisis for middle-class families at home. The leading public schools sent a high proportion of their boys to the Indian civil service or the colonial service, which provided prestigious and well-paid work. Once abroad, many men who came from relatively modest middle-class backgrounds discovered that one could live a charmed existence equivalent to that of the aristocracy in Britain. In the twentieth century the colonies were, in fact, increasingly to serve as a dignified dumping-ground for distressed landed figures, as well as for politicians to whom the government owed a debt or whom it wished to remove from the domestic scene.

Patriotism, Nationalism, and Militarism

During the nineteenth century the British regarded themselves as a patriotic but not an aggressive nation. Although foreigners, noting the remarkable expansion of the Empire and use of force it entailed, felt understandably unconvinced, this self-image was not without foundation. What contemporaries really meant was that Britain, in contrast to other European powers, was not *militarist*. Gladstone's attacks on Disraeli reflected the view that a reckless use of force to win territory was alien to British traditions. Not only did Britain's army remain comparatively small, but she showed very little enthusiasm for it. Below officer rank army life was barbarous, the men being confined to dismal barracks and becoming vulnerable to all kinds of immorality. They invariably joined up out of desperation or to escape something in their civilian life.

Naturally, dislike for the army as an institution did not indicate that the British were a pacific people; their militarism simply took different forms. For example, as a result of the military problems of the 1850s – the Indian Mutiny, the Crimean War, and the fear of an attack by Napoleon III – the authorities cast around for ways of improving Britain's capacity for home defence that would not lead to great expenditure or a large standing army. In 1859 an expedient was found in the Volunteer Movement. This involved part-time training for voluntary rifle corps. Volunteering brought with it some pride and satisfaction in defending one's country, an attractive uniform, and the social prestige conferred by the patronage of the Lords-Lieutenant – without the drawbacks of the regular army. Between the 1860s and the 1880s recruitment fluctuated between 224,000 and 288,000. Initially, middle-class and lower-middle-class men came forward, but by 1900 70 per cent were working men, mainly artisans. These 'Saturday soldiers' gradually helped to change attitudes towards the army in general in the late Victorian and Edwardian period. Their influence spread to youth organizations such as the cadet corps of the public schools, which were attached to Volunteer Units. Members of the Volunteers also founded the Boys' Brigade in 1883 and the Church Lads' Brigade in 1891.

After 1870 the renewed sense of an external threat also began to make an impact; the relatively relaxed British patriotism gave way to a more shrill and aggressive nationalism. The shock of Prussia's swift defeat of the French triggered the publication of a famous fictional account of the 'Battle of Dorking', based on the idea of an enemy landing on the south coast which caught the British unprepared and led to an advance on London. The Battle of Dorking was to be fought many times on paper and on the platform in the years before 1914; it served to keep the subject of national defence high up the agenda, and prepared several generations of British people for the need to defend the country against a surprise invasion.

What changed gradually during this period was the perceived source of the external threat. Most authors of invasion literature took it for granted that the enemy would be the French or the Russians. This made sense in the imperial context rather than in the European. For some years British attitudes towards Germany had been relatively favourable. The two countries largely operated in different spheres; moreover, British statesmen regarded it as perfectly proper for so civilized a state as Germany to assume a number of colonial responsibilities. The two countries were seen to share a common culture and religion, and to have connections through the royal family. There was much admiration in Britain for German achievements in philosophy, music, science, and education. Many German Liberals saw the British parliamentary system as a model. Unfortunately Bismarck did not, and the steady decline of Liberal forces in Germany was one sign that the sympathies between the two countries were ebbing. Relations deteriorated because of fears that the Germans were dumping goods at below-cost prices in the British market. However, the significance of economic rivalry should not be exaggerated: the commercial relations between the two countries were highly profitable for Britain, and Germany was only one of a number of industrial competitors. By the 1890s Germany was still a lesser irritation than France or Russia, though incidents like the Kaiser's telegram of support to President Kruger of the Transvaal over the Jameson Raid in 1896 severely damaged relations. Matters deteriorated around the turn of the century with the realization that, for the first time, Germany was building a fleet capable of threatening Britain in the North Sea. The combination of the German army and a first-class navy represented a nightmare, and Germany's transformation into Britain's national enemy was rapidly accomplished in the Edwardian period.

Meanwhile, the Boer War had the effect of exposing Britain's lack of friends in Europe and her military vulnerability; any serious challenge during the South African conflict would have been extremely difficult to meet. Thus, in spite of the disputes over the origins of the war and the methods adopted by Lord Kitchener to overcome the Boers, there was widespread agreement that it had exposed alarming deficiencies in

Britain's military, political, and social systems. This provided the stimulus to the movement for National Efficiency and, after 1903, to Chamberlain's tariff-reform campaign. The reformers attacked British party politics as incompetent in the planning and management of war. One modest innovation was the establishment of the Committee of Imperial Defence, which was designed to bring more expertise and less politics to bear upon strategic problems. Critics also looked for more professionalism in the army; the Officers' Training Corps (1902) and the General Staff (1904) were steps in this direction. More generally, they looked to more extensive forms of military training, higher educational standards, and improvements in the physical condition of children; hence the discussion of the registration of births, compulsory training of midwives, the provision of school meals, and the medical inspection of schoolchildren, much of which was recommended by the Inter-Departmental Committee on Physical Deterioration which reported in 1904.

Although National Efficiency as a movement was soon overtaken by normal party warfare, both the ideas and the general sense of external threat remained; they left a legacy in the form of several right-wing pressure groups whose object was to pressurize the government and to arouse popular support for a more robust national defence. One of these, the Navy League, had been in existence since 1895, but it had greater success after 1900 in pushing the government into extra naval rebuilding programmes. The most interesting yardstick of public concern over an invasion was the National Service League, formed in 1901 to campaign for universal male conscription. Although it became a substantial body in British terms, the NSL was never comparable in size or influence to similar organizations in Germany at this time. In spite of the Boer War, or perhaps because of it, the NSL had only 4,000 members by 1906: it took the prestige of Britain's foremost military hero, Lord Roberts of Kandahar, who became its President, to push up membership to 98,000 by 1912, though the mounting fear of invasion must have helped a good deal. Moreover, the NSL failed to convince the political parties of the need for peacetime conscription. Even the Conservative leaders thought this a bridge too far. It was widely believed that working men would simply not accept compulsory training, and the distaste for the Prussian approach to national defence was clearly deep-rooted. The British took pride in their own distinctive militarism, based on the voluntary system of recruitment. The huge popular response in 1914–15 proved the wisdom of this policy, and underlined the extent of patriotic sentiment.

Part II

The Reorientation: The Emergence of the Interventionist State, 1902–1918

7

The State, Social Welfare, and the Economy

Government intervention in social reform was by no means a novelty; several of the Edwardian reforms essentially extended the Victorian tradition of encouraging and requiring local authorities to provide certain services and minimum standards. However, in several respects the Edwardian innovations also broke with the localist and permissive principles of the Victorian era, they took the state into areas of its citizens' lives hitherto avoided, and they involved a major increase in the central bureaucracy. During the twenty years before 1906, the accumulating evidence about the extent and causes of urban poverty had gradually increased the pressure on governments to involve themselves more closely. None of this, however, had produced any startling changes in the methods of dealing with poverty. Local authorities lacked the financial resources, while Whitehall felt it had already increased grants considerably. The alternative options became clearer, but the political will had not yet been found. Thus the Fabians made detailed and well-researched proposals but remained on the political periphery. The New Liberals were not yet entrenched in the local organizations, and while they became influential at national level, their party remained largely out of office. In the Conservative ranks, Joseph Chamberlain had attempted to force the pace in the 1890s by advocating old-age pensions, but there could be little progress under the dead hand of Salisbury's premiership.

Eventually, short-term shifts in party politics created more favourable circumstances. The domestic crisis engendered by the South African War helped to discredit Salisbury, and positively encouraged some of those on the right to see the necessity for state intervention as a means of raising the physical efficiency of the population. Consequently, Balfour's premiership saw several initiatives, including the 1902 Education Act and the 1904 Inter-Departmental Committee on Physical Deterioration. A worsening economic climate led the government to introduce an Unemployed Workmen's Act in 1905. In a sense this represented the last fling of late Victorian experiments with public-works schemes by local authorities. The interesting thing about it was the way in which contemporaries interpreted it as a commitment by the government to provide employment for respectable working men. But Balfour intended nothing so radical; the 1905 Act was purely a temporary expedient. He dealt with the political problem by setting up a Royal Commission on the Poor Laws, which reported in 1909 and gave further publicity to the Webbs' case for breaking up the poor-law system and shifting some of its responsibilities to central government.

Consequently, by the time of the 1906 general election expectations were clearly high, both among politicians and in the working-class community. Though the Liberals returned to office without a formal programme of social reform, their MPs were undoubtedly committed to a number of innovations (see p.123). The emergence of a substantial Labour Party in the House of Commons simply strengthened the pressure. This became clear when scores of Liberal back-benchers joined with Labour MPs in support of amendments to the King's Speech demanding old-age pensions and the Right to Work Bill. In fact, the new Chancellor of the Exchequer, H. H. Asquith, had already begun to set aside resources for the introduction of pensions, so the reformers were pushing at a half-open door.

This is a reminder of the importance of accidental factors such as political personnel in explaining the timing and character of social reform. Asquith played a vital role, first as Chancellor in preparing pensions and taxation reforms, and then as Prime Minister by giving support to Lloyd George's controversial budget. The Local Government Board, in some ways the most likely vehicle for social reform, failed to contribute much, largely because of the negative and conservative attitude adopted by the President, John Burns. Not until 1914, when he was succeeded by Herbert Samuel, did the board help to promote welfare policies. In contrast the Board of Trade emerged as a key ministry under the leadership of a succession of radical Presidents, Lloyd George, Winston Churchill, and Sydney Buxton. Though Lloyd George's contribution to taxation and social policy is often seen in terms of his comparatively humble origins, this is clearly far from sufficient explanation, as the record of Burns, a working-class figure, suggests. Lloyd George's characteristic asset lay in his unorthodox methods. Lacking the patience for the usual committee-led route to legislation, he preferred to use advisers and assistants in whom he had confidence and with whom he could talk through the solutions unhindered by conventional civil servants. A typical decision was the dispatch of William Braithwaite to Germany to draw up a quick report on the operation of state welfare schemes there. He noticeably ignored the proposals in the reports of the Royal Commission on the Poor Laws, much to the chagrin of the Webbs, and relied on politicians like Charles Masterman who were outside his ministry but politically sympathetic.

Winston Churchill emerged as another unorthodox collaborator in these discussions. In 1905 Churchill's only Liberal credentials were his support for free trade. When offered a post at the Local Government Board he evidently saw it as a backwater. But Churchill was a man of sudden enthusiasms. By 1908 he was, in Masterman's words, 'full of the poor whom he has just discovered', and with Lloyd George he mapped out the series of reforms that became the centrepiece of the government's programme. Churchill brought to government a Tory paternalist's confidence that the power of the state should be freely employed to improve the efficiency and promote the stability of society. Asquith deliberately promoted Churchill in the face of jealousy amongst other Liberals towards this rather suspect newcomer to the party; but after 1911, when he had gone on to the Home Office and the Admiralty, Churchill found new fields for his restless energy.

It is also noteworthy that, as in the nineteenth century, social reform owed something to the role of civil servants and experts. Though Sidney and Beatrice Webb enjoyed much less influence than they expected to have, they managed to get the young William Beveridge taken on at the Board of Trade to prepare the scheme for unemployment insurance. At the Board of Education Sir Robert Morant, a more orthodox figure, worked successfully, under three transient ministers who were preoccupied with the politics of education, to introduce a scheme of school medical inspection in 1907. Morant understood that this would generate so much evidence of ill health amongst children that it would be comparatively easy to extend inspection into a school medical service.

Edwardian Social Reform

The early reforms of the Liberal government concentrated on children: school meals, medical inspection, compulsory registration of births, and the Children's Act of 1908. This reflected the current consensus expressed in the 1904 report of the Inter-Departmental Committee on Physical Deterioration – a mixture of humanitarian concerns and Bismarckianism. However, the measures did not entirely fail to provoke controversy. Even the modest bill which allowed local authorities to feed necessitous schoolchildren was interpreted in some quarters as subversive. For whereas those who received poor-law relief were disfranchised, there was to be no such penalty for parents whose children enjoyed free school meals. 'Why a man who first neglects his duty as a father and then defrauds the State should retain his full political rights is a question easier to ask than to answer,' complained A. V. Dicey, the eminent jurist. The more important change actually came in 1914 when the government made the provision of school meals *compulsory*, one of the several indications that the reforming impulse was still going strong when war broke out.

The scheme for old-age pensions, introduced in Lloyd George's first budget in 1908, was also modest and widely supported, but infringed traditional thinking in several ways. Elderly people aged seventy years with under ten shillings weekly income were eligible. The radical aspects in the scheme were the absence of any contributory element and the retention of voting rights by the recipients. In the event, the original estimate of 572,000 eligible pensioners turned out to be too low, for 668,000 qualified in the first year. Initially those in receipt of poor relief had been excluded, but this restriction was abandoned in 1911. The weekly pension was set at 5 shillings and, after some hesitation, the full rate was paid to married couples. Thus in several ways the scheme rapidly became more costly than originally anticipated.

Subsequent reforms proved much more complicated, largely because so many vested interests had to be squared or outfaced. Lloyd George's health insurance programme, which was embodied in the 1911 National Insurance Act, took a long time to produce because of opposition by commercial and professional bodies, and was less radical than he would have liked. In one sense the scheme was rooted in the Victorian practice of making contributions in return for benefits at a later stage. In return for weekly payments of three pence by the worker, two pence from the state, and four pence from the employer, the insured person received 10 shillings for each of 13 weeks and 5 shillings for each of 13 subsequent weeks in any one year; in addition, he enjoyed free treatment from a doctor. The boldness of the scheme lay in the fact that it replaced the multiplicity of local and optional policies with a compulsory, national one. This is why it attracted criticism from trade unions, whose members already subscribed to their own forms of insurance, and worried the Friendly Societies and commercial insurance companies, who feared competition from the state. Eventually many of them were appeased by the option of being incorporated into the health insurance provisions as approved societies through which benefits could be paid. The British Medical Association also objected to state intervention. As the spokesman of wealthy professionals it instinctively disliked the prospect of doctors becoming employees of the state; the health of the people came low down the BMA's priorities. It threatened to make the Act unworkable by refusing to treat insured people whose income was under £2 a week. However, it transpired that the complacent BMA leadership was out of touch with ordinary GPs, who commonly received only 4 or 5 shillings per patient per year under existing schemes. Lloyd George offered 6 shillings. Doctors agreed to participate and the BMA had to climb down. In a way this was unfortunate, for it meant that Lloyd George did not carry out his threat to create a state medical service by offering a salary of £350 per annum to 12,000 young doctors to run his scheme. The obvious weakness of health insurance was that it did not include children and the majority of women. Lloyd George was acutely aware of this, though he did introduce a maternity benefit of 30 shillings which was at least a recognition by the state of the vital contribution of women as mothers, and of the hardship endured by them.

Unemployment, an even more complicated problem, was tackled in several stages. The first came in 1908 with the introduction of labour exchanges, and the second in 1911, in the form of unemployment insurance. The latter was, again, based on contributions from worker, employer, and the state in return for benefits of 7s. 6d. for each of fifteen weeks in any one year. This was a distinctly cautious scheme, both in the sense that the unemployment fund was intended to be self-supporting and in the sense that only engineers, shipbuilders, and building workers were covered. These groups included some highly paid men, who were relatively well placed to afford the contributions. Also, their occupations were notoriously subject to sudden cyclical slumps. The motive here was to save skilled, respectable working men from being forced onto the poor law. Because these proposals for the unemployed ignored the recommendations of the minority report of the Royal Commission on the Poor Law they provoked criticism from the Webbs, who considered that attendance at labour exchanges ought to be compulsory, not voluntary, in order to enable the idlers to be distinguished from those genuinely seeking work. But Lloyd George and Churchill understood working-class politics better than the Webbs. They knew that, as it would take time to overcome the suspicion of labour exchanges, a voluntary approach was wiser. Nor was it feasible for the state to take over wholesale responsibility for the unemployed from the poor-law authorities, as the Webbs advocated. The potential costs were huge and the political complications considerable. Neither workers nor the government were keen to adopt a complete system of state regulation of labour.

Taken as a whole, the Edwardian social reforms were in no sense a welfare state, although they enjoyed an important link with the post-1945 system in the shape of the insurance principle. The Liberal measures were not intended as a comprehensive or uniform system of welfare provision. Rather, they involved targeting certain discrete parts of the problem of poverty; those not included continued to require a safety net, which meant that it was necessary to leave the poor law in being, though it clearly had a diminishing role to play.

Any weaknesses in detail do not detract from the overall, long-term significance of the innovations. They represented a marked departure from the principles governing welfare and poverty in the Victorian era, and they led the state into a position of responsibility from which no political party was subsequently able to remove it. At the time, the Conservatives became anxious to deny any intention of abandoning old-age pensions; and in spite of their attacks on the 1911 National Insurance Act they claimed only that they could improve the schemes if returned to office. By 1911 a Unionist Social Reform Committee had emerged, with a view to recovering the initiative for the party in this field. They favoured a break-up of the poor law and transfer of its functions to the county councils; they advocated the *compulsory* adoption of slum clearance, school meals, and medical treatment by local authorities; and they believed the necessary finance should take the form of grants-in-aid from central government, supported by direct taxation. This was a synthesis of Fabian socialism and Toryism, at once interventionist and nationalist. Although the party leadership avoided taking up these proposals, the fact that the committee included much of what was to be the inter-war leadership of the party – Stanley Baldwin, Edward Wood, Samuel Hoare, F. E. Smith, L. S. Amery, and William Joynson-Hicks – was an indication that social policy was to occupy a central place in British politics in the future.

Finally, it is now clear that the Liberal programme was not a brief expedient but part of a strategy whose momentum was being maintained right up to the outbreak of war. Ministers were very conscious of the shortcomings of their programme. For example, it was always intended to extend unemployment insurance once the viability and acceptability of the scheme had been established. The war and its aftermath influenced the timing of this, and it was 1920 when the majority of workers were incorporated. Similarly, Lloyd George had been frustrated in his wish to include widows' pensions in his 1911 reforms, and the money could not be found until 1925. Other innovations were, however, being introduced in 1914. Herbert Samuel pioneered a new policy for women in the shape of local-authority maternity clinics which were backed by a 50 per cent grant from the government. Subsequently this policy became obligatory for local councils, as did school meals in 1914. The Liberals were also anxious to tackle housing, especially as some Conservatives began to take up the issue. In 1914 Lloyd George made loans of £4 million available for local-authority building, and under the Land Campaign he proposed direct state intervention to build cheap rented housing. Edwardian reform was thus an ongoing programme, required by the political logic of the situation and made possible by the new approach to public finance adopted by the Liberals.

Popular Attitudes to the State

In 1912 Hilaire Belloc denounced state welfare on the grounds that the workers were being made to give up political freedom for servility. In one sense this was absurd; as we have seen, it was a characteristic of the Edwardian reforms that they involved no loss of political rights for the recipients. Pensions, in the words of the *Nation*, were 'a right conferred by citizenship, rather than a boon conferred by poverty alone'. As the extensive reforms of 1918–19 were to show, wider political rights now began to go hand in hand with social legislation. Thus commentators like Belloc and Dicey were fighting in the last ditch to resuscitate the ideas of mid-Victorian individualism. It was no longer seen as enough to exhort the working classes to stand on their own feet, for many who lived industrious, thrifty, and abstemious lives were still liable to fall into poverty through circumstances beyond their control.

However, if contemporary propaganda designed to discredit the social reforms carried little significance, we should not, on the other hand, assume that the innovations were automatically popular. We know that working-class communities traditionally regarded state intervention in their lives with suspicion and hostility. The Edwardian years are interesting for the way in which they reflect the transition from negative Victorian attitudes towards the post-1945 views of the state. The response seems to have varied according to the way in which each particular measure affected families and individuals. Old-age pensions were already widely advocated in the working class by 1908, and Asquith's scheme attracted emphatic support. This was because it involved no cost to the beneficiaries and none of the indignity associated with seeking relief from the poor-law guardians. The only bureaucratic impediment was the necessity to establish one's claim by producing a birth certificate; but once this stage had been reached an elderly person simply collected a pension at the post office, which came as a great relief to those who dreaded the onset of old age.

At the other extreme, reforms involving children continued to create some friction and resentment. The 1908 Children's Act allowed local authorities to take children into care, and made parents liable to prosecution for neglect; medical inspection exposed parents to embarrassment and criticism; and even school meals were problematical because it was open to the authorities to try to recover the costs from those parents deemed able to pay. Clearly, not every reform was welcomed at first. But some writers have followed contemporary propaganda by concentrating on the initial criticism of, for example, the 1911 National Insurance Act, and failing to see how this reform won acceptance in time. Though some employers and union leaders attacked the contributory element in the 1911 Act, this was to overlook the fact that most families were already accustomed to the idea of contributory insurance in one form or another. The question was whether one got value for money, hence Lloyd George's 'ninepence for fourpence' claim. Part of the problem lay in the fact that contributions began to be collected in 1912, though the benefits were not paid until 1913. But by that time the controversy had subsided; for most workers the state scheme marked an improvement over the failing benefit schemes traditionally available.

It is fair to say that one potential drawback to state welfare was the involvement of the recipients with an unsympathetic official bureaucracy. For example, unemployment insurance benefits would be paid at labour exchanges. However, the government was quite well aware, when launching its schemes in 1908, that many working men regarded labour exchanges with suspicion because in the past they had been used to supply labour to break strikes or to offer work at below union rates. These problems were effectively tackled partly by making attendance voluntary and allowing men to reject any jobs which seemed to undermine the terms and conditions for which the unions were contending. Churchill sought to reassure them further by employing trade unionists in many of the minor posts in the labour exchanges. This practice extended higher up the civil service too. Indeed, some Conservatives accused the Liberals of trying to shackle the labour movement by jobbery. *The Times* claimed in 1913 that 374 known activists in socialist and labour organizations were employed in government departments, notably in National Insurance, the Board of Trade, and the Home Office. The best known was David Shackleton, formerly of the Weavers' Union and a Labour MP, who was taken on by Churchill as a labour adviser at the Board of Trade. In this context, it is not surprising that the new labour exchanges attracted rapidly rising business, and the fears expressed about the men's attitudes were at least partly overcome by familiarity and the tangible benefits available.

Another important qualification to be made about the social reforms is that they were naturally of more relevance to some sections of the working class than to others. Historians have often underestimated their value because they have neglected women's interests. Yet nearly three in five pensioners were female. Women had a compelling interest in improvements in health and Lloyd George was alive to this, as is indicated by the introduction of the maternity benefit and his losing fight for widows' pensions. Certainly these were the reforms advocated by the Women's Co-operative Guild, which sought to focus attention on the poor health endured by working-class women as a result of excessive child-bearing. The introduction of local maternity clinics in 1914 was the next step in this direction and helped to generate support for further expenditure on subsidized milk, family allowances, and even birth control during and after the war.

Attitudes towards the state also varied according to the employment experience of different groups. The unskilled and low-paid were less likely to enjoy trade union membership benefits or to be able to maintain Friendly Society policies, and so incorporation in a state scheme could represent a real gain. Similarly, in very poorly paid occupations where surplus labour held down wages, trade union action was never very effective, and the intervention of the state offered the prospect of improvement. Thus, for example, the Trade Boards Act affected men and women in the 'sweated trades' after 1909; in chainmaking there followed a 50 per cent wage rise between 1911 and 1914, which still left wage levels comparatively low but none the less represented a gain that was unattainable by any other means. Women stood to benefit because of their concentration in low-paid occupations. Trade unions also derived some indirect benefits from state intervention. The effect of trade boards in tailoring and chainmaking was to stimulate membership recruitment in these traditionally difficult sectors. There were similar advantages for unions, which acted as approved societies for payments under the National Insurance scheme, notably among railwaymen and shop assistants.

Ultimately, popular reactions to state interventionism were determined by the perceived effectiveness of the new schemes. Some clearly took time to make an impact. By 1911, for example, only four in ten local authorities were actually providing school meals; hence the scheme was made obligatory in 1914. However, some of the criticisms made by historians about the shortcomings of the reforms are misleading, for example the claim that health insurance did nothing for those most in need, namely women and children. The key purpose of the scheme was to replace the income lost when the family's breadwinner became too ill to work, so that the indirect advantage for wives and children was very important. It is probably true that the free medical treatment proved less attractive, largely because working-class families were unconvinced about the value of doctors and hospitals and preferred to rely upon patent pills purchased at chemists. Unemployment insurance was, initially, limited to relatively well-paid men, though the 2.5 million covered represented a considerable part of the workforce. Four-fifths of these men had previously enjoyed no insurance against unemployment, and in the first year of the scheme 23 per cent made a claim. However, the fact that national insurance operated for only a year before the outbreak of war disrupted the pattern of employment makes it difficult to assess the impact. The official figures issued by the poor-law authorities offer an inadequate measurement of poverty because they excluded the majority of the poor. However, they do reveal something of the impact of the reforms. Between 1910 and 1914 those assisted by the guardians diminished from 916,000 to 748,000. This was largely due to the disappearance of men and women over seventy years old, previously on outdoor relief and now receiving pensions. Workhouse visitors commented on the sudden absence of elderly people, though some 50,000 who needed institutional care did remain.

One should not exaggerate the material benefits arising from state welfare. For many families, the cheap food which was perceived to be the result of free trade was the most tangible advantage of Liberal policy. Moreover, many working men naturally placed full employment and high rates of pay at the top of their agenda. But they also appreciated how vulnerable they were to the vicissitudes of the economy. Unemployment fluctuated from only 3.7 per cent in 1907 to 7.8 per cent in 1908, but was down to 3.0 per cent by 1911 as the Edwardian economy enjoyed its last boom. Social welfare could only be a supplement to the 'family wage' to which trade unionists aspired. But it was one that even the best-paid workers needed at several stages in their lives. Increasingly they could look to the state, in preference to the anachronistic poor-law system which had cast such a long shadow over the lives of their Victorian predecessors.

The Taxation Revolution

The Boer War had had the effect of increasing the National Debt to £640 million by 1901 and to £800 million by 1904. Orthodox economists and politicians advocated retrenchment in government spending to enable this sum to be reduced. But the conflicting pressures to service the debt and to cut taxation meant that Chamberlain's hopes for the introduction of old-age pensions were strangled once again. Asquith, the new Liberal Chancellor, managed to redeem £41 million of the National Debt; but despite this concession to orthodoxy, his chief role was to lay the foundations for a drastic shift in the basis of government finance. Faced with evidence of widespread evasion of income tax, in 1907 he introduced compulsory returns of all classes of taxable income, and obliged employers to make returns of their employees' wages and salaries. Also in the 1907 budget he adopted the principle of differentiation by setting the tax on earned incomes at 9d. in the pound and that on unearned incomes at 1s. Asquith was coming round to the idea that taxation should be graduated according to ability to pay, the case for which had been recently argued by Leo Chiozza Movey, a Liberal MP, in his book *Riches and Poverty* (1905). In particular Asquith committed himself to the introduction of a supertax levied on a rising scale on all incomes above £5,000. He also insisted that it was no longer realistic to 'treat each year's finances as though they were self-contained; the Chancellor ought to budget, not for one year, but for several years.' This heralded a long-term commitment to fund new social-welfare policies, which made inevitable some broadening of the basis of taxation and a greater reliance upon direct taxation.

Lloyd George's celebrated budget of 1909 was built on the foundations laid by Asquith, though the pressures of an anticipated budget deficit of £16 17 million drove him to adopt an unexpectedly large catalogue of measures: 'I have got to rob somebody's hen roost next year.' In all he proposed to raise an extra £13–15 million revenue. Income tax on unearned incomes over £700 was increased from 1s. to 1s. 2d.; on earned incomes up to £2,000 it remained at 9d. but went up to 1s. in the £2–3,000 range. A supertax was introduced on incomes over £5,000, payable at 6d. at £3,000. Additional sums were levied on death duties, stamp duty, and liquor licences. There were two novel sources of revenue: a tax on petrol and a motor car licence proved excellent ways of taxing conspicuous wealth. And, most controversially, Lloyd George proposed a half-penny in the pound levy on the value of undeveloped land, excluding agricultural land, and a 10 per cent duty on benefits gained by a lessor on the termination of a lease.

The key feature of these measures was that they affected only a small number of rich people. Income tax liability, which began at £160, included around one million people in all. But most of them were unaffected by the new levies: for example, there were only 25,000 taxpayers above the £3,000 level. In addition, the Cabinet persuaded Lloyd George to assist those on modest middle-class incomes by introducing tax relief of £10 for each child under sixteen years to those below £500 annual income. The only aspects of the budget which affected the working class were the extra purchase taxes on spirits and tobacco.

The 1909 budget proved to be of very considerable significance, both immediately and in the long term. It established the principle that taxation ought to be related to capacity to pay. This was reflected both in the system of graduation and in the shift from indirect taxes on consumption, which were largely paid by the poor, to direct taxes on income and wealth. By 1914 around 60 per cent of all government revenue was derived from direct taxation, compared to 44 per cent in the late 1880s. Much of the extra revenue was now spent on social welfare. Old-age pensions alone cost £8.5 million; and total spending on social services stood at £33 million in 1913 as against £5 million in the late 1880s. Thus the effect of this combination of social and taxation policies was unquestionably to redistribute the nation's income from rich to poor, albeit very slightly. In fact, the Edwardian reforms inaugurated a trend in British public finance which lasted for many decades. The modest redistribution of income continued until the 1979–92 period, when reactionary governments reversed the pattern by transferring income from poor to rich via the tax system.

Moreover, the People's Budget created a momentum which was still carrying the Liberal government forward in 1914. Despite some internal dissent, Lloyd George was actually extending his innovations in the last peacetime budget. Incomes above £2,500 were subject to a tax rate of 1s. 4d. in the pound. The supertax limit was lowered from £5,000 to £3,000 and a maximum rate of 2s. 8d. was applied. Death duties were again raised to 20 per cent on estates worth over £1 million. And the tax relief for children was doubled to £20. Clearly, the vigorous application of the redistributory elements in the 1909 budget prepared the way for even more drastic measures during the war.

The State and the Economy

In the heat of the Edwardian political debate, Conservatives sometimes accused the Liberals of introducing socialism. This usually implied the expropriation of individual wealth and ownership and the creation of a huge central bureaucracy. Government was undoubtedly becoming a bigger employer: between 1901 and 1911 the number of jobs in the civil service, local administration, and the armed forces rose from 960,000 to 1,270,000. The administration of the new schemes of national insurance and land valuation created a further 5,000 permanent and 10,000 temporary posts up to 1914. However, while happy to justify a proper role for the state, the Liberals indignantly rejected the imputation of socialism. Their tax policies still left private wealth and individual ownership intact; the economy would continue to be governed largely by free-market forces – it could hardly be otherwise for a party wedded to free trade.

Indeed, it is only comparatively recently that historians have recognized that the Liberals had discovered, perhaps by luck as much as by judgement, a viable mixture of collectivism and individualism. The balance in policy reflected, in an approximate fashion, the blend in their support in the country. For, vital as their working-class vote was, the Liberals clearly remained a party with significant backing among the middle classes too. Although a number of employers jibbed at the radical taxation policies, it seems that 37 per cent of Liberal MPs were still businessmen in 1906, which represents only a modest reduction over the 1890s. Indeed, some of the country's most successful entrepreneurs in manufacturing and finance, including William Lever, Sir John Brunner, George Cadbury, Joseph Rowntree, and Lord Swaythling, were major contributors to Liberal funds as well as supporters at elections. Yet their role should not be exaggerated. They were not typical businessmen; rather, they were sufficiently well established in business to be able to indulge their own interest in political and social causes. Also, apart from Walter Runciman and J. A. Pease, there were no businessmen in Asquith's Cabinet, though this is not particularly significant since the same is true of the late Victorian Cabinets of both parties.

Why did a number of employers remain Liberal supporters when the party's policy was moving to the left? For some, free trade continued to retain its importance; it helped to maximize export markets, maintain cheap raw materials, and reduce the pressure for higher wages. Some clearly remained Liberal for non-economic reasons or simply out of habit. But a number of the more successful and enlightened businessmen broadly accepted the diagnosis of the progressives. They were increasingly concerned about Britain's failure to educate and train her labour force, and to improve the health and housing of her people. Some employers had already pioneered welfare schemes which were subsequently overtaken by the state. Above all, they recognized that, if simply left to itself, private enterprise did not work efficiently in all areas; in particular it failed to generate the necessary investment for the utilities and services on which all producers depended. Thus, if the adherence to free trade and the involvement with capitalists placed limits upon Liberal collectivism, it did not stop, and in some sense actually promoted, moves away from an unqualified *laissez-faire* approach.

State intervention was seen as appropriate by the Edwardian governments in three broad areas. First, and most simply, there were problems or sectors in the economy where an acute national interest overrode a private interest. For example, since 1870 the government had enjoyed the legal power to take over the operation of the railways, which it did promptly on the outbreak of war in 1914. During the Edwardian period ministers intervened drastically, following the outbreak of foot-and-mouth disease, to force farmers to cooperate in checking the spread of the disease. But the most striking case of intervention, which involved a change of ownership, not simply of control, came as a result of Churchill's decision to authorize a switch from coal- to oil-fired battleships. Oil was more efficient but brought the danger of dependence on foreign supplies. Consequently the Cabinet agreed to his request to purchase the Anglo-Persian Oil Company for £2 million. This proved a highly profitable investment for the government and the taxpayer over many decades until, in the 1980s, it was disposed of by the Thatcher government. Churchill, of course, was no socialist, but regarded the state's participation as an owner as justified where an important national interest was at stake.

The second area of state intervention consisted in promoting the rights and the efficiency of the labour force. This proceeded from the belief that trade unions had a legitimate role to play in the marketplace in seeking improved wages and conditions. This led governments beyond efforts to promote arbitration and conciliation to direct interference with the rights of employers and owners. In 1909 the Trade Boards Act imposed minimum wages in a number of 'sweated trades' where the surplus of labour, the presence of women workers, or the lack of skills had the effect of holding pay at very low levels. In 1912 the government eventually resolved the industrial dispute in the coal industry by legislating for an eight-hour working day. Even as an *ad hoc* measure, this was not to be undertaken lightly by a government committed to private enterprise. During 1913–14 the momentum was maintained by Lloyd George's Land Campaign. It emerged that six out of ten agricultural labourers earned less than 18s. a week, which put them below the poverty line as defined by Rowntree. The Chancellor proposed to tackle this by appointing regional wage tribunals, which would impose minimum wages in each area.

The third aspect of the government's interventionism centred around the problem of unemployment and inadequate growth rates of the economy. A first step towards a solution had been the labour exchanges, which were seen as a way of improving the mobility of labour and, eventually, reducing the incidence of casual employment. During the first five months of the scheme, the exchanges filled 900 vacancies a day, but by the first six months of 1914 they were filling over 3,000 a day. Thus, although the economic boom sharply reduced unemployment, men were increasingly being drawn into the system. But in his 1909 budget Lloyd George pointed the way to a more radical approach to unemployment and investment in the form of the Development Commission. He proposed to set aside £200,000 a year, plus any surplus revenue, for a Development Fund which would promote scientific research, exploit resources not presently being used, and finance land reclamation, afforestation, transport, and experimental farming. The bill actually defined the scope of the Commission's work to include 'any other purpose calculated to promote the economic development of the United Kingdom'. Conservatives attacked this chiefly on the grounds that it allowed governments to bribe voters or industrialists with Exchequer funds, was likely to be wasteful of public money, and would result in a swarm of new officials.

In one sense the Development Commission was a recognition of the need to invest more in certain sectors of the economy where private enterprise failed to serve the national interest vigorously enough. Progressive industrialists believed there to be a role for government to play by complementing private business rather than competing with it. But the fund was also a sign that government expenditure might usefully be regulated according to the state of the labour market. The minority report of the Royal Commission on the Poor Laws called for public spending to be raised in those years when the economy was depressed so as to stabilize the level of employment. In the event, little was done to realize this objective. During the last years of peacetime the economic boom largely removed the need to stimulate employment. Significantly, the Land Campaign included proposals for state-subsidized rural house-building which would have been a useful counter-cyclical measure if applied. Even in the Treasury the notion of public works designed to modify the trade cycle was under discussion. In this way the ground was prepared to some extent for wartime interventionism, and for the economic strategy associated with J. M. Keynes between the wars.

8

The Liberal–Labour Alliance

The Edwardian era represented a watershed in the scope and substance of British politics which set the pattern for the twentieth century. The performance of the economy, the level of unemployment, the standard of living, and social welfare all began to occupy a central role in political debate and in the work of governments. Gladstone and Salisbury had always feared such a development, but contrived to avoid it by concentration on the traditional politics of the constitution, the law, the Church, and foreign policy. After 1906 each party contributed to the shift of emphasis. The New Liberals and their ideas came to the fore partly as a by-product of the rise of certain individuals – Asquith, Lloyd George, and Churchill – to key positions, but also because the traditional Liberal causes began to run into the sand, victims of the House of Lords; by contrast, social and financial reform restored the populist cutting edge to Liberalism. For the Labour Party the need to tackle social and economic questions was all of a piece with its claim to represent working-class interests. As such issues rose up the agenda the party had to move with them for fear of being squeezed between New Liberalism and Socialism. Conservatives also found themselves drawn into social politics by Chamberlain's plunge into tariff reform after 1903. He claimed for the government some ability to raise the level of employment by excluding foreign-made goods, and a capacity to pay for social reform with the additional revenue generated by tariffs. Either way the role of the state was being emphasized and Conservatives, often to their dismay, became involved in a debate which their left-wing opponents seemed likely to win.

The Politics of the People's Budget

In spite of the changes in Edwardian politics, it would be an exaggeration to suggest that the new agenda dominated throughout the period. Indeed, between the end of the Boer War and the general election of 1906 some very traditional issues held sway. The effect of the war in increasing the national debt and taxation gave the Liberals an opportunity to steal the Conservatives' clothes and advocate retrenchment. A. J. Balfour's 1902 Education Act reactivated Liberal Nonconformity because it abolished the school boards and obliged Nonconformists to contribute towards the cost of Anglican schools in some areas. This led to a passive resistance campaign involving non-payment of rates, and to a selective application of the new legislation by county councils in Wales under Liberal control. As this coincided with a religious revival in Wales, it was as though the spirit of Victorian politics had been restored. Above all, the tariff-reform campaign allowed Liberals to sink their differences in defence of free trade. Although debilitated by external attack and internal division, Balfour's government hung on because it enjoyed a large majority. A number of Tory free-traders, such as Winston Churchill, who saw how the rank-and-file was falling into the protectionist camp, decided to quit in order to join the Liberals. But by December 1905 Balfour had decided to resign. The new Premier, Sir Henry Campbell-Bannerman, won a landslide victory in January 1906 with 401 Liberals against 157 Conservatives, 83 Irish Nationalists, and 29 Labour.

In the light of the Liberals' decline and downfall during and after the First World War this election has sometimes been seen as a freak result, the last fling of Victorian radicalism. Clearly the high profile of free trade, retrenchment, Nonconformity, and 'Chinese slavery' in South Africa gives some grounds for this view. However, this only demonstrates that the best tactic for an opposition is to concentrate on purely negative arguments. The rhetoric of 1906 could not hide the fact that the Liberals were a changed party. Half of the MPs had never sat in Parliament before. The majority of them were committed not only to the traditional causes but to old-age pensions, poor-law reform, and graduated taxation. In any case, if the election had been no more than the last gasp of Victorianism, 401 seats would have been an extraordinarily robust death rattle.

The defeated Tory leader professed to see in the rejection of his party and the emergence of a substantial Labour Party an echo of the revolutionary currents affecting some European societies. This seems a far-fetched analogy for Ramsay MacDonald, Philip Snowden, Arthur Henderson, or Keir Hardie, all men deeply committed to the parliamentary system. But their election in 1906 was a portent of change. Labour's initial breakthrough resulted from some careful tactics designed to make the most of the new party's limited resources. The alarm amongst trade unionists over the Taff Vale decision encouraged more affiliations to the Labour Representation Committee, which boosted its funds and its authority. At the same time the success of its candidates in by-elections at Clitheroe, Woolwich, and Barnard Castle in 1902–3 raised expectations. But the lesson MacDonald drew was that Labour could do well in partnership with the Liberals because its candidates could expect to be elected only by winning over the votes of Liberal working men. This was the logic of the 1903 electoral pact between MacDonald and the Liberal Chief Whip, Herbert Gladstone. Thus, Labour fielded 50 candidates of whom 31 were not opposed by the Liberals; since 24 of the 31 were elected in 1906, MacDonald's tactics appeared to have been fully vindicated.

From the Liberals' perspective the pact also had advantages, for they stood to lose from any split in the non-Conservative vote. In the short run Labour's willingness to finance candidates, especially in areas where the Liberals were weak, saved Gladstone a good deal of money. In addition, Liberals like Gladstone regarded the Labour politicians as similar to the existing Lib-Lab trade unionist MPs, that is, as essentially working-class Liberals, not socialists. By maintaining co-operation with them, it would be possible to preserve the balance within Liberalism which was otherwise liable to be tipped to the right by the growth of Liberal Imperialism and by the recruitment of Tory free-traders.

In retrospect it is easy to see the 1903 pact as an historic error for Liberalism: it had taken a cuckoo into its nest. But although the Liberals would have won the 1906 election even without the pact, this was not obvious in 1903. Even after the landslide, it was wise to avoid alienating Labour and unrealistic to think that the new party could simply be crushed. In by-elections in 1907 Labour won Liberal seats at Jarrow and Colne Valley, while also making extensive gains in municipal elections. For a time it appeared as though the rising expectations in the working class might give Labour a sustained boost in popularity. Briefly after 1906 the party began to influence the political agenda with its bills for school meals and trade union reform, which were taken up by the Liberal government. Keir Hardie also embarrassed ministers with the 'Right To Work' Bill, based on the ideas of the Webbs, which made central government responsible for providing either a job or relief for the unemployed worker. As this could not be accepted, the campaign highlighted the shortcomings of Liberal policy on unemployment.

Pressure on the left during 1906–7 coincided with the deflation of Liberal hopes in Parliament. Refusing to be overawed by the huge Liberal majority in the Commons, the House of Lords amended a series of bills on education, land, and plural voting. For a time it seemed as though the Liberals were back in the dilemma faced by Gladstone and Rosebery in the 1890s. There was a danger of disappointing the expectations of both the traditional party activists and the working-class voters. However, by 1908 the tide had begun to turn and the government regained the initiative. The introduction of old-age pensions heralded this recovery. When Asquith took over the premiership and Lloyd George became Chancellor, it was inevitable that the government would tackle the questions of finance and the House of Lords. Lloyd George struck a blow at both with the 'People's Budget' of 1909.

Behind this famous budget one may detect three distinct motives. The immediate one was the need to meet a budgetary deficit resulting from falling tax revenues during the trade depression, and the extra expenditure of £8 million on pensions and £3 million for the naval estimates. Second, the budget was designed to deal with the loss of morale consequent upon rejection of Liberal legislation by the peers. Banking on the convention that the upper chamber could not amend financial legislation, Lloyd George planned his budget as a Trojan horse which incorporated increases in licence duties and proposals for land valuation which had been thwarted as ordinary legislation. This would amount to an effective repost to the arrogant peers, and was an indication that Lloyd George did not anticipate outright rejection. The third motive was more fundamental. For some years politicians had faced growing pressure to tackle the issue of poverty, but had backed away from the financial implications for local and national government. Chamberlain's advocacy of old-age pensions in the 1890s threatened to outflank the Liberals, and the tariff campaign made this more likely. Conservatives calculated that the Liberals' commitment to free trade would leave them unable to finance social reform and thus squeezed between protectionism on the one side and Labour on the other. The 1909 budget demolished these calculations at a stroke by demonstrating that one could retain the advantages of free trade while raising additional tax revenue to pay for pensions and new Dreadnoughts. This freed the Liberals from their dilemma and forced their opponents back onto the defensive. Since Labour was bound to welcome measures damned as 'socialism' by the Conservatives, the effect of the budget controversy was to reinforce the loyalty of Labour to the electoral pact and thus maintain the government's base.

Though Lloyd George had believed his budget to be politically shrewd, he did not at first appreciate just how central it was to become. This is because he underestimated the strength of Tory reaction. Within days the protectionists, seeing how the new taxes had checkmated them, began to assert the right of the peers to reject the budget. Many peers felt tempted to do so because of the implications of the scheme for land valuation; though Lloyd George expected to raise little revenue from the land taxes in the first year, it was clear that once he possessed comprehensive information about landownership he would be in a position to realize a long-standing aim of Victorian radicals: effective taxation of the vast landed wealth of Britain. Soon the opposition organized a Budget Protest League, and Lloyd George set about provoking the Conservatives into adopting an intransigent position. At famous meetings held in Limehouse and Newcastle upon Tyne, he derided the Tory peers as selfish rich men who wanted Britain to rebuild the navy but refused to contribute a fair share towards it. He resurrected the old radical attack on landowners as idle parasites living off the hard work of both employers and workers:

> The question will be asked whether five hundred men, ordinary men chosen accidentally from among the unemployed, should override the judgement . . . of millions of people who are engaged in the industry which makes the wealth of the country. That is one question. Another will be: who ordained that a few should have the land of Britain as a perquisite? Who made ten thousand men the owners of the soil, and the rest of us trespassers in the land of our birth?

Even before these words had been uttered, Balfour had decided to acquiesce in the wish of the extremists to reject the budget in the House of Lords. Though this was an immensely dangerous step in view of Lloyd George's challenge to the class interests of the peers, to back down at that stage would have been divisive and demoralizing for the party and exposed the leadership to attack. As a result, the peers threw out the budget by 350 votes to 75. 'At last, with all their cunning, their greed has overborne their craft,' declaimed Lloyd George; 'we have got them at last.'

He was essentially right. Faced with no alternative but to seek a popular mandate to override the peers, the government dissolved Parliament. At the election in January 1910 the Liberals lost many seats, but that was inevitable after the landslide of 1906. The return of 275 Liberals, 82 Irish, and 40 Labour MPs meant that the Conservatives had been heavily defeated.

Not only were they obliged to swallow the budget, but they had also put reform of the House of Lords at the top of the Liberals' agenda. Reform could only be accomplished if the new King, George V, agreed to create hundreds of new Liberal peers. Obviously reluctant, he appealed for an attempt to find a compromise, which led to a constitutional conference between representatives of the two sides in the summer of 1910. But the Conservatives were still too angry or too arrogant to see that they would have to back down, and the conference failed. The King could not afford to reject the advice of his elected government without involving the Crown in party-political controversy, and so he undertook to create peers on condition that Asquith won a second election. The December campaign

THE LITTLE DOTARD.
REGISTRAR JOHN BULL (*to bearer of venerable infant*). 'WELL, WHAT CAN I DO FOR IT – BIRTH CERTIFICATE OR OLD-AGE PENSION?'

Lloyd George's 1909 budget passed

concentrated on the House of Lords issue and produced a repetition of the January result. Thereupon Asquith introduced the Parliament Bill, which removed the power of the peers over financial legislation; other bills could be delayed for up to two years, but would become law if approved in the Commons in three successive sessions; finally, Parliament's lifetime was reduced from seven to five years. Eventually enough peers backed down for the bill to pass.

In detail the Parliament Act was something of a mistake for the Liberals. They had left the hereditary membership of the upper chamber intact, and undermined the power of the Commons by shortening its life. But the whole episode clearly represented a triumph for the Liberals. Through serious misjudgements the Conservatives had put themselves in the position of opposing financial and social policies that commanded popular support; as a result they had largely destroyed the bulwark against radicalism used by Salisbury and opened the way to further innovations. Not surprisingly, a by-product of all this was the infliction of bitter internal divisions on the party for some years.

The Conservative Dilemma

To some extent all the parties found themselves divided and perplexed by the social and political issues of the early twentieth century; but the Conservatives were the most deeply troubled. The predicament of being excluded from office and unable to find the way back exacerbated tensions that had been simmering for years. Tariff reform provided the immediate source of dissension. In fact this issue had been developing for many years. In the 1880s, when some manufacturers began to suffer from German competition, the protectionist sentiment that had hitherto been confined to agricultural interests began to revive. As a result, the National Union of Conservative and Unionist Associations debated tariff reform at its annual conferences, and swung in favour of protectionism. The ground was thus well prepared when Chamberlain gave a lead in 1903. His immediate motive was to lift the party out of the toils of the Boer War and give it fresh momentum. In theory, tariff reform hit a number of targets all at once. By offering protection to British producers it guaranteed more jobs for British workers. It appealed to Empire enthusiasts because the policy could be applied so as to favour British colonies and thus foster trade amongst them. And it offered a source of revenue to a government otherwise trapped by a swollen National Debt and inflated income tax. At first the protectionists seemed to carry all before them. The Tory free-traders were reduced to a small minority in 1906, partly by the election but also by defections and bitter attacks on them from the tariff-reformers in the constituencies. As a result the parliamentary party became largely committed to tariffs, though in many cases this represented opportunism more than depth of conviction.

However, the strategy suffered from many flaws. Much of British commerce benefited too much from free trade to want to abandon it; even manufacturers were divided between those who prospered by the free exchange of goods and those in steel or cutlery who suffered from foreign competition. The fortunes of the tariff cause were also vulnerable to the broad pattern of the economy. In the trade depression of the early Edwardian years its appeal to the unemployed naturally strengthened; but by 1909, as the economy began to boom again, the momentum was lost.

Above all, popular fears of higher food prices deterred the voters and gradually sapped the resolve of the Conservatives themselves. Defeat in 1906 had seemed unavoidable, but the failure to recover sufficiently in 1910 caused Conservatives much more concern. Balfour decided to try to take the sting out of tariff reform by offering to hold a referendum before abandoning free trade. Subsequently it was decided to exclude food from the new policy. But such concessions failed to rescue the party from its position in opposition. Increasingly tariff reform appeared to be too great a handicap. Worse, it had led the party into the decision to reject the 1909 budget, thereby precipitating a constitutional crisis and the loss of safeguards against radical attack upon the Church Establishment, the Union with Ireland, and private property. Consequently, the more pragmatic Conservatives began to water down the commitment to tariff reform after 1910. This left the party in disarray over its future economic policy and casting around desperately for alternatives.

The obvious expedient was to find a new leader. After losing three elections, Balfour had become highly vulnerable. While the traditionalists felt he had failed to stand up to Chamberlain, the reformers blamed him for prevaricating over tariffs and for failing to back the majority line. Eventually the parliamentary party asserted itself, and in 1911 Balfour became the first of a long line of twentieth-century Tory leaders to be driven out by their own followers. When the party divided evenly between Austen Chamberlain and Walter Long, a compromise candidate emerged in Andrew Bonar Law. With his modest middle-class background, Bonar Law initiated the twentieth-century tradition of taking Conservative leaders from outside the ranks of the landed elite. Even Asquith – sprung from a West Riding woollen manufacturer, though polished by Balliol College – referred to the Tory leader as 'a gilded iron-merchant with the mind of a Glasgow baillie'. Many Conservative dignitaries were equally dismissive of the bourgeois and relatively inexperienced new leader. In the event, Bonar Law proved to be a success in party terms. A crisper and more aggressive leader of the opposition than Balfour, he adopted a simple, negative solution to the Conservative dilemma. He concluded that it was futile for Conservatives to attempt to compete with the Liberals over social reform as some wished to do; the more the debate concentrated on such issues the more advantage the radicals would enjoy. Instead, Bonar Law determined to rally the party around opposition to Irish Home Rule. This at least had the merit of uniting the fractious Conservatives, and allowed them to portray the Liberals as mere tools of John Redmond, desperately clinging to office by buying the votes of the Irish MPs.

However, the Orange card had been played before. If there were no votes for the Liberals to win on the issue, neither were there many for the Conservatives. The British electorate was no longer moved by the Irish question, and the Conservatives privately admitted that their campaign fell flat in the country. Even in the event of a civil war breaking out in Ireland it is not certain who would have been blamed; the Conservatives' irresponsible role in encouraging armed resistance to the Home Rule policy made them highly vulnerable. In the event, the Conservatives continued to be frustrated by the Liberal–Labour alliance up to 1914, and saw little prospect of winning the next election. The party was also unnerved by the growth of trade unionism and militancy among the working class; many doubted its ability to control the industrial situation in the event of its return to office. In short, the Edwardian period left the Conservatives badly shaken and demoralized. It is, in this light, an historiographical eccentricity that the period was characterized as 'The Strange Death of Liberal England' in the 1930s. For it is clear that it was nothing of the sort. Rather, the world that had collapsed was that of the *Conservatives*, who had clung too long to the political strategies of the Salisbury era. It was to take a world war to rescue the party from its dilemma.

Labour's Turning-Point

It was traditionally assumed that well before 1914 the Liberals were on the way to being ousted by Labour as the alternative party to the Conservatives; the disruptive effects of the war on the Liberals and its stimulating effect upon Labour were thought simply to have accelerated the process. However, this view was the product of hindsight. Historical research on the Edwardian period has cast much doubt on the strength of popular support enjoyed by Labour, and on the assumption that it outflanked Liberalism by offering a more radical alternative. As we have already seen, the policies of the two parties largely converged on a progressive formula of social welfare and taxation reform; moreover, as the Labour leaders were the products of the same Victorian era as the Liberals, they were equally enthusiastic about the traditional radical programme of free trade, Home Rule, and temperance. It was therefore difficult for Labour to establish a distinctive position except by advocating socialism; but that was neither particularly popular with the working class nor supported by most of the Edwardian Labour MPs.

However, policy was only one aspect of the question. The other was the electoral strategy. The growing organization and political awareness of the working class inevitably threatened to some extent the Liberals' ability to maintain a cross-class alliance, more especially when a rival party could offer working men much more direct representation. Yet the elections of 1910 give grounds for thinking that it was indeed feasible for the Liberals to maintain their electoral position. Although they lost many of the seats gained in 1906, these did not go to Labour; rather, they were rural and residential middle-class seats recaptured by the Conservatives. The Liberals held onto their industrial, working-class strongholds in the Midlands, the north, Wales and Scotland. In fact the social-geographical pattern of modern elections was already in place by 1910. Most of the territory north of a line from the Wash to the Bristol Channel was Liberal and Labour; south of it the Conservatives predominated except in parts of London, East Anglia, and the West Country. Class-based politics had substantially arrived in Britain before 1914, and thus pre-dated the rise of Labour as one of the major parties.

Within the confines of the electoral pact the Labour Party performed strongly. But while their candidates usually won if the Liberals stood down, in three-cornered contests they invariably came third or, at best, second. When a local pact broke down the Liberals regained seats from sitting Labour MPs, and they also recovered seats lost in by-elections. This electoral weakness on Labour's part was obscured by gains and losses between 1906 and 1910. The party lost eight of its existing seats and gained three. But the decision of the miners' federation to affiliate to the party meant that technically most of the Lib-Lab MPs were now Labour. Hence the official total rose from 29 to 40. In practice, however, many of these MPs remained Liberals both at Westminster and in their constituencies.

This picture is corrroborated by the weakness of Labour's organization in the country. By 1909 the number of affiliated constituency organizations stood at 155, and it was only 158 by 1913. Thus, the party's modest total of parliamentary contests – 50 in 1906, 78 in January 1910, and 56 in December 1910 – was not simply the result of the pact but a reflection of limited local organization. Even in the seats that were fought, the candidates sometimes had to rely upon the Liberals or a trade union to run an election. The federal structure of the party represented a weakness here, for membership was largely indirect, that is, it consisted in the membership of unions or socialist societies which affiliated to the party. There was little direct membership as such.

However, some of the local activists in the ILP felt that the party would grow if the leadership adopted a bolder strategy. MacDonald and Snowden were accused of collaborating with the government in order to save their own seats. In 1908 Bell Tillett wrote a famous denunciation characterizing them as 'flunkeys to Asquith' in *Is the Parliamentary Labour Party a Failure?*. Victor Grayson, briefly the MP for Colne Valley, argued that by adopting a real socialist policy they would attract the working-class vote from Liberalism. But the party leadership believed this to be nonsense, and they largely disapproved of attempts by ILP activists to increase the number of candidates. They could not entirely prevent wildcat or propagandist candidates in by-elections because it was usually possible to bring in enough outside help to make good the deficiencies in local organization. As a result, the 1911–14 period saw many three-cornered by-election contests. But in the process Labour lost all four of the seats it was defending and made no gains. MacDonald appreciated that excessive provocation of the Liberals would only result in the loss of Labour seats at the next general election; but he also feared the wider effect of splitting the non-Conservative vote, which would simply restore the Tories to power. In 1910 the Conservatives won 46 per cent of the poll, more than enough to give them a majority under the first-past-the-post system in the absence of the electoral pact. Thus, by 1914 Labour was expected to contest 37 seats already held and 18 others in which the national executive had agreed to a contest. In addition, 22 constituencies had selected candidates without approval by the NEC. Even this modest total of 77 was not out of line with the December 1910 candidacies, and was likely to be cut back as the election drew near.

Understandably, the activists were partly inspired by the mood of militancy among working-class communities which manifested itself in the strikes and expanding union membership of these years. From 2 million in 1900, the unions had grown to over 4 million by 1914. Ever since the formation of the ILP, the goal of tapping the funds and resources of trade unions for socialist politics had proved rather elusive. However, during the last years of peacetime success appeared to be much closer. In 1908 a Liberal trade unionist official of the Society of Railway Servants, W. V. Osborne, had successfully taken his union to court to stop it using its funds for party political purposes. As a result, Labour suffered a loss of income estimated at £30,000 during 1909–14. Many working men regarded this as patently unfair, since the wealthy were free to spend huge sums on political causes.

Eventually the Liberal government tackled the financial grievances of the labour movement. In 1911 MPs received a salary of £400 per year for the first time. In 1913 the Trade Union Act made it legal to raise a separate fund for political purposes subject to two qualifications; every union must first hold a ballot to obtain members' consent to a fund, and any member might opt out of the political levy while retaining his membership. There was something oddly illiberal in a system that put the onus on the individual to refuse the political levy. In the long run the scheme proved advantageous to Labour, since many men paid the levy without realizing it. The ballots conducted by most unions in 1913 produced a majority in favour of the political fund, though the minority was substantial and the turnout low. Here was at least a potential gain of major proportions for those who wanted Labour to break out of the electoral pact. But as the new funds were only being established when the war broke out, it is difficult to estimate their effect. During the spring and summer of 1914, MacDonald seemed intent on working with the Liberals up to the next election. There was not sufficient evidence either of Labour strength or of Liberal weakness to justify any major change. Pressure from the activists might well have split Labour itself into a socialist minority and a Lib-Lab majority, but such problems were largely put into suspension by the war.

While Labour had not reached the point where it could replace the Liberals, Asquith's party seemed entrenched in office, if battered by a succession of controversies. The real flaw in the Liberals' position lay in their neglect of some aspects of traditional radicalism – the stubborn resistance to women's suffrage, and the drift of foreign policy towards the Continental commitment to France and Russia, for example. But in social and financial policy progress was impressive. The claim that reform was checked by 1914 because of the fear of alienating the middle class is simply not borne out by the facts. In 1914 new social reforms were being introduced, graduated taxation and income redistribution was being pushed further ahead, and the Land Campaign brought fresh forms of interventionism onto the agenda. As long as the Liberals held the initiative over social reform the alliance was likely to continue.

9

Crisis and Controversy in Edwardian Britain

From the perspective of the 1930s, the Edwardian era was sometimes seen as a society already in the first stages of breakdown and about to be engulfed in the greater chaos of world war. The violent challenges to authority posed by strikes, suffragettes, Ulstermen, Irish Nationalists, and even the antagonists in the constitutional crisis over the House of Lords, were seen as aspects of a single crisis that threatened to make Britain ungovernable. In this view the war further undermined conventional ideas and institutions, thus leading to the decline of Britain in the inter-war period, when liberal democracy was in headlong decline in several European societies. However, though undeniably marked by unusual controversy, the Edwardian years have been misunderstood, and an examination of the component parts of the crisis suggests a less apocalyptic assessment.

Class Struggle and Class Collaboration

A characteristic of the Edwardian period was the growth of trade unions and the assertiveness of the working-class movement. We have some measurement of this in the pattern of annual stoppages of work, which rose from 300–400 a year in 1902–6 to 800–900 a year in 1911–14; the peak came in 1913, with 1,459 separate strikes, though the largest number of working days – 41 million – was lost in 1912, when the miners struck. Since many of these strikes proved to be successful they gave a fillip to union membership, which had doubled to reach 4.1 million by 1914.

However, whether this phenomenon amounted to a threat to parliamentary government depends less on the size than on the character and motivation of the labour movement. Do the workers' struggles with their employers indicate greater class consciousness and a rejection of the parliamentary system? There is some evidence for a qualitative change in the attempts made to unite men divided by skills or jobs into single unions. There was also a certain amount of sympathetic strike action. For example, in 1911 the seamen's strike led to supportive stoppages and thus to the emergence of the National Transport Workers' Federation under Ben Tillett. During the London dock strike of 1912 the men agreed to stay out until the grievances of every group of workers had been resolved. The culmination of this trend came with the formation of the so-called Triple Alliance of miners, transport workers, and railway men in 1914. A joint stoppage by these groups would have paralysed the economy and thus been tantamount to a general strike – a challenge to the government as much as to the employers. Emotions were also heightened by several clashes between workers and the police and troops. In 1910 two miners lost their lives at Ton-y-Pandy in South Wales, when troops were ordered out to restore order by the Home Secretary, Winston Churchill. The authorities misjudged the situation when they arrested several syndicalist leaders: for example, Tom Mann was prosecuted for encouraging the troops not to fire on strikers in 1912, and in the following year Jim Larkin of the Irish Transport Workers received a seven-month sentence for sedition.

In fact, though Mann and Larkin were inspiring orators and figure-heads they were not really leading the labour movement; syndicalism remained a minority creed. Official heavy-handedness only gave them more prominence and credibility than they would otherwise have had. Ordinary working men came out on strike for largely pragmatic reasons rather than from ideological motives. In the late 1890s falling prices had given way to a renewed period of inflation, and for several years money wages failed to keep pace. Inevitably the men wished to recover lost ground when circumstances changed. After 1908 the economy began to expand again, and by 1911 unemployment had fallen to 3 per cent, which gave workers more bargaining power. In addition unions had become legally liable for the costs of strike action to their employers as a result of the Taff Vale decision of 1901; but after 1906 the Trades Disputes Act removed this restraint, and the accumulated grievances of the men inevitably prompted more widespread industrial action.

Of course, it is possible that, however moderate and pragmatic the original objectives behind the Edwardian strikes, in the course of struggling against stubborn employers the men's sense of class consciousness increased, and that they began to become more politically motivated. This period certainly saw a good deal of interaction between workers and government. But it does not follow that they were necessarily antagonistic. We have already noted how working men benefited from government intervention over minimum wages and working hours, from the employment available in labour exchanges, and from the indirect advantages of the national insurance scheme for the unions. Government also impinged directly upon the industrial situation. Since 1896 it had been empowered to offer arbitration in disputes if the two sides agreed. Under the Liberals, the Board of Trade soon became very active in all major strikes. Lloyd George, Churchill, and Sydney Buxton were all keen on intervention, and from time to time other ministers, including Sir Edward Grey, John Burns, Herbert Samuel, and Charles Masterman, also participated. Their mediation was not entirely even-handed, for they frequently felt exasperated at the refusal of employers to recognize the union leaders and exerted pressure on them to engage in regular collective bargaining.

Taken together, the various government measures dealing with industrial relations, union law, wages, hours, and social security amounted to a formidable catalogue of concessions. Not surprisingly, most union leaders and Labour MPs recognized this. For them, the lesson of the Edwardian years was that, by a combination of organization in the country and steady pressure in Parliament by both Liberal and Labour MPs, it was now possible to use the power of the state to the advantage of working men. This trend was to be taken considerably further during the First World War. This does not mean that the unions were wholly satisfied with official policies, but it made it highly improbable that the labour movement in general would become alienated from the British parliamentary system in these years; on the contrary, the underlying significance of events, both industrial and political, was that they gave working-class men a foothold within the apparatus of the state that they had not hitherto enjoyed.

The Challenge of Feminism

To some upper-class Edwardian men, the campaigns of the suffragettes represented one of the most alarming symptoms of the underlying malaise in British society. Although founded in 1903 by Emmeline and Christabel Pankhurst, the Women's Social and Political Union failed to attract much public notice until 1905, when it adopted militant tactics. Militancy began mildly enough, with the questioning and heckling of Cabinet ministers at public meetings over their reluctance to introduce a women's suffrage bill. Thereafter it gathered pace, with attempts to rush into the lobby of the House of Commons and interrupt debates in the chamber itself, though the WSPU also used traditional methods, including huge rallies and marches, to demonstrate their support, and organized campaigns in by-elections designed to demoralize the government by securing the rejection of its candidates. However, the Pankhursts found that, despite the interest the newspapers took in these activities, the novelty quickly wore off. Since publicity was the best way of attracting new funds and members, it was necessary to keep the movement in the public eye. This meant finding new forms of militancy capable of evading the successful counter-measures taken by the authorities. Rough treatment of the suffragettes by both the police and the public led them to resort to attacks on property, including smashing the windows of West End clubs and setting fire to pillar-boxes and country houses, and the pursuit of Cabinet ministers on golf courses and at railway stations. The imposition of prison sentences resulted in a number of the suffragettes going on hunger strikes to which the authorities responded by forcible feeding. This so endangered the lives of the women that in 1913 the government resorted to the so-called 'Cat and Mouse' Act, which allowed them to release prisoners but rearrest them later to complete their sentences.

To the more extreme anti-suffragists, these escapades were shocking proof of their claims about the unfitness of women to be voters and also about the unrepresentativeness of those women who were demanding the vote. In a notorious letter published in *The Times* in 1912 a leading doctor, Sir Almroth Wright, resurrected the commonly held but unspoken assumptions that women were prone to nervous hysteria as a result of the menstrual cycle:

> these upsettings of her mental equilibrium are the things a woman has most cause to fear; and no doctor can ever lose sight of the fact that the mind of woman is always threatened with danger from the reverberations of her physiological emergencies.

In the case of the militant suffragettes Wright alleged that they were 'sexually embittered', that is, unmarried, women 'in whom everything has turned into gall and bitterness of heart and hatred of men. Their legislative programme is licence for themselves or else restrictions for men.' To some extent Christabel Pankhurst played into the hands of the anti-feminists by publishing her pamphlet *The Great Scourge and How to End It* (1913), which urged that, since the majority of men had contracted venereal disease, women should avoid marriage altogether. However, just as Wright's prejudices no longer commanded much serious attention, so Christabel Pankhurst's wilder claims tended to discredit her cause. Both distracted from the substantial question of whether women should have the vote, and their contribution was thus purely negative.

In fact, the significance of Edwardian militancy is clear only when it is placed in perspective. The Pankhursts' claim that militancy was justified and necessary because the women's movement had made no progress by relying on respectable methods since the 1860s was quite erroneous. As we have seen, late Victorian feminists won a series of improvements for women, and by their successful participation in public roles, both in local government and in party politics, they had steadily undermined the traditional case against women's enfranchisement. As a result, by the turn of the century political opinion had swung in their favour; the 1906 Parliament in particular contained a large suffragist majority. Even the League for Opposing Women's Suffrage admitted from its own private surveys that this was the case. Of course, many of the MPs were lukewarm suffragists, and were not prepared to treat the issue as a political priority; but broadly the argument over the principle of women's enfranchisement had been won by the time the Pankhursts joined the campaign. The outstanding difficulty was now to find a precise formula, defining how many women would have the vote and on what qualifications, which could be embodied in a parliamentary bill.

In this situation the relevancy of militant tactics was questionable. Its most valuable contribution lay in attracting publicity and thus stimulating women hitherto uninvolved in the women's movement to become active. This was important, because the anti-suffragists had always claimed that the majority of women were apathetic about the vote. Up to 1906 both constitutionalists and militants represented a mere stage army of activists. From 1909 this changed, however, in that, though few joined the Pankhursts, growing numbers rallied to the constitutional movement led by Millicent Fawcett, the National Union of Women's Suffrage Societies, whose membership exceeded 50,000 by 1914.

On the other hand, militancy clearly damaged the cause. It alienated a number of Liberal and Labour MPs, who felt it unreasonable to attack them rather than anti-suffragist Conservatives at by-elections. Also, men in all parties were antagonized by the suffragette violence. The effect became obvious in the House of Commons when a pro-suffrage vote of 255–88 in 1911 turned into an anti-suffragist vote of 222–208 in 1912. Moreover, the Pankhursts were a highly divisive element within the women's movement; this was not so much because women disliked their militancy as because of their autocratic style of leadership, which was so much at odds with the democratic cause for which they were fighting. The Pankhursts forced no fewer than three splits from the WSPU, first the Women's Freedom League in 1907, then the United Suffragists led by the Pethick-Lawrences in 1911; finally Sylvia Pankhurst was driven out, to establish her own group in East London. This left the WSPU as a dwindling band of Pankhurst family loyalists based on a scattering of groups in London and the Home Counties. The idea that such an organization possessed the moral influence to compel a well-entrenched government to back down was totally unrealistic. Thus, in spite of its embarrassment, the government continued to use the full force of the law against the suffragettes. It did not believe that women's suffrage had become a mass movement. The Pankhursts themselves threw away their opportunity when they left Manchester for London and severed connections with the Lancashire working-class women in order to concentrate on cultivating well-to-do ladies in south-east England. This, ultimately, explains why the suffragette movement never posed a serious threat to parliamentary government. It was entirely dwarfed by the big battalions of trade unionism and Ulster Unionism.

Politicians did, in fact, show themselves more susceptible to female pressure where it seemed to reflect the social and economic grievances of ordinary working-class women. Thus, the greatest progress in the Edwardian period was actually made by conventional methods. Several trade unions were persuaded to drop their opposition to women's suffrage, and the National Union of Women's Suffrage Societies began to forge links in working-class communities. Their pact with the Labour Party in 1912 contributed strongly to this, for it brought middle-class and working-class radicals together and put pressure on the Liberals by fostering new Labour candidacies in seats held by anti-suffragists. Combined with the alienation of many women Liberals and the defection of some of them to Labour, this sounded a warning to Asquith about the consequences of his obduracy. On the eve of war he had been forced by a mixture of internal and external pressures to accept the need to enfranchise women on terms that would include the wives of working men – the solution that was in fact to be adopted in 1917.

The Irish Question

The first act of defiance towards Parliament during the Edwardian period took the form of the 'passive resistance' to the 1902 Education Act offered by some Nonconformists who refused to pay rates to maintain Anglican schools. However, the campaign was not widely supported by

Nonconformists and soon petered out. The central controversy of the period developed out of the dispute over the 1909 budget. This was complicated by the absence of a written constitution. The Liberals took their stand on the *convention* that the peers did not amend money bills, and the lower house pronounced the House of Lords action to be 'a breach of the constitution and a usurpation of the rights of the Commons'. The problem was, of course, wider than this, for the peers undoubtedly enjoyed the right to amend and reject ordinary legislation. Yet there was no satisfactory way of resolving a dispute when the peers rejected a bill passed by a large majority in the Commons, which had become a regular occurrence since the 1870s, when the peers had largely become Conservatives. In practice, Liberal governments simply hoped that the huge increase in the electorate after 1867 would overawe the Tory peers. Lord Salisbury's response to this had been to invent the notion of the mandate: if the elected government was held to lack a popular mandate for a particular measure the Lords were entitled, indeed morally bound, to delay it until the people had spoken at a general election. This preposterous principle was applied only under Liberal governments, and it threatened to make general elections an annual occurrence if the government were to pass its legislation.

The Parliament Act of 1911 only partly resolved this problem. Though dealing with the veto powers of the House of Lords, it left intact the anomaly of its hereditary membership and its extraordinary political bias. As the preamble to the Act had included a promise of further reform, the Conservatives insisted that the British constitution was in suspension while the pledge went unredeemed; the country, they argued, was now subject to single-chamber government. After 1914 the role actually played by the House of Lords certainly diminished, and the constitutional question languished on the margins of political debate for decades. Although Tory peers produced a plethora of schemes for further reform, their leaders steadfastly declined to live up to their own professed intention of restoring the upper chamber to a proper role in the system. In the long term they acquiesced in what the Liberals had done.

However, the immediate ramifications of the Parliament Act proved more dramatic. As a result of the 1910 elections the Liberals became dependent upon the 82 Irish MPs for an overall majority. In these circumstances the Nationalist leader, Redmond, naturally expected a new Home Rule bill, more especially now that the removal of the peers' veto made it practical politics. Thus in April 1912 the third Home Rule Bill began its passage through the Commons, and by May 1914 it had been passed three times as required by the Parliament Act. However, the Conservatives and Unionists exploited the two-year interval to promote resistance to the measure in Ulster. Led by Captain James Craig and Sir Edward Carson, the Unionists argued that a Dublin parliament for the whole of Ireland would be oppressive towards the Protestant minority, politically radical, likely to jeopardize the advantages Ireland enjoyed through free trade with England, and certain to open up a weakness in Britain's defences. The truth was that, as the dominant elite in Ulster, they feared a loss of power and status.

However, the government refused to be overawed. They saw the weakness of the Unionists' position. In one sense their cause had already been abandoned by the English Unionists, who had taken government money and surrendered their territorial interest in Ireland. In spite of their concentration in parts of Ulster, the Unionists could hardly expect to veto indefinitely the wishes of the majority. Four-fifths of the Irish constituencies had consistently voted for a Home Rule parliament, and from 1910 onwards their representatives held the balance of power in the House of Commons. On the other hand, Asquith underestimated the depths of Ulster resistance and the lengths to which British Conservatives were prepared to go in order to thwart the policy of the elected government. The campaign began with a dignified declaration of adherence to the Union known as the Ulster Solemn League and Convenant, to which a quarter of a million people put their signatures. But the rebels also established an Ulster Volunteer Force of some 90,000 men, which gained a measure of respectability as a result of the irresponsible pledge given by Bonar Law to support the Ulstermen even if they resorted to violence.

The government continued to regard this as bravado until January 1914, when a number of gun-running episodes occurred. Before long the Ulster Volunteers possessed 35,000 rifles, landed from Germany, and the Nationalists countered by organizing the Dublin Volunteers. The emergence of these private armies raised the prospect that civil war would errupt when the Dublin parliament was set up. That seemed even more likely after the notorious Curragh Mutiny of March 1914. Some fifty-seven officers, stationed in Ireland, offered to resign rather than enforce the Home Rule policy in Ulster. They had been encouraged in this action by senior army officers in London, who placed loyalty to party above their duty to the government. The failure to court-martial those responsible suggested that the government could not wholly rely upon the army to carry out orders. Although this clearly represented the most serious challenge to parliamentary government, it remains a matter of speculation as to what the consequences of civil war would have been. Asquith could not retreat from the Home Rule Bill because he would have lost his parliamentary majority, and because to do so would have caused chaos in the three nationalist provinces. A civil war in Ireland might have led to some violence in Liverpool and Glasgow, but most of the population of the mainland would almost certainly have remained indifferent. They did, after all, subsequently witness something like a civil war in the early 1920s. These events undoubtedly aroused the political elite, especially the far right, but there is no indication of a similar degree of concern in the country at large.

Invasion Scares and Plans for War

Irish developments were steadily eclipsed by the far more serious conflict abroad. The Anglo-Japanese Alliance had set in train a series of events that exacerbated the existing tensions in Europe. It immediately alarmed the French because of the possibility that, if Russia and Japan became involved in a war, they might involve their respective partners, France and Britain, who had no interest in such a conflict. Russia's stunning defeat by Japan in 1904–5 put a different perspective on the alliances. The weakness of the Dual Alliance had been dramatically exposed. If this was France's problem, it was one that increasingly concerned Britain. For not only did she feel threatened by the German navy, she was equally worried about the prospect of another French defeat at the hands of the German army. As the security of the British Isles loomed larger in the minds of the strategists and politicians, it seemed necessary to downgrade Britain's extensive colonial obligations. If these could not be met through force, the alternative was diplomacy and compromise. The first step in this direction came in the form of the Anglo-French Entente of 1904. While this dealt with certain specific disputes over Egypt and Newfoundland fishing rights which had complicated relations for years, the underlying rationale behind the entente was the appreciation that the two countries faced a common threat in Europe.

Although the entente fell far short of an alliance, Germany chose to interpret it as more than simply a settlement of disputes. She therefore tested the strength of the new relationship, putting pressure on the French by challenging their attempt to take over Morocco. The Germans calculated that in a crisis the British would abandon France, leaving her aggrieved and isolated again. Up to a point this proved correct, for the French were checked in North Africa. But in reality the German government had seriously miscalculated, for its clumsy intervention proved to London and Paris how necessary it was to cooperate against the bully. The immediate effect was to prompt Sir Edward Grey, the new Liberal Foreign Secretary, who had entertained suspicions of German ambitions since the 1890s, to move closer to the French. This took the form of conversations between the two navies. The move took place in the context of an existing redeployment of the British fleets. Squadrons in the Pacific and Canadian waters had already been withdrawn, and others were reduced or amalgamated in order to concentrate on an Atlantic Fleet based at Gibraltar, and a Channel Fleet. As the relationship with the French developed, this resulted in further concentrations of British ships in the Channel and the North Sea, where Britain undertook to defend France's northern coast, while in the Mediterranean Britain could rely upon French naval support in protecting her trade routes.

As yet all this was not widely known in the country, or in the Cabinet for that matter. What caught the public attention was the launching of the Dreadnought battleship in February 1906. With a speed of twenty-one knots and ten twelve-inch guns, the Dreadnought outclassed all existing competitors. On the other hand, it rendered Britain's existing fleet obsolete and vulnerable if another power were to build more rapidly to Dreadnought standard. In this situation scares about invasions, spies, and internal subversion flourished. Erskine Childers (*The Riddle of the Sands*, 1903), William Le Queux (*The Invasion of 1910*, 1906), and H. G. Wells (*The War In The Air*, 1903) were among those who pandered to the fashion. Following the precedent of the 'Battle of Dorking', Le Queux collaborated with Lord Roberts of the National Service League and Lord Northcliffe of the *Daily Mail* to focus attention upon the likely route chosen by an invading army; it followed towns in which Northcliffe was anxious to boost sales of the

Daily Mail, thereby enabling him to mix profit and patriotism in a rather satisfying way! It soon became difficult to separate fact from fiction. It was well known that thousands of Germans and other aliens lived in Britain working as waiters, servants, barbers, and clerks. They were now alleged to be largely spies, plotting acts of subversion to facilitate the landing of an enemy force. Reports of suspicious activities from members of the public found a welcome in the pages of the *Daily Mail* and other papers. The excitable mood prompted the Navy League to foment panic over the navy in 1909, on the basis of information that the Germans planned to build fourteen Dreadnought-type ships. At that time Britain possessed six, and the Admiralty planned to add to these at the rate of four a year. The League, the Northcliffe press, and the opposition in Parliament demanded eight new ships immediately.

Though the scares about spies, invasions, and the German navy were baseless propaganda, they were not without some political influence. Many of those in high places wanted to believe the rumours, as did the general public. In addition to becoming amateur spy-spotters, many Edwardians undertook first-aid training, especially under the Voluntary Aid Detachments organized by the Red Cross, so as to be useful when the invasion came. The Cabinet could not afford to be outflanked either. It established a regular secret service, introduced registers of aliens to enable the authorities to control their movements, and made arrangements to have strategic buildings properly guarded. In 1910, on a quiet Friday afternoon, the government squeezed a new Official Secrets Bill through the House of Commons; this was intended to enable the police to check subversion by prosecuting any person who was considered to be suspicious but without producing any evidence that he was a spy or had actually spied. Though this was a scandalous piece of legislation in a liberal society, the Act was accepted without serious question because of the climate of fear. Once war had broken out the authorities immediately arrested twenty-one 'spies'; and it was widely believed that the British forces would have done much better had they not been betrayed by the network of spies ensconced in Britain. John Buchan's famous novel *The Thirty-Nine Steps* (1915) caught the mood very neatly.

Meanwhile Asquith also defused the naval scare in 1909 by undertaking to build eight new Dreadnoughts if the rate of building in Germany appeared to warrant it. In fact, the Germans never regarded an invasion of Britain as feasible. By 1911 they had only six Dreadnoughts, and Britain retained a comfortable lead right up to 1914. However, the scares helped to lock the two countries into the naval race, and obliged the Chancellor of the Exchequer to fund the higher naval estimates. Although Lloyd George offered some resistance to this, he had come to agree fundamentally with the new trend in Britain's foreign and defence policy; consequently he always found the money for new Dreadnoughts.

While such signs of determination on the naval front were reassuring to the French, the situation in Europe remained very worrying. After all, Britain had given no firm or public commitment to support them. In the event of war, the Royal Navy would play no part in resolving the immediate strategic problem; for the German armies were expected to cross the frontiers of France and Belgium in a series of arcs which would envelope Paris. In a matter of weeks the French government would collapse, and the war might be effectively over before the Russian forces had swung into action. Increasingly, therefore, the French wanted to know whether Britain would put any real weight behind her new friendship in the early stages of a war.

Reluctantly Grey concluded that Britain must do so because a further defeat of France could lead to her being seriously weakened as a great power; once the Germans gained control of northern France they would be in a strong position to threaten the British Isles. The plans designed to give effect to this intention emerged from the discussions of the Committee of Imperial Defence and the Army's General Staff, which had been created as recently as 1904. Faced with Britain's chaotic efforts to raise a substantial army for the South African War, the new Secretary of State for War, Richard Haldane, chose to reorganize his forces by amalgamating the regulars, the militia, and the volunteers into two. The Field Force, basically the professional army, was to be capable of mobilization within fifteen days and thus to be available for use abroad in an emergency, while the Home Force or Territorial Army, consisting of partly trained men, would be ready for action six months after the outbreak of war.

At the time the authorities avoided saying that the Field Force, or British Expeditionary Force, was to be deployed in Europe. Since the memory of the Boer War was still vivid and apprehensions about the Russian threat to India's North-West Frontier were very much alive, the army was seen in terms of colonial emergencies. Indeed, as the Russian railway system reached Tashkent in 1904 it became feasible for the first time to convey a substantial army to within striking distance of the Indian Empire. But the British government had no intention of giving the Commander-in-Chief in India the extra forces he believed he required. It followed that a Russian threat must be tackled and defused by diplomatic means, as the French had been. This was accomplished by the Anglo-Russian Entente of 1907. The Russians agreed to communicate with the Amir of Afghanistan only through Britain, and to keep their agents out of his country – a remarkably favourable deal considering how vulnerable Britain's position was. In Persia the two powers agreed to confine themselves to separate spheres of interest, the Russians to the north around Teheran, the British to the south-east around Bandar Abbas. The rationale was identical to that underlying the Anglo-French Entente: both powers considered the threat of war in Europe as the greater priority.

The 1907 Entente virtually completed the diplomatic revolution. It allowed the Committee of Imperial Defence to plan for the dispatch of 120,000 men of the BEF to reinforce the left flank of the French forces in Europe, with a view to checking the German advance towards Paris. For a time the Admiralty prevaricated over this, because it did not wish to be downgraded to become a transport service for the army. But the navy's own ideas about the war simply failed to address the key strategic problem, and by 1911 it had been overruled; Britain was effectively, though not formally, committed to fighting a Continental land war.

As a result, Britain was better prepared for the First World War than for almost any war in her modern history. She had diagnosed the strategic threat, rationalized her forces and adapted her planning to meet it, and in the event successfully implemented her policy: Paris was saved, and the Allies won vital time in which to mobilize both Russian manpower and British industrial might. Yet this has not been generally recognized in received accounts of the period. This partly reflects the force of Edwardian propaganda on the theme of Britain's unpreparedness for war. It also reflects the fact that, in spite of her planning, Britain gave no binding promise to join France and Russia in war with Germany. This was to some extent due to political inhibitions. In the Cabinet, half the ministers tended to oppose participation in a Continental war. Much of the Liberal and Labour forces in the country and in Parliament strongly disapproved of the naval race, felt that more should be done to maintain good relations with Germany, and abhorred the entanglement with the oppressive Tsarist regime in Russia. Yet even if such pressures had not existed, it is not clear that an *alliance* with France would have been wise. Grey rightly felt that any firm pledge to the French would have made them all the more willing to risk a war with Germany. Since above all things he hoped to maintain peace, Grey saw the wisdom of keeping the other powers in some doubt about Britain's course of action in a crisis.

However, none of the criticisms had much effect on the development of government policy. The radicals found it impossible to deny the necessity for some naval building, especially since, in spite of several overtures from the British, Germany refused to modify her own programme. In addition, the government took some of the sting out of the criticism by saving money on the army reorganization and by managing to finance its social policies in spite of the swollen naval estimates. After 1910, when the Liberals lost their independent majority and party passions were inflamed by the House of Lords and Home Rule issues, the critics felt reluctant to upset the boat over foreign policy. This still left a serious charge against Grey over the secrecy in which British policy was being made; the radicals correctly suspected that the Foreign Secretary was committing Britain to more than he was prepared to admit in public. This concern was shared by Cabinet ministers, who were largely excluded from policy-making. The issue came to a head as a result of the second crisis over Morocco in 1911, when the German government again tried to bully the French and demanded compensation for the French occupation of North Africa. This provoked a famous warning to Germany not to treat Britain as though she were of no account in international affairs delivered by Lloyd George in his speech at the Mansion House. In Cabinet John Morley spoke for many when he complained that Grey seemed to be giving unauthorized promises of support to the French. As a result the ministers formally decided that no communications should take place between British and French forces that committed Britain to intervention, and that any talks must first be approved by the Cabinet. This, however, was shutting the door after the horse had bolted.

During the next three years fears among the political elite were, if anything, diminished by Grey's successful diplomacy. When a crisis erupted in late July 1914 following the assassination of the Austrian Archduke at Sarajevo, the initial expectations were of yet another *regional* war in the Balkans, albeit one involving Austria and Serbia. However, such expectations were dashed by the decision of the Russians that they could no longer stand by and allow their interests in the Balkans to be trampled upon. Since Germany intended to back up Austria, France was unavoidably involved too. Faced with the prospect of war between the great powers, the British Cabinet initially declined to commit itself to either intervention or neutrality. But on 2 August the ministers took a crucial half-way step by authorizing Grey to assure France that Britain would not allow Germany to use the Channel for hostile purposes. The doubters in Cabinet appreciated that Britain had entered into some obligations to the French for mutual naval defence; it was also tempting to see Britain's role in the forthcoming conflict as primarily a naval one. News of the violation of Belgian territory helped to bring the waverers into line behind an ultimatum demanding Germany's withdrawal. Eventually all but two Cabinet ministers, John Burns and John Morley, accepted this decision. But British entry into the war was not essentially bound up with the Belgian issue, even though this provided a respectable legal and moral course. Rather, Britain calculated that her own security was now closely linked with the maintenance of France as a great power.

An Ungovernable Society?

Was the Cabinet's fateful decision to take Britain to war a desperate ploy to save a turbulent society sliding into chaos? It is fair to say that for Asquith the international crisis was a timely diversion from domestic troubles. War had the immediate effect of producing a party-political truce and the suspension of the controversial bills to give Ireland Home Rule and to disestablish the Church in Wales. Strikes and suffragette campaigns swiftly ceased. John Redmond gave a remarkable pledge of loyalty to the British cause and the warring parties in Ireland withdrew from confrontation.

However, the deep loathing with which the Liberal ministers committed the country to war leaves no room for suspicion that they opted for it as a way out of domestic difficulties. Indeed, they saw war as the ultimate triumph of Toryism which would probably check, if it did not destroy, all the progress that had recently been made. This may well represent a contrast with the situation prevailing elsewhere. For the unstable regimes of Germany, Austria, and Russia there was always a temptation to use external conflict to manufacture unity and silence domestic opposition.

But the British parliamentary system remained strong and viable, and the Liberal government was well entrenched in spite of the controversies surrounding it. At worst Asquith had been weak towards the Ulster Unionists, whose leaders ought to have been arrested, and he had been foolishly stubborn in resisting the claims of women to enfranchisement. But bit by bit the various challenges facing the government were all being surmounted. Asquith had consolidated his position by twice winning elections in which first 86 per cent and then 81 per cent of the electors had voted. Above all, the particular configuration of parties meant that social class was never the source of acute disunity and instability that it became in some other countries. To Lloyd George's opponents it naturally appeared that he employed the language of class war against the dukes, but in fact the rationale behind his radical rhetoric was to shore up the alliance of working men and the middle class. The economic boom of the last years of peacetime made this all the easier to achieve. Neither the labour movement nor the women's movement was really anti-parliamentary; their aim was to join the system rather than to overthrow it. Nor were there any significant links between the various extra-parliamentary movements. They were coincidental controversies rather than symptoms of a common malaise.

If a sense of frustration and alienation existed at all in these years, it was characteristic of the right wing of the political elite. Conservatives found it hard to accept their protracted exclusion from office, and the landed class could see that it was being marginalized politically. Above all the Edwardian period was characterized by bitter attacks by the 'radical right', including such men as Leo Maxse, Sir Henry Page Croft, and Willoughby de Broke. Much stirred by Britain's poor showing in the Boer War, they espoused simplistic conspiracy theories to explain the country's decline. They diagnosed an enemy within in the form of Irish Nationalists, Liberals, Socialists, Jews, aliens, and 'radical plutocrats'. The latter group comprised wealthy cosmopolitan businessmen who remained outside the Conservative Establishment and chose to promote Liberal causes such as land reform. The radical right also felt betrayed by its own party leaders, who now attracted criticism as an effete, upper-class elite inherited from the Salisbury era. Balfour was the obvious victim of this mood. The outbreak of war made these Tory rebels even more neurotic and suspicious of their own party leadership, and even the return to office failed to heal the divisions that had developed in the party.

10

Politics and Society in the Great War

In Britain the conflict that broke out in August 1914 was quickly dubbed 'the Great War'. From the perspective of the Second World War this seems an exaggeration, for it was essentially fought in the European theatre with a few peripheral campaigns in the Middle East and Africa. Yet the British people had experienced nothing on this scale since the Revolutionary and Napoleonic Wars a century earlier. Since then war had been a matter of minor colonial skirmishes; even the Crimean War, which did involve three great powers, was peripheral by comparison, and far less was at stake than in 1914–18. This was the sense behind Sir Edward Grey's one famous remark: 'The lamps are going out all over Europe; we shall not see them lit again in our time.'

In fact, by comparison with the regimes in Germany, Russia, Italy, and even France, Britain was to emerge remarkably unscathed by the ordeal. None the less, contemporaries and many later observers judged the Great War to have made a far-reaching impact on almost every aspect of life in Britain. During the 1920s and 1930s contemporaries naturally concentrated on the purely negative aspects, notably the loss of some 750,000 British males. But after 1945, influenced by the aftermath of a second war, historians increasingly emphasized the constructive effects of the First World War. It began to be seen as a new kind of war – 'Total War'. Previous conflicts, especially for Britain, had usually involved quite small numbers of fighting men and made only a limited impact on society. During 1914–18, on the other hand, over 5 million British men enlisted in the armed forces; moreover, the support and maintenance of the troops required a huge civilian effort by both men and women. The necessity for mass participation led some historians to argue that the political elite had no option but to grant a succession of concessions or rewards of both an economic and a political nature which had the effect of levelling some of the inequalities in British society. The war was also seen to have discredited many prevailing beliefs and institutions, and to have had a liberalizing effect on social behaviour.

The Continental Commitment

At first the war was widely expected to resemble the Franco-Prussian War; it would involve swift mobilization, rapid flanking movements, decisive battles, and thus reach a speedy conclusion. In a famous phrase, it would all be over by Christmas. Britain's role would be primarily naval and economic; her industrial and financial resources would enable the French and Russian armies to overcome the Central Powers. In order to prevent an early knock-out blow against Paris by the advancing German armies, the British Expeditionary Force, commanded by Sir John French, was conveyed across the Channel to reinforce the left flank of the French – an important but limited role. At Mons the BEF, which found itself directly in the path of much larger German forces, quickly joined in the general retreat south. A counter-attack at the Battle of the Marne in September checked the German advance, but movement rapidly ceased. The rival forces then raced to the coast to prevent flanking attacks, and dug elaborate lines of trenches protected by rolls of barbed wire. By January 1915 the prospect of rapid advances and frontal attacks had nearly disappeared and four years of stalemate on the western front had begun.

Although British strategy had already scored a major success – the defeat of the Schlieffen Plan – this was not widely recognized at the time. Though Paris had been saved she remained, throughout the war, only a few precarious miles from enemy forces and thus vulnerable to a sudden attack. The man who did most to develop British strategy beyond the first stage was Lord Kitchener, whom Asquith appointed Secretary of State for War. Kitchener quickly perceived that the struggle would in fact be prolonged, and that Britain would have to play a greater role on the Continent than had been anticipated. Ignoring the pre-war Territorial Army, he proceeded to create an entirely new British army from the mass of eager volunteers.

The popular response to the crisis is one of the most remarkable features of the whole war. The authorities were swamped by the numbers of men who came forward voluntarily to serve in the forces. Recruiting figures ran at 300,000 in August, 450,000 in September, 137,000 in October, 170,000 in November, 117,000 in December, and 156,000 in January 1915. Explanations for this response vary. Even though there was no specific territorial issue at stake, people had grown to expect a war and to regard the Germans as a threat to British interests. For many, the most concrete issue was Belgium, which was presented by the press and government propaganda as the innocent victim of the German bully. Tales of German atrocities – almost wholly unsubstantiated – and the arrival in Britain of Belgian refugees helped to reinforce this view and to give a high moral tone to the British cause. The Liberal government enlisted liberal writers like H. G. Wells and John Galsworthy to present Britain's case in idealistic terms. Those who sought comfort from the appalling situation were attracted by the idea that this was a 'war to end war', not one for territorial aggrandizement.

However, most people were more moved by fear and hatred of Germany. Economic motives for volunteering were not particularly important; unemployment was not high at this time, and the rush of recruits clearly went far beyond any particular class or industry. Many young men were swept up in the excitement, felt vaguely patriotic, were attracted by the status conveyed by a uniform, and anticipated a brief adventure in foreign parts in the company of fellow workers. Hence the popularity of the 'Pals Battalions' based on particular occupations and companies.

Consequently, by early 1915 an unusually large British army was undergoing training for service abroad. The question arose: exactly what use should be made of it? Britain's planning did not go beyond the original commitment to France. Kitchener, in common with almost all other generals, adopted the view that, since the war could be lost very easily on the western front, Britain had no option but to concentrate her forces there and seek a decisive victory over Germany. This view gained force from the demands made by the French that Britain must relieve the pressure on them by taking over more of the front line.

Against this were ranged a number of arguments and interests. The Royal Navy, never happy at being downgraded by the Continental commitment, enjoyed a good deal of political support, especially from those Liberals who wanted a limited war. Up to a point the traditional naval strategy was implemented. Enemy ships were soon cleared from the oceans, Germany's foreign trade was largely stopped, the British Isles protected from invasion, and the navy prepared for a decisive battle with the German High Seas Fleet. However, this was far from being a recipe for overall victory in the war. Apart from the Battle of Jutland in 1916 the Germans avoided major naval actions, and it was far too dangerous for Britain to attack close to the enemy's home waters; battleships were very vulnerable to mines, submarines, and torpedoes. After all the controversy of the Edwardian years, the navy proved something of a disappointment and was by no means the key to Britain's war effort.

But critics of the army's strategy like Lloyd George and Winston Churchill contended that over-concentration on the western front simply dissipated Britain's precious manpower for no strategic gain. Instead, they wished to use the flexibility offered by the navy to transfer the new armies to other theatres in which the central powers were more vulnerable. To some extent this strategy was implemented in the campaigns in the Dardanelles, Salonika, Mesopotamia, and East Africa. By thus 'knocking away the props' in the shape of Turkey, Austria, and the German colonies, Britain would isolate Germany. Unhappily these territories turned out not to be props for Germany at all. In any case, it proved almost as difficult to achieve victories in these campaigns as on the western front, and the military authorities were never willing to release enough resources to attain decisive results.

The politicians were generally reluctant to overrule expert military advice, even when it failed to produce positive results. Both wartime prime ministers found themselves exposed to charges that they had failed to provide either munitions or troops in sufficient quantity and thereby hampered the army's efforts. This was to have important political consequences. For example, when Sir John French's spring offensive of 1915 failed he blamed Asquith rather than admitting his own incompetence, arguing that if he had had more high-explosive shells the breakthrough would have been achieved. This situation contributed to Asquith's decision to replace his Liberal government with a three-party coalition in May 1915. It also led to the creation of the Ministry of Munitions under Lloyd George, which generated huge quantities of armaments, albeit at great cost, and made his reputation as a war minister when others were being discredited.

As the generals lost one excuse they sought another. By the start of 1916 debate had come to focus on the supply of men necessary to sustain an offensive on the western front. Now that the number of volunteers had dwindled, the generals, supported by most Conservatives and some Liberals, argued that the time had come to adopt conscription like the other belligerents. This was resisted in Cabinet partly on the grounds that compulsion was morally wrong, and more effectively on the grounds that it was essential for Britain to maintain her industrial output so as to sustain the allied war effort for a long period; frittering men away on futile offensives would achieve nothing except a German victory. In 1916 the government eventually opted for conscription first of single and then of married men. This made possible Sir Douglas Haig's disastrous offensives of 1916 and 1917, first the Battle of the Somme, which produced 60,000 British casualties on the first day alone, and then the prolonged Passchendaele campaign. Neither achieved any significant territorial or strategic gain.

In spite of Haig's repeated assurances that his offensives had worn down and demoralized the enemy, in the spring of 1918 the Germans launched a major attack which very nearly brought them victory. It broke up the British Fifth Army, forced the allies into a headlong retreat, and almost divided the British forces from the French. This military setback created a political crisis for Lloyd George, who was accused of deliberately withholding troops from Haig as a means of deterring him from embarking on fresh offensives. This came close to destroying the government, but Lloyd George managed to bluff his way out, largely because even his critics thought his government the best one available. When the allied forces, now strongly reinforced by American troops, resumed the attack in the summer of 1918, they were surprised to find the Germans in retreat. Although never comprehensively defeated, the German leaders agreed to an armistice in November and the war came to an expectedly sudden end. At last it was the turn of the politicians to seize some advantage from military events. Lloyd George lost no time in holding a general election in December 1918, while his prestige as the 'man who had won the war' was at its height.

State Intervention in the Economy

The adoption of a Continental policy involving the recruitment of mass armies had major implications for wartime economic policy because of the strains it imposed upon industry. There was to be no 'business as usual'. From the very beginning, substantial numbers of skilled men in engineering, mining, and steel, who were essential to the war effort at home, joined the armed forces. To some extent this loss was checked by government action to exempt key workers from military service, and in 1917 Lloyd George introduced a National Service Department with the intention of distributing manpower rationally between the various industries and the armed forces. But this proved impossible, largely because the army had already distorted the labour market and it was politically impossible to force it to disgorge its recruits.

Instead, the chief expedient was simply to introduce large numbers of unskilled men, women, and boys into the jobs done by the men at the front. At first many war workers felt grateful to have a full week's work and to escape occupations like domestic service which suffered from low pay and status. In time, however, they realized that they were being exploited by the government and by the employers, who were making huge profits. Not surprisingly, trade union membership rose from 4 to 6.5 million by the end of the war. The number of strikes, which had been running at a high level, dropped sharply in 1915 and 1916, but then rose again in 1917 and 1918. Workers, angered by the rising cost of food and housing, appreciated increasingly that they were in a strong bargaining position. Many companies were making assured profits on war orders and paying unskilled wages for skilled work. Since the government itself was desperate to avoid any interruption to the flow of production, it encouraged the owners to grant a substantial increase in money wages during the latter half of the war and in the immediate aftermath.

The other process at work was inflation. Since Britain depended heavily on imported food some substantial price rises were inevitable, especially when German submarines attacked merchant shipping indiscriminately. An equally serious problem arose over accommodation, which was almost

Registrar of Women Workers. "WHAT CAN I DO FOR YOU?"
Applicant. "YOU PROBABLY WANT A FOREWOMAN: SOMEBODY WHO IS USED TO GIVING ORDERS AND WORDS OF COMMAND. I'VE BROUGHT MY HUSBAND TO SPEAK FOR ME."

Novel wartime developments in employment

wholly rented at this time. The diversion of workers from the building industry meant that there could be no improvement in the housing stock during the war. Moreover, the major migration of workers to such areas as Clydeside, Newcastle, and Sheffield, where munitions factories were concentrated, created severe shortages and led to much higher rents.

Wartime governments took these economic and social problems very seriously. Originally the Liberals had expected the war to have the effect of disrupting industry and creating mass unemployment. However, they soon realized that the huge demands of the war machine would lead to a shortage rather than a surplus of labour. They particularly feared that, as the initial euphoria wore off, civilian discontent would offer opportunities both to syndicalists and to pacifists, who were campaigning for a negotiated peace, to disrupt industry. The result was a barrage of interventionist measures to regulate the economy and ensure the free flow of vital goods. From the beginning of the war the government began to buy food supplies on the world markets and release them so as to check price increases. Propaganda designed to achieve the more economical use of food and appeals for voluntary rationing largely failed, and by 1917 a compulsory rationing system had been introduced. Key items like bread and potatoes were subsidized, and 80 per cent of all food was purchased by the Food Controller. Draconian steps were taken, in the form of the 1917 Corn Production Act, to impose precise output targets upon farmers. As early as 1915 a Rent Restriction Act was passed to avert serious discontent in the industrial districts.

Industry also came under extensive official control. In August 1914 the government invoked its powers under an act of 1870 to take over the operation of the railways during the war. Thereafter the state intervened as and when shortages, bottlenecks, or industrial unrest created a threat to production. Between 1915 and 1917 the coalfields were brought steadily under official control. Under the auspices of the Ministry of Munitions, the government became the owner of 250 factories and supervised the work of a further 20,000 involving 2 million employees. Above all, the cost of the war meant that government expenditure accounted for nearly 60 per cent of gross national product by 1917 compared to 7 per cent in 1913. Although 70 per cent of the cost of the war was met by loans, the remaining 30 per cent was financed by taxation, which meant an increase in income tax from a maximum prewar rate of 2s. 8d. to 6s. in the pound. In addition the tax net was greatly extended because the threshold was lowered from £150 to £130 and rising wage rates made millions of working men liable for income tax for the first time.

Of course, all this did not mean that Liberal and Conservative politicians had suddenly become Socialists. By and large, wartime economic innovations resulted from *ad hoc* responses to a succession of crises; they were expedients rather than considered strategies. Consequently, most were quickly reversed, though there was to be no return to low income tax. But an important precedent for sweeping state intervention had been set, and the inability of unfettered private enterprise to meet national needs had been demonstrated. Both trade unions and employers appreciated how much they had to gain from the closer involvement of the state in the economy. For example, wartime experience strengthened the pressure among the miners for the nationalization of the coal industry. But some socialists noticed how beneficial state control could be for the owners too. It helped them to reduce strike action. Although limits were placed upon profits, this was a small price to pay for the certainty of large profit margins resulting from government contracts. Industrialists, now increasingly organized in such bodies as the Federation of British Industry, determined to maximize their influence on government in the future so as to obtain a range of policies – from lower taxation to tariff protection – to help maintain their wartime grip on the domestic market. In this sense the war marked the start of a 60-year retreat from traditional capitalist enterprise in favour of a more corporatist approach.

Political Transformation

Military and economic policies inevitably had the effect of expanding the size of government during the war. Under the Defence of the Realm Act the authorities enjoyed broad, ill-defined powers to interfere in the lives of their citizens. This coercive power was supplemented by the official propaganda machine and by the unofficial cooperation of the press, which largely abandoned any responsibility to report truthfully to the public about the war. Under Asquith's premiership the innovations in government machinery were confined to the new Ministry of Munitions, cabinet committees to look into issues such as postwar reconstruction, and successive War Cabinets which debated military strategy. In December 1916, however, the traditional twenty-member Cabinet was bypassed by a five-man War Cabinet, which met daily so as to make quick decisions on vital war issues. It was made more efficient by the Cabinet Secretariat under Sir Maurice Hankey, and backed up by Lloyd George's personal secretariat of unofficial experts and advisers. These two elements substantially survived the transition to peacetime, and contributed to the growing dominance of the Prime Minister at the expense of the Cabinet in British government.

A conspicuous feature of Lloyd George's premiership was the increase in his powers of patronage, especially the creation of new departments and ministries: Labour, Pensions, Food, Health, Shipping, Air, Information, and National Service. This naturally involved a considerable increase in the number of civil servants centrally employed, and although several of the new departments were purely wartime expedients, Health, Labour, and Pensions outlasted the war and reinforced the trend towards a larger role for government. They were suspect in Treasury eyes because of their association with costly social policies; but it seemed inexpedient to abolish them, given the pressure of a mass electorate in 1918 and the rise of the Labour Party. These were the key underlying developments on the political front after 1914.

Although Asquith's Liberal Government successfully took Britain to war with only two ministerial resignations and bolstered its standing by the addition of Lord Kitchener, its fortunes soon began to deteriorate. This was due not only to the absence of any conspicuous naval or military successes but also to its vulnerability when accused of lacking enthusiasm for the war effort. Many Liberals certainly had misgivings about wartime policies – the Defence of the Realm Act and the infringement of free trade in the 1915 budget, for example. However, this would never have become a serious problem but for a major miscalculation in May 1915, when Asquith invited the Conservatives and Labour to join him in a coalition. In the short run this had the advantage of enabling him to avoid holding a wartime general election. But it was seriously flawed because it reversed the logic of prewar political strategy which pointed to a coalition with Labour and with the Irish Nationalists who, under John Redmond, were notably loyal to Britain's cause after the outbreak of war. A coalition with the Conservatives was destructive of the rationale that had underpinned the Liberals' position since 1906. It immediately alienated the Irish who, though invited in, stayed out of the new government. As the implementation of the Home Rule Bill seemed increasingly unlikely, the Nationalists' own position in Ireland began to be undermined. The Easter Rebellion of 1916 greatly accelerated the process, so that by 1918 Redmond's party was largely ousted by supporters of Sinn Fein, exasperated by the failure of the parliamentary strategy for achieving Home Rule.

More immediately damaging to Liberal morale was the perception that alliance with the Conservatives had led to the adoption of conscription and committed the country to the total defeat of Germany – the 'knock-out blow', in Lloyd George's phrase. Many Liberal and Labour politicians who had had misgivings, especially over cooperation with Russia, now began to conclude that the prewar policy, including the arms race and the entanglement with France, had been a mistake; in view of the horrendous casualties suffered in 1915 and 1916 it seemed wiser to seek a negotiated peace. The attractions of this admittedly minority view were strengthened when President Woodrow Wilson expressed similar sentiments; and the evident inability of the government to define Britain's war aims in anything but vague terms only exacerbated fears that the government, influenced by Tory imperialists, would drift into a vindictive peace and a protectionist policy.

The alternative lay in a compromise settlement which respected the wishes of the various national groups, general disarmament, and a new method of conducting international diplomacy – the League of Nations. Such policies were advocated by the Independent Labour Party and the Union of Democratic Control, a new pressure group which included Socialists like Ramsay MacDonald and such Liberals as Charles Trevelyan and Arthur Ponsonby. By 1916 such groups were gaining recruits among Liberals who felt that the war was no longer being pursued by liberal methods or for liberal objectives; consequently, there seemed less necessity to maintain Asquith as Prime Minister. Although many Liberals greatly disliked the Lloyd George coalition which replaced Asquith in December 1916, there was little attempt to organize an effective opposition to it, partly because this would have appeared unpatriotic and might well have provoked a wartime general election. Consequently the Liberals found themselves divided between two leaders, each of whom was inadequate, acting as Prime Minister and leader of the opposition; the party inevitably lost its sense of purpose once it had abandoned the strategy that had served it so well up to 1914.

Conversely, the war gave a renewed sense of purpose to the Labour Party. The attainment of several ministerial positions in both the coalition governments raised the party's status, and the growing influence of the trade union movement made it conscious of its central role in the war effort. In this way Labour succeeded on the one hand in being patriotic and helpful in the war effort and on the other in adopting a critical view of government policies. By concentrating on defending the interests of working-class families over wages, pensions, wartime allowances, prices, and rents Labour achieved concrete results. Moreover, the enactment of new social policies such as state-subsidized housing after the war enabled the party to seize the initiative from the demoralized Liberals at the grass roots.

The party also sensed that, with the collapse of Liberal organization in the country, the time had come to abandon the Edwardian electoral pact and run candidates of its own in a majority of constituencies. As the unions gained members and built up their funds, as was permitted under the 1913 Act, this strategy became increasingly feasible. Above all, the 1918 Representation of the People Act gave further encouragement by enfranchising almost all working men; it also helped Labour by reducing some of the costs of fighting elections. Prompted by these favourable circumstances, the party took a series of initiatives between the autumn of 1917 and the autumn of 1918. It produced new statements of party policy in both domestic and external affairs. It adopted many extra candidates and recruited members from middle-class and Liberal backgrounds in addition to its indirect trade union membership. The party also devised a new constitution which included the famous Clause 4, committing it to socialist objectives for the first time. Most important in this scheme were the arrangements for individual party membership and the establishment of a local Labour Party in every constituency. The 1918 constitution effectively placed central power in the hands of the trade unions by allowing their block vote to determine the membership of the National Executive Committee at each annual conference. This had the effect of undermining the role of the small socialist societies and the ILP, who had been so influential under the original party constitution. In this way Labour equipped itself to challenge the Liberals for their role as the chief alternative to the Conservatives in 1918.

The outbreak of war found the Conservatives in a highly fractious condition as a result of their prolonged spell out of office. In collaboration with their allies in the press, they accused their opponents of lacking patriotism and being unqualified to run the war. Yet at the same time the Conservative leaders also sensed that, with the rise of the labour movement, conditions were still moving against them. They proved reluctant to risk forming a Conservative government in 1915 and in 1916, for fear that they would be unable to obtain the cooperation of the working class in maintaining the war effort. Coalition, however uncomfortable, was the best they could hope for.

On the other hand, the crisis of war and the threat to the empire helped to reunite the Conservatives and did at least restore them to office. Particularly under Lloyd George's premiership, Bonar Law, Balfour, Austen Chamberlain, and Lord Milner found a very satisfying role to play. Victory, however, only raised in an acute form the underlying problem of the party's future strategy: it had not won an election since 1900. Now the concatenation of working-class enfranchisement, the growth of the Labour Party, and the prevailing mood of radicalism which seemed to be affecting all European countries by 1918, threatened to create a left-wing landslide at the postwar general election. In these circumstances the Conservatives decided to swallow their dislike of Lloyd George, maintain the coalition, and fight an election under his leadership. In this way they could hope to benefit from association with military victory, benefit from Lloyd George's personal popularity, and also keep the Liberals divided.

In the event, this plan worked even better than expected. At the election of December 1918 the supporters of the Coalition won 541 seats, including 478 Conservatives and 136 Lloyd George Liberals, on the basis of 54 per cent of the vote. Against them were ranged 73 Sinn Feiners (who did not take their seats), 63 Labour, and a mere 28 Liberal MPs. This was the result partly of the way in which Asquith had been discredited and partly of the split between Labour and the Liberals which divided the non-Conservative vote. But it also reflected the unexpectedly emotional circumstances in which the election was conducted. It was fought hurriedly within a few weeks of the armistice while wartime feeling ran high. Politicians discovered that voters had not yet reorientated themselves to domestic politics; rather, they showed themselves anxious to make Germany pay for the war and to hang the Kaiser. The coalitionists simply pandered to the hysteria. One government minister made the notorious promise: 'We will squeeze Germany like a lemon; we will squeeze her until you can hear the pips squeak.' It thus proved all too easy to condemn the more restrained Liberal and Labour candidates as weak and unpatriotic, and as responsible for obstructing Britain's victory in the war. By 1918, then, the war years had effectively destroyed the left-wing alliance that had dominated Edwardian politics, and ushered in an era of Conservatism.

Social Reconstruction

Although the war did not fundamentally change the Edwardian system of state-financed social welfare, it helped to expand it and indirectly strengthened the political foundations, at least in the long term. As a result of the mass enlistment of young men, many of whom left wives and children behind, it became necessary early in the war to pay allowances and, as casualties mounted, pensions to civilian dependants. This experience had some long-term significance in that it convinced some feminists that payments made directly to mothers were a highly cost-effective means of relieving poverty. At the time the idea was referred to as the 'endowment of motherhood', though subsequently it became better known as the family allowance. Under Eleanor Rathbone and others, a campaign began to persuade the state to recognize that the role of the wife and mother was of equal importance to other occupations and should therefore receive remuneration in the form of a direct payment. During the Edwardian period, politicians had become more susceptible to pressure on behalf of married women, with the result that in 1914 Samuel had offered local authorities grants to set up maternal health clinics. The huge loss of young men during the war made governments even more concerned to promote the health of infants and to make motherhood less burdensome. The result was a considerable extension of existing policy, but not the adoption of the more radical ideas of Rathbone. The number of clinics increased from 350 to 1,290, and health visitors from 600 to 1,350, by 1918. This culminated in the 1918 Maternity and Child Welfare Act, which compelled local authorities to implement nationally determined standards of care. This was one of the few enduring wartime reforms.

Another byproduct of wartime concern about the younger generation was H.A.L. Fisher's 1918 Education Act, which made secondary education compulsory to the age of fourteen, proposed a complete system from nursery education to higher education, and offered a 50 per cent grant to local authorities. But the main reconstruction debate concentrated on the deterioration of the housing stock during the war. It was officially estimated that 300,000 extra houses would be required after the war. The result, much delayed, was Dr Addison's 1919 Housing Act. This measure was based on the idea that each local authority should prepare a house-building programme reflecting local needs, be responsible for carrying out the scheme, and offer homes to rent to working-class families. It was housing that assumed the highest profile among social reforms, then and subsequently, because the Lloyd George government promised to build half a million 'Homes For Heroes' in 1918. However, the shortage of labour and the uncooperative attitude of builders made the housing targets difficult to meet. Eventually, additional government subsidies helped to achieve a total of 170,000 new houses. Although this fell a long way short of the promises made, it represented a major advance on previous housing policies, and was important not least for setting a higher standard for working-class housing. Unfortunately, the higher cost made the housing programme very vulnerable to postwar expenditure cuts in 1920; by 1921 the Addison Act had been abandoned and its author driven from office. Similarly, the education reforms were severely curtailed by financial retrenchment. This speedy reversal of large parts of the reconstruction programme clearly suggests that it is at least an exaggeration to claim that the mass participation in the war effort had a fundamental impact on official thinking; in many ways the rhetoric of reconstruction proved an ephemeral feature, designed to serve a temporary political need.

In some ways the most solid concession granted in wartime – because it could not be withdrawn even when circumstances changed – was political reform. Up to 1914 only six out of every ten men, and no women, enjoyed a parliamentary vote, and the Liberal government had been severely embarrassed by the consequences of its failure to effect some reforms. The dilemma deepened during the war simply because many men lost their place on the electoral register as a result of leaving home for war work in the armed forces. Fearful of holding an election from which large numbers of patriotic war workers were actually excluded, the political parties devised a compromise scheme of reform in 1917. This enfranchised virtually all males over the age of twenty-one and women over thirty years who were themselves, or were married to, local government voters. This transformed the pre-1914 electorate of between 7 and 8 million into one comprising 13 million men and over 8 million women.

However, it is very doubtful whether this reform can be explained in terms of fundamental changes in attitude brought about by popular participation in the war. Politicians certainly paid tributes to patriotic war work, implying that the vote was a reward or, in the case of women, a recognition that they had proved their capability as war workers. But in fact most politicians continued to disapprove of women as paid employees. They had no hesitation about forcing them out of their new jobs, and they deliberately withheld the vote from the young women who had worked in the vital munitions factories. War had clearly failed to alter traditional ideas about the role of the two sexes. The reforms of 1918 represented the conclusion of the attempts and manœuvres of Liberal and Labour radicals for many years before 1914; it was passed more to meet party political objects than because of a radical change of heart. Though the Conservatives had given most away in the all-party compromise, they took comfort from the patriotism shown by the working classes, the redistribution of the constituencies, and their alliance with Lloyd George.

The Lost Generation

In the 1920s and 1930s the most pervasive view of the war, fostered by the anti-war literature, was that it had swept away the cream of British youth and manhood, thereby blighting the lives of a whole generation. Yet without in any way minimizing the physical and psychological damage caused by the conflict, subsequent historians have largely revised this bleak view of the social impact of the war period. This is especially true for the majority of the population who occupied civilian roles. In spite of high taxation, the payment of death duties, and the loss of domestic servants which detracted from the comfort of wealthier families, most people maintained or even improved their material condition. Graduated taxation and social welfare speeded up the modest redistribution of income in favour of the poor that had been initiated in the Edwardian period. More importantly, the economic boom which produced full employment and plentiful overtime for both sexes boosted the income of working-class families; it also accelerated the move out of low-paid occupations. The rationing system went some way to ensuring that the diet of poorer people did not deteriorate. The results of this were improved life expectancy for civilians, and in particular the sustained reduction in infant mortality rates during the war.

On the other hand, the war left an indelible mark in the shape of three-quarters of a million male deaths and a large number of men too crippled or shell-shocked to be able to readjust to a normal, civilian life. Careful study of the casualty rates among the social classes has gone some way to corroborating the idea of a Lost Generation. The losses were highest among junior army officers because they were required to be first out of the trenches to lead their men into the attack. Since the junior officers were largely drawn from men in their late teens and twenties from public schools and universities, they were clearly concentrated in a very limited social range. It is this loss that was to be reflected so vividly in postwar literature.

However, the broader implications so famously depicted in such works as Vera Brittain's *Testament of Youth* (1932) are not borne out by the evidence. At the time it was natural to assume that thousands of young women would be deprived of husbands, which would not only blight a generation but also undermine the British state by weakening family life and reducing its supply of manpower. In fact this proved not to be the case. While war led directly to male deaths, it also had the indirect effect of checking the emigration which had removed each year several hundred thousand, largely male, persons from the British Isles. Consequently the supply of marriageable men did not diminish in relation to the female population; in fact, it improved because after 1918 emigration never returned to its traditional levels. Moreover, the desire to marry appears to have grown quite markedly. During the 1920s a higher proportion of men, and especially of women, were marrying, and the trend gathered pace throughout the 1930s.

This is a reminder that the significance of the war for women was complex and contradictory. On the one hand, there is some evidence that it had a liberating effect. The number of women in paid work outside the home increased by 1.2 million. For middle-class women in particular this did open the way to a freer social life. They had more money, fewer chaperones, and opportunities to mix with the opposite sex in cinemas and at dances. Hence the contemporary criticism of the so-called 'flappers', who were thought to be promoting an irresponsible and immoral lifestyle, while capitalizing on their new political and economic power to avoid marriage and motherhood.

In fact there was more continuity than change in women's lives in this period. Before the war had come to an end they were already being pushed out of their new jobs, and in a short time there were *fewer* women in the labour force than there had been before 1914. By and large women's work continued to be a matter of temporary, low-paid employment, undertaken as necessity and opportunity dictated. For the majority of women domesticity determined the pattern of their lives just as it always had. In all social classes girls aspired to marriage and motherhood. The fact that family size was falling only helped to make domesticity more attractive than hitherto. Indeed, the most significant portent of this period was not women's war work but rather the publication in 1918 of Marie Stopes's book *Married Love*, which evidently caught the mood of the ordinary woman perfectly. In this Stopes argued that, if married couples deliberately spaced pregnancies rather than allowing them to follow rapidly after one another, this would lead to an improvement in the health of both babies and mothers. By freeing sexual relations from the perpetual fear of pregnancy, she claimed, husbands and wives would enjoy married life much more. For many couples this was the advice they wished to hear, and the widespread adoption of birth control led to a sharp fall in family size during the 1920s.

The Impact of War

On investigation, many of the trends and innovations attributed to the Great War turn out to be not so much the direct product of war as the outcome of long-term developments whose origins lie in the pre-1914 period. The long-running campaign for reform of the parliamentary franchise is perhaps the foremost example of this. The war clearly continued existing patterns of change, in some cases strengthening or accelerating the process. The extension of the Edwardian innovations in taxation is a case in point. Several social reforms such as the 1911 unemployment insurance scheme and the provisions for infant and maternal welfare were also extended as a result of the war. Similarly, the cause of tariff reform gained force from the war years because the seizure of German markets generated demands for the protection of several growth industries.

But in several major respects the war interrupted the pattern of development, and even caused a reaction against change which persisted for many years. For example, the war created a huge increase in the National Debt which severely limited the scope of government policy in the 1920s. The great increase in government resources engendered during 1914–18 only stimulated a backlash in favour of lower taxation and retrenchment which seriously disrupted existing social policies and delayed the implementation of others. State economic intervention was very rapidly reversed, so that by 1921 industry had been restored to private control and proposals for nationalization of coal and the railways were abandoned. For a time the object was simply to return to unfettered free enterprise and remove government from the economy; it was some time before it was generally appreciated that this was a wholly unrealistic objective. Finally, as we have seen, the ephemeral liberation for women during the war years had the effect of making men anxious to force women back into domesticity. During the 1920s women workers were invariably sacked when they became married on the grounds that, as they were supported by their husbands, they ought not to deprive a man of a job.

On the other hand, not all the innovations of wartime perished in the postwar reaction. Both the Ministry of Labour and the Ministry of Health, whose responsibilities included housing, survived to keep alive the interventionist spirit in Whitehall. Amongst the social policies, those for mothers and children's welfare did continue to gain resources even in the 1920s. And although Addison's Housing Act was abandoned, the policy of state-subsidized building was revived in 1924 by the first Labour government. Equally important in the long term was the fact that rent restrictions were never entirely given up, and from 1918 there was a steady decline in the role of rented accommodation in favour of private ownership and council housing.

There is, finally, little evidence that changes actually began during the war years as opposed to growing out of earlier developments. One or two infant industries were coaxed into life. The introduction of duties on some imports in the 1915 budget marked the first real retreat from free trade. However, the most important effect of the war lay in the way in which it disrupted the Edwardian pattern of politics. By destroying the progressive alliance it paved the way for two decades of Conservative predominance in Britain. Yet the huge expansion of the electorate placed definite limits upon the kind of policies the Conservatives felt able to pursue; indeed, in many ways it forced them to embrace the interventionist Edwardian social policies they had once criticized. There was certainly a measure of continuity, not only in terms of detailed policies but also in the survival of the parliamentary system and the constitutional monarchy in Britain. The very fact of the country's victorious conclusion to the war served to strengthen her institutions and to foster pride in her empire; in this way war helped to create a somewhat unrealistic impression that Britain could retain her traditional role in the world, an impression which only another great war decisively destroyed.

Part III

The Period of Confusion: Collectivism versus Capitalism, 1918–1940

11

The Failure of *Laissez-Faire*

The Edwardian period had ended in a boom enjoyed by the traditional staple industries – coal, textiles, shipbuilding, and steel. But concern about Britain's underlying weaknesses had led to extensive experimentation with interventionist policies: public-works schemes, redistributive taxation, state welfare, and regulation of wages. All of this was submerged in the greater but ill-considered state interventionism during wartime. When the dust had settled however, the underlying problems of the Edwardian economy resurfaced in more acute form, and the question had to be faced – what was the proper role for the state in economic affairs? This became the persistent theme of the 1920s and 1930s.

The Legacy of War

The war exacerbated the existing imbalance of the British economy by shifting resources away from consumer goods into sectors whose output was desperately needed for the war effort: coal, steel, engineering, shipbuilding. Consequently Britain ceased to supply many of her foreign markets with textiles and coal – key elements in the Edwardian balance of trade. Other countries, notably the USA and Japan, stepped in to supply Britain's former customers, while several of the less industrialized states, such as Canada, India, Brazil, and Argentina, became more self-sufficient during the war. India for example doubled her cotton textile production, which contributed to a huge fall in British postwar exports. Admittedly, the war did give a boost to chemicals, electrical goods, and motor vehicles, in which production had lagged for some time. But overall the effect of the war was to bequeath a deteriorating balance of trade in the 1920s.

Of course, a deficit on visible exports and imports was by no means new, but it had previously been covered by invisible earnings from foreign investments, shipping, and insurance. However, these were much more elusive after 1918. The prevailing depression between the wars restricted the scope for such income, which is why many economists and politicians believed so strongly in the necessity of restoring international trade to its prewar level as soon as possible. Unfortunately, this was now largely beyond Britain's power. In the first place, the war triggered the spread of protectionism – even Britain had infringed her free-trade principles in 1915 – and many of the fragile, newly created states adopted tariffs.

Moreover, during the war Britain had liquidated a high proportion of her overseas assets, and she had borrowed extensively from the USA to whom she owed debts of £1,150 million. As a result she found it difficult to recover her former role as the leading creditor nation and could not, therefore, effectively stimulate the general level of economic activity. The USA, which might have occupied this role, tended to recall her loans and retreat into protectionism. Finally, a severe fall in the price of food and raw materials between the wars impoverished many of the less developed countries, who were consequently unable to purchase manufactured goods from Britain as they had traditionally done. Hence, in spite of periods of growth, especially in the later 1920s, the world economy was never fully restored to the buoyant condition which had been the key to Britain's former prosperity.

At home the war also bequeathed problems for the government's finances. Even when the huge military costs of the war had been scaled down, governments were still left with expensive commitments in terms of expanded social policies. In the early 1920s government expenditure consumed 24–9 per cent of gross national product, compared with around 12 per cent before 1914. This put postwar governments in a strong position to influence the level of output and income, but they generally saw the negative rather than the positive side of state expenditure. They concerned themselves chiefly with the need to balance the budget and to deal with the national debt, which had risen from £650 million to £8000 million. Annual interest payments on the debt increased from £20 million in 1913 to £325 million by 1920, and remained around £300 million until the 1930s. Naturally these payments absorbed a substantial part of the government's expenditure: 24 per cent in 1920 and as much as 40 per cent later in the decade. Consequently, governments felt that they had little room for manœuvre; while servicing the national debt, reducing income tax from the 6s in the pound levied during the war, scaling down military estimates, and curtailing social spending without endangering their political position, they must somehow continue to end up with a balanced budget. This was not an insuperable problem, but it was tackled with a lack of vision and imagination. For much of the time the government's budgetary policy was so deflationary as to have an inhibiting effect on economic growth. The policy amounted to a rather traditional mixture, in which academic economic ideas were often overlaid by fundamental moral beliefs and by simple political prejudice. For example, the restoration of the pound to its prewar value could scarcely be justified by economic considerations, but really reflected an emotive and irrational belief in sterling as a patriotic symbol of British strength; as such, it was adhered to when it no longer made much sense. Similarly, after 1918 ministers often felt they had been led morally astray by the financial irresponsibility of Lloyd George, both before and during the war. The reaction against his interventionist and inflationary style was accelerated by the political necessity for the Conservatives of giving their supporters some relief from income tax. On the whole they attempted to return to Victorian policies, only to be frustrated eventually by the scale and persistence of the interwar depression.

The Return to Gold

British policy-makers were overwhelmingly anxious for a return to the pre-1914 era, when Britain had been central to the world's trade and the pound the leading medium of exchange. But this involved backing the currency with gold. Unfortunately, during the war the government had been forced to abandon the gold standard and to issue paper pound and ten-shilling notes for the first time. After 1918 it was seen as a matter of urgency to return to the gold standard. Indeed, the Cunliffe Committee of 1918 pronounced this to be the overriding object of British policy, a view fully accepted by the governments of the period. This reflected their essentially internationalist, Victorian view of British interests; to them it seemed self-evident that the country's prosperity lay in the traditional staple industries on which the industrial revolution had been based; and since these relied upon export markets it was necessary to restore the smooth functioning of the prewar world system. Until this was accomplished, it was inevitable that unemployment would remain high.

However, there were several practical flaws in this strategy. The first has already been noted – that Britain lacked the capacity to influence the international economy that she once had. Second even when the level of world trade did begin to return to pre-1914 levels, as it did in the mid-1920s, Britain proved unable to recover her former share of it. Third, there was a crucial judgement to be made as to what value should be put on the pound when she did return to the gold standard. The Cunliffe Committee had not so much as considered anything lower than the prewar rate; and most orthodox economists, civil servants, and politicians made exactly the same assumption. This meant valuing the pound at $4.86. Yet since 1914 the actual value commanded by the pound had fallen far below this level. The politicians understood fully that to maintain the higher level required deflation and wage cuts in Britain. This is why they hesitated to return to gold until 1925, fearful of popular unrest over falling wages.

However, in the deflationary climate after the war the advocates of a return to gold felt encouraged by the upward trend of the currency. From $3.40 in February 1920, the pound rose to $4.63 dollars by the end of 1922. After the return of Baldwin to power in 1924 it rose to $4.79. Treasury officials felt convinced that the time for a return to gold had come. The decision, taken in 1925, was probably the most important single act of economic policy between the wars – and the worst mistake. The Chancellor, Winston Churchill, who was somewhat out of his depth in economics, made a point of hearing both sides of the argument. J. M. Keynes and Reginald McKenna, the former Liberal Chancellor, explained why a return to gold at $4.86 was undesirable, but Churchill followed the Treasury's advice – and later admitted he had been wrong.

The drawbacks to the policy were twofold. The obvious one was that by increasing the value of the pound the government immediately made British goods more expensive abroad, which checked economic growth and employment. Keynes put the overvaluation of the pound at 10 per cent, but it is nowadays thought to have been 20–5 per cent. The decision reflected the excessive influence of City of London financiers, who stood to benefit from an overvalued pound, at the expense of manufacturers whose interests suffered; the latter were of much more importance for the overall health of the economy.

Even worse were the domestic implications of the return to gold. In order to defend the pound at its new, high level it was necessary to attract and retain foreign funds; this meant raising the bank rate and keeping interest rates high. Meanwhile industry had to cope with its uncompetitive prices by reducing its domestic costs, especially wages. Thus the government's policy imposed a phase of deflation on the economy in which businessmen were inhibited from investing and in which domestic demand was restricted. One immediate effect was to arouse trade-union fears about general wage cuts, a situation which produced the General Strike in 1926. Moreover, while other industrial countries began to enjoy substantial economic growth in the second half of the decade, Britain's economy continued to be sluggish. By throwing away the chance for a real economic recovery, the Baldwin government of 1924–9 proved itself the most incompetent of the interwar period.

To make matters worse, the government's persistence with the policy of deflation and the gold standard ultimately failed even on its own terms. For in spite of the debacle of the General Strike, wages did not generally fall to any significant extent. Consequently, as the 1920s wore on the attempt to hold the pound at its 1925 level, in the face of a balance-of-payments deficit, was increasingly unrealistic. Yet politics and national pride prevented both Conservative and Labour governments from devaluing the currency, even in the face of evidence that France and Belgium had successfully stabilized their currencies by accepting a lower rate. Fortunately for Baldwin, his government was defeated at the 1929 election, largely as a result of his failure to reduce unemployment, thus leaving Ramsay MacDonald to face the consequences of his misjudgements. But policy continued as before. Philip Snowden, the Labour Chancellor, fully accepted the Bank of England's right to determine interest rates with a view to maintaining foreign investments in Britain. 'Parliament', he declared, 'is not a competent body to deal with the administration of such highly delicate and intricate matters.' As a result of this attitude the Labour government eventually broke up over its difficulties in balancing the budget, to be replaced in 1931 by a National Government. Though established essentially with a view to maintaining the pound, the new government was rapidly driven to abandon the gold standard and to adopt devaluation. A decade had been wasted in a vain attempt to sustain an unrealistic objective.

Unemployment

It would be misleading to think of the interwar economy as a period of unrelieved depression. In the immediate postwar years, for example, the high family incomes engendered by wartime produced a considerable demand for consumer goods that had been unavailable for four years. Businessmen enthusiastically abandoned war production and attempted to switch resources into domestic goods once again; there was rapid restocking and some unwise speculation. But in the summer of 1920 the boomlet collapsed suddenly, as industry realized it had put too many goods on the market and had not recovered its foreign customers. By 1921 unemployment had leapt to 2 million or 17 per cent of the insured labour force. The subsequent recovery proved to be slow, leaving unemployment between 1 and 1.5 million throughout the 1920s. After the 1929 Wall Street Crash the economy deteriorated again, and unemployment reached 3 million by 1932. A period of modest and uncertain economic growth from 1933 to 1937 brought unemployment down to 2 million in the mid-1930s.

The national figures conceal considerable fluctuation, over time and also geographically; indeed, the concentration of certain depressed industries in the north and west of the country tends to make unemployment a regional problem. In 1921, for example, when national unemployment stood at 17 per cent, it rose to 27 per cent in engineering and 36 per cent in shipbuilding. In 1932, London and south-east England reported 13.7 per cent out of work compared to 27 per cent in the northern counties of England and 36 per cent in Wales. By 1934 St Albans, typical of the towns around London attracting light industry, had only 3.9 per cent unemployment while Jarrow, dependent on a collapsed shipbuilding industry, had 68 per cent. The regional concentrations of unemployment were held by the Treasury to be proof that the problem was structural. This was true to the extent that the staple industries – cotton, coal, and steel – had been experiencing declining productivity for decades. However, much of the unemployment reflected cyclical problems; even in the depressed areas, unemployment fell in response to a general improvement in the economy. With a less deflationary policy and a more competitive exchange rate, far fewer jobs would have been lost.

Insofar as the governments of the 1920s accepted that unemployment was cyclical in nature, they believed that it reflected a purely temporary

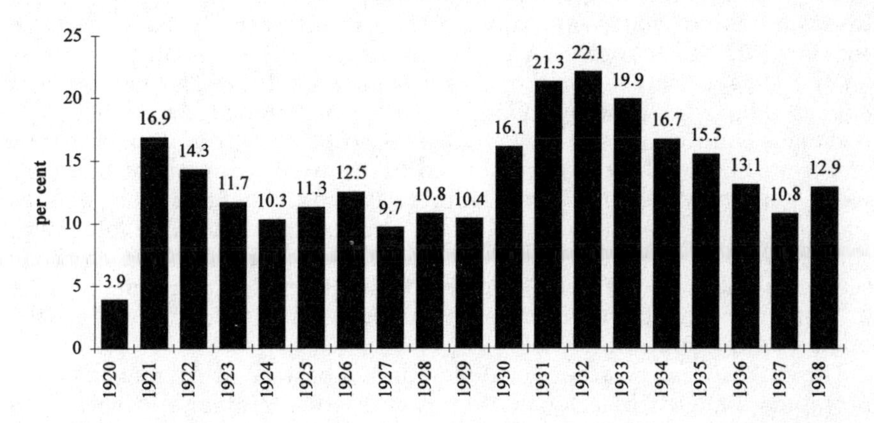

Figure 11.1 Interwar Unemployment as a Percentage of Insured Workers*

* Unemployment was lower amongst uninsured workers; if they are included, the annual figure for unemployment is 2%–3% lower.

imbalance in the economy. After a short period of depression, certain adjustments were thought to operate so as to restore full employment. Businesses, for example, would take on more workers. However, in the 1920s this process was evidently not working properly. For one thing, the government itself inhibited business activity by maintaining high interest rates to support its exchange rate policy. But at the time politicians, businessmen, and civil servants widely believed that the crucial adjustment – falling wages – was being checked; wages had to drop significantly before the economy would recover.

In addition, the government's budgetary policies contributed to depressing the level of domestic demand for the products of industry. It worked on the traditional assumption that each year its revenue and its expenditure ought to be in balance; any surplus should be used to redeem the national debt or reduce taxation. As a result of huge wartime spending, the balanced budget had been temporarily abandoned. But now the problem was firmly taken in hand. In 1921 a committee under Sir Eric Geddes recommended sweeping cuts in government expenditure, and as a result the budget deficit of £300 million was turned into a surplus of £200 million. This made possible a cut in income tax from 6s. to 4s. in the pound. In subsequent years there was invariably a budget surplus which was devoted to the national debt. In this way, money that could have been spent so as to boost the demand for goods and services was diverted to relatively wealthy people who had loaned money to the government. Government deflationary policies, in short, helped to keep the economy sluggish and to delay recovery.

In spite of this the government continued to believe it was spending too much on social welfare. In particular, it regretted the cost of unemployment insurance. The original scheme of 1911 had been extended to cover 12 million workers in 1920; domestic servants and agricultural labourers were the only major groups still omitted. Lloyd George's system had worked well enough while unemployment remained low. But in an era of mass, long-term unemployment, more was paid out by the Unemployment Insurance Fund than was paid in; thus the Treasury was obliged to subsidize it. Much rhetoric was expended by politicians and the press on the alleged extravagance of this system after 1918. The government appeased its critics by withdrawing benefits from women, even when they had made contributions, if they refused to accept jobs as domestic servants; and from 1921 applicants were obliged to meet the requirement that they were 'genuinely seeking work' before receiving benefit. But in spite of such efforts the costs remained high, because genuine unemployment continued to rise. In 1930 the Labour government withdrew the 'genuinely seeking work' condition. This, combined with the rapid deterioration in the economy, pushed the fund further into deficit, so that its debt stood at £100 million by the end of the year.

The dual threat of a budget deficit and a balance-of-payments deficit prompted Ramsay MacDonald to resort to the traditional device of a special committee to investigate government spending. Sir George May's Committee estimated the budget deficit for 1931–2 at £120 million. It advised £97 million of expenditure cuts, including lower unemployment benefit and salary reductions for public employees. There were, in fact, alternatives, including the use of a revenue tariff and devaluation of the pound. Even the object of balancing the budget need not have created a crisis; for no less than £60 million in the 1931 budget was devoted to repayment of the national debt, a wholly unnecessary item which no other major country paid at this time. But the Labour government accepted the need to balance the budget by retrenchment; it disagreed only on the means to that end. Its successor, the National government, eventually imposed cuts, thereby weakening the level of demand for goods and worsening the depression.

Capitalism, Socialism, and Keynesianism

From the tone of political debate between the wars one would expect policies to have fluctuated sharply between *laissez-faire* and state socialism. Conservatives warned about the threats to private wealth, and demanded that individual enterprise be freed from government interference. Labour politicians advocated a levy on capital, coal nationalization, and welfare reforms, while some socialists even praised the centralized economic planning of the Soviet Union.

Yet on both sides much of this was empty rhetoric, designed essentially to arouse party supporters. In practice, as we have noted, bipartisan policies prevailed on all fundamentals during the 1920s and 1930s. Philip Snowden was every bit as orthodox a Chancellor as Austen or Neville Chamberlain. Conservative governments maintained historically high levels of income tax and social-welfare spending; Labour retreated from radical ideas about the capital levy and land reform, and never prepared a precise scheme for nationalization of industry. They disagreed only over tariffs, where Labour remained loyal to free trade, and the level of unemployment benefits. But this hardly separated a party of capitalism from a party of socialism.

This, however, does not mean there were not real divisions over economic issues. The alternatives to the Treasury orthodoxy adhered to by both Conservative and Labour governments were propagated by an assortment of individuals and groups, including J. M. Keynes, Lloyd George and his Liberal inquiries, the Independent Labour Party, Sir Oswald Mosley, younger Tory MPs such as Robert Boothby, Oliver Stanley, and Harold Macmillan, and the 1930s pressure groups, Political and Economic Planning and the Next Five Years Group. Although they cannot be equated with one another, their remedies had a great deal in common. Essentially they shared a scepticism about the orthodox belief that the solution to unemployment lay in a restoration of the international market and the role of sterling. Instead, they looked to the domestic market and emphasized the capacity of governments to stimulate production and consumption at home. To this end they advocated, in varying combinations, tariff protection, public works, cheap money, central control over banks and investment, and devaluation of the pound.

Keynes offered much the most comprehensive and well-thought-out alternative economic programme. It began in 1918 with his warning about the effects of reparations payments in *The Economic Consequences of the Peace*. By 1923 he had begun to argue against deflationary policies in his *Tract on*

Monetary Reform, and he totally condemned the return to gold at $4.86 in *The Economic Consequences of Mr Churchill* in 1925. His famous collaboration with Lloyd George and others involved abandoning attempts to balance the budget and to reduce wages, and instead using the government's resources constructively to raise the level of demand for British-made goods. Keynes found it frustrating that the government actually maintained a fairly high level of expenditure, but wasted too much of it on the national debt instead of boosting the incomes of those who would increase spending. In 1929 the scheme propagated by Lloyd George involved an expenditure of £251 million on public works over two years. This was a realistic approach, given the extent of Britain's problems; between 1929 and 1932 gross domestic product fell by 30 per cent in the United States, but by only 5.5 per cent in Britain.

However, the Treasury rejected the Keynesian/Lloyd George programmes on the grounds that public-works schemes would take much longer to put into operation than they anticipated, that the budget must be balanced, and that since there were no unemployed financial resources any further state investment would merely have the effect of reducing the funds available to private businesses. There was no validity in these objections except for the first one; Keynes' policies would have been slower to achieve their cumulative effect than he realized, though this was hardly an argument for not applying them.

It is often suggested that Keynes's ideas came too late to have any real impact between the wars. He himself once described his work as 'the croakings of a Cassandra who could never influence the course of events in time'. But although his most celebrated work, *The General Theory of Employment, Interest and Money*, did not appear until 1936, it is clear that he had subjected conventional economics to a continuous attack since 1918. The case for countercyclical public-works schemes was widely accepted in the 1920s among both economists and left-wing politicians. By the early 1930s the TUC and key people like Ernest Bevin had been won over to Keynes's programme. And when in 1938 Harold Macmillan published *The Middle Way* it was clear that he, too, had now absorbed Keynesian thinking. The post-1945 agenda had begun to take shape.

The other novel feature of interwar economic policy was the bipartisan move towards state collectivism in industry. Labour had put nationalization in its 1918 constitution, but did little to turn this into a practical policy until the 1930s. More remarkably, the Conservatives showed a growing appreciation that it was not always in the national interest to leave industry to the free play of market forces. Several Tory ministers in the Lloyd George coalition had been ready to support nationalization of the railways – an obvious example of a key industry losing efficiency because it was divided into too many competing private companies. In the event, nationalization was killed in the postwar reaction and the companies were amalgamated into larger units instead. There was a similar but even stronger case for nationalization or rationalization in the coal industry, which was divided into 1,400 separate colliery companies, many of them small and inefficient. Although the majority on the Sankey Commission accepted the case for a change of ownership, the report was ignored for political reasons.

Eventually in 1930 the Labour government intervened to set total output, apportion quotas for each pit, and fix prices, thereby propping up the less efficient parts of the industry. Subsequently the Conservatives continued this policy, and took a further step in nationalizing the royalties of the coal-owners in 1938, a policy long advocated by radical critics of the landowners.

However, Conservative collectivism went much further than this. It really began under the postwar coalition with the establishment of the Forestry Commission. This was a recognition that, as Lloyd George and the radicals had long claimed, private enterprise simply neglected to invest in forests and woodland. The war had underlined the danger to national security of being too dependent on imported wood, and in the 1920s extensive investment was undertaken to repair the deficiency. An even more important example was electricity. In 1919 the government allocated £20 million for new investment and divided the country into districts, each under a Joint Electricity Authority with responsibility for extending the supply. In 1927 Baldwin created the Central Electricity Generating Board to own and operate the national grid: the national interest required an efficient electricity supply which could only be achieved by economies of scale.

Subsequently the National government also recognized the inadequacies of private enterprise in several sectors. In 1933, for example, it adopted a bill prepared by Herbert Morrison to establish a London Passenger Transport Board, thereby creating a state monopoly over all the transport services in the capital. By 1938 it had decided that the competition between Imperial Airways and British Airways had been disastrous, and so the two were merged into a state company, British Overseas Airways Corporation, under Sir John Reith. But in some ways the most significant, and certainly the most expensive, aspect of interventionism under the Conservatives was in agriculture, even though it involved no change of ownership. As a result of wartime food shortages the farming industry had expanded, and in the 1920s many men were assisted to establish themselves in smallholdings. But in 1921 the government withdrew the subsidies that had been designed to maintain a minimum price for corn; and during the 1920s and 1930s food prices fell drastically, driving many farmers out of business. The second Labour government offered a solution in the shape of Dr Addison's 1931 Agricultural Marketing Act. The idea was to impose quotas on imports of food, to offer subsidies to stimulate domestic producers, and to establish marketing boards to buy up and resell the farmers' crops. By 1933 the National government had extended the policy to cover milk, potatoes, pigs, wheat, sugar, and hops. The effect, as with coal, was to protect the inefficient, but it certainly stimulated an increase in agricultural output of around 16 per cent between 1931 and 1937, though at an annual cost of between £30 and £40 million. It can, of course, be argued that each case of state interventionism was a special one, not part of an overall plan or philosophy. But by 1939 the list of special cases had become rather too long for such a view to be satisfactory. State collectivism was becoming a habit for both parties. The Conservatives in particular had breached *laissez-faire* principles so deliberately that there could be little doubt that their leaders, at least, had abandoned the old confidence in an unfettered free-enterprise economy.

Economic Recovery in the 1930s

For several years after the formation of the National government the slump grew even worse. But by 1934 clear signs of recovery manifested themselves as unemployment began to fall and output exceeded the 1929 level. Indeed between 1932 and 1937 unemployment fell from 3 million to 1.5 million while industrial production increased by 46 per cent. This improvement has given rise to a much more optimistic view of the 1930s than has traditionally been held. As with all claims in economics, much depends on which periods are chosen for comparison; but the basis for the optimistic view rests on the evidence that, whereas national income increased by only 10 per cent from 1921 to 1929, it rose by 17 per cent from 1929 to 1937.

Of course, this growth was rather patchy in geographical terms. It developed within a limited range of industries largely dependent on the domestic market for consumer goods rather than on a recovery in exports. The building industry was the major example. Whereas in the 1920s 150,000 houses were built on average each year, after 1934 well over 300,000 were being built annually. This meant a rapid increase in employment, since small firms were able to take on extra labour quickly, and a stimulus to companies supplying bricks, pipes, glass, and paint. Typical of the expanding consumer-goods industries was vacuum cleaners, whose production rose from 37,000 in 1930 to 409,000 in 1935, and motor cars, where annual output increased from just over 200,000 in the late 1920s to 400–500,000 by the late 1930s. Finally, the aircraft industry, transformed by rearmament after 1935, generated (on one estimate) around a million extra jobs by 1938.

How far can the limited recovery from the depths of depression be ascribed to the policies of the National government? This is a complex question; many different pressures were at work simultaneously. However, it is clear that the government was by no means as orthodox as Neville Chamberlain liked to pretend. In several ways it abandoned the policies it had been established to defend in 1931; as we have seen, it was responsible for major infringements of *laissez-faire*, and it can be seen as having laid the foundations for economic management in the post-1945 era.

Unquestionably the greatest gain was that the National government, in spite of itself, abandoned the gold standard in September 1931 and effectively devalued the pound. By March 1932 the pound had dropped from $4.86 to $3.40. Though the effect was temporary, devaluation stimulated exports, checked the fall in Britain's share of world trade, and restricted imports, thereby boosting domestic employment. But the greatest benefit was that monetary policy had been freed from external pressures, and it thus became possible to cut the bank rate from 6 per cent to 2 per cent in 1932. Interest rates remained low for the rest of the decade. All this represented a victory for Keynesian thinking and a defeat for the Treasury and the government. However, the effect of cheap money was to boost industry, especially house-building. Unfortunately, the government failed to use cheap money as a countercyclical weapon as Keynes wished; the Treasury's reason for favouring low interest rates was simply that this reduced the cost of the National Debt.

Other policy innovations of the 1930s were of more marginal significance. For example, the government passed an Abnormal Importations Act in November 1931 which empowered it to levy 50 per cent tariffs on certain goods, and an Import Duties Act in 1932 which imposed a general 20 per cent duty, with exemptions for British Empire products. Free trade had come to an end at last. Certain industries benefited from protection; imports of vacuum cleaners fell off sharply, for example, and fresh investment in steel led to new plant at Ebbw Vale, Corby, Shotton, and Workington. But other run-down industries were not easily restored. The government's attempts to foster Empire trade by means of the Ottawa Agreements had only a marginal effect, because India and the Dominions insisted on protection for their own industries.

The government recognized the depth of the problems in the old industrial districts by a belated and half-hearted attempt at a regional economic policy. The first stage was the appointment of commissioners to investigate the 'Distressed Areas' – South Wales, Tyneside–Durham, West Cumberland, and Scotland. They identified schemes including improvement of water and sewage supplies, harbour repairs, and hospital building, but the £2 million allocated for this scarcely reflected the extent of the problem. From 1936 the government offered to remit rent, rates, and taxes for companies which moved into distressed areas, which fostered the development of trading estates. But even in 1938 only 17 per cent of newly opened factories were in the special areas, compared to 40 per cent in Greater London, though this at least represented a major improvement on previous years. If the practical effects were marginal in the 1930s, they were a pointer, if not to full-scale economic planning, then to further state intervention after 1945.

As Chancellor of the Exchequer, Neville Chamberlain obscured the innovatory character of economic policy by proclaiming his commitment to orthodox financial principles. Certainly, the government began with £70 million of expenditure cuts and increases in taxation, designed to balance the budget and restore confidence. But the effect was only to inhibit the recovery. Soon, however, Chamberlain departed from the straight and narrow, if partly for political reasons. From 1932 his budgets were mildly inflationary. By the 1935 election, income tax had been reduced and the expenditure cuts abandoned. By this time Chamberlain was also coming under pressure to increase spending on rearmament, and though he resisted, he eventually lost the struggle.

The overall record is thus rather mixed. In spite of the recovery, unemployment remained at nearly 11 per cent by the outbreak of war in 1939. The growth industries had failed to make a major impression on the worst of the problem. This was partly because growth was concentrated in the south-east and Midlands. Factors such as coal, iron, or water that had influenced the original location of industry no longer applied. Electricity was available everywhere, and the London market made the south attractive to businessmen. It has been claimed that the unemployed workers failed to migrate to the new industries because they were cushioned by unemployment benefits. This is doubly untrue. In fact considerable movements of population took place; boom towns like Slough in Buckinghamshire absorbed large numbers of men from South Wales, for example. But there were real obstacles too. Most of the jobs lost in the old industries were for skilled men, whereas in the growth industries employers wanted women, and particularly young workers, for unskilled production-line work because they would accept low wages and were unlikely to be union members. Older men with families could not easily move to areas of low wages and relatively expensive housing.

Apart from being limited in extent, the recovery also proved to be fragile. In 1937 it collapsed into another brief slump from which it was saved largely by the rearmament programme. Even in 1939, only 68 per cent of cotton-weaving capacity and 76 per cent of spinning capacity was being employed. Britain had a balance-of-payments deficit each year from 1937 to 1939. The truth was that the growth industries flourished in the home market but were unable to compete abroad. The structural problems of British industry remained – there was simply less of it than in 1918. Such success as the National government had enjoyed reflected the adoption of parts of Keynesian economics more by force of events than out of conviction. Consequently the politicians never succeeded in deriving the full benefit from their innovatory policies.

12

Mass Democracy in an Age of Decline

By the end of 1918 it seemed that the war had obliterated many of the familiar landmarks of Edwardian politics. Each of the four political parties had suffered some division since 1914, and two of them were now in headlong decline. The country was now firmly in the grip of a coalition government from which, it was widely predicted, a new centre party would shortly emerge. Yet in the end most of the expectations of radical change failed to materialize, or at least proved to be exaggerated. In spite of the severe social problems generated by the economic depression and the enfranchisement of millions of new voters, the political system remained remarkably stable; the forces making for change were somehow absorbed and assimilated.

Structural Changes in Politics

Tremendous potential for change was embodied in the 1918 Representation of the People Act, which created an unprecedentedly large electorate of 21 million people. In social terms the new voters differed from the prewar voters by age, class, and sex. Politicians felt apprehensive because the bulk of the new male voters were relatively young; they were assumed to be ignorant of politics, unattached to party, and easily swayed. It remained unclear how their experience in the armed forces had affected them.

This concern was compounded by the fact that many newly enfranchised voters were working-class. Indeed, so far as men were concerned the post-1918 electorate was now representative of the working-class majority of the population whose interests the Labour Party claimed to represent. Under the guidance of the Party Secretary, Arthur Henderson, Labour built up a formidable organization during the 1920s. As many as 3.5 million trade union members paid the political levy to the party in this decade. A former union official of moderate Liberal views, Henderson was ideally placed to win the full cooperation of the unions; in 1920 he even obtained a 50 per cent increase in the political levy. The trade unions also sponsored large numbers of parliamentary candidates by offering financial subsidies to the constituency associations that adopted them. The number of local Labour

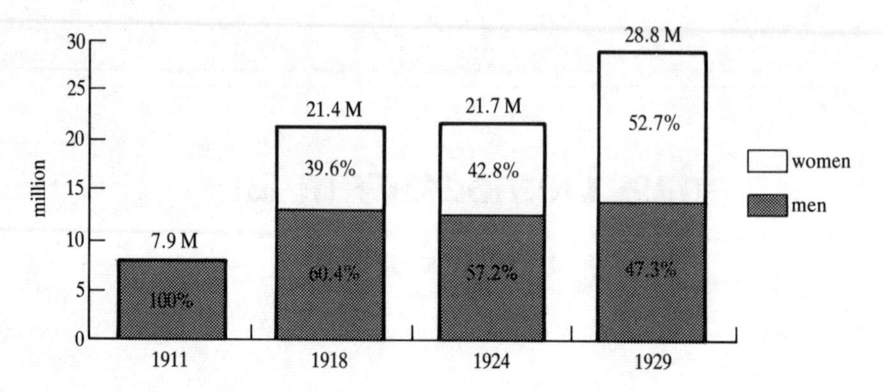

Figure 12.1 Parliamentary Voters in the UK, 1911–1929

Parties affiliated at headquarters increased from 397 in 1918 to 626 in 1924; a series of general elections in 1922, 1923, and 1924 helped to accelerate the spread of the organization and the nomination of the candidates, even in strongly Conservative seats. By 1928 the party's individual membership stood at 215,000; but much of the local work was done by the 1,800 women's sections, who claimed approximately 300,000 members by 1929.

There were, however, limitations. Labour was by no means as wealthy as the Conservatives, financed by business and the honours system. Also, Labour's resources remained concentrated in a limited band of very safe constituencies, not in the marginals where they were needed. One sign of this was the deficiency of full-time paid agents, who numbered little more than 100 in the 1920s. Nor did the individual membership seriously rival that of the Conservatives. However, the result of the organizational improvement was a steady increase in the party's vote to 37 per cent by 1929. Clearly, Labour had not yet mobilized the majority of the working-class vote, otherwise it would have done even better than this. Yet this performance was quite sufficient to establish Labour as the second party from 1918 onwards, and as the Opposition in Parliament. This, though it ruffled feathers at the time, did not pose any fundamental challenge to the political system, for Labour politicians were dedicated to parliamentary methods and traditions. But Labour was, of course, different; the product of extraparliamentary forces, it had virtually no representation in the House of Lords: and as an avowed class party it at least seemed to threaten to polarize politics between the representatives of capital and those of labour.

The novel element after 1918 consisted in the 8.4 million female voters. In 1928, when Stanley Baldwin enacted the Equal Franchise Act, a further 5 million women became enfranchised, and as a result women became a substantial majority of the total British electorate (see Figure 12.1). However, the impact of this was not as great as contemporaries anticipated. Women proved slow to come forward as parliamentary candidates, and the parties showed reluctance to give them winnable seats. By 1935, after seven general elections for which women had been eligible, only 5 per cent of all candidates were female. In 1919 Nancy Astor became the first woman MP to take her seat; like several of the early members, she owed her success to family connections. Between the wars the highest number of women MPs was 14, elected in 1929, and 15 in 1931. Moreover, almost all of those who were elected were not motivated primarily by feminist concerns, but stood out of loyalty to their respective parties.

How far did the new voters influence the character and the agenda of politics? Paradoxically, the larger electorate proved to be much quieter and better-behaved than its Victorian predecessor. Women's participation contributed to this change. Fewer people attended big public meetings, and politicians increasingly communicated with the voters by radio broadcasts and cinema newsreels. In the Edwardian period, political debate had already begun to focus on social-economic issues and Labour had some reason to feel that unemployment, pensions, housing, and health would dominate the political agenda after 1918. This was correct inasmuch as the Conservative leaders also concluded that the combination of a new electorate and the rise of Labour made it dangerous for them to neglect questions of social welfare and standards of living. The Conservatives also gave much less emphasis to imperial issues than they had traditionally done, preferring a bipartisan approach to a polarization of views. This, however, did not mean that the party was simply obliged to echo its rivals in social reform. It took the initiative in tackling questions such as widows' pensions in 1925. On the negative side, the Tories sought to discredit Labour by equating it with Socialism, which in their view implied atheism, hostility to the family, state appropriation of private property, and Stalinist autocracy. All this was so far-fetched that it made little impression on the working class, though it probably helped to frighten some of the middle classes. The Conservatives also emphasized the need to protect British industry, and thus British jobs, with tariffs. It also proved advantageous that mass enfranchisement had coincided with the movement of many working-class men into the income-tax net. This gave them a direct interest in keeping taxation low, and thus limited their willingness to support expensive social policies.

The enfranchisement of women reinforced the existing trend for politicians to concentrate upon economic management, living standards, and welfare. Most politicians chose to believe that women's interests were largely confined to domestic matters. As one party agent predicted in 1921:

> The women's vote is having a narrowing effect upon politics, making them more parochial and is, at the moment, reducing them to bread and butter politics and the cost of living . . . their votes will probably be given on purely home questions . . . while Imperial and foreign issues will leave them cold.

The Liberals and Labour felt sure that women's role as household managers made them keen free-traders, fearful of the effects of tariffs on food prices; and when the Conservatives lost heavily over this issue in 1923 they certainly blamed the volatile female electors. What is more certain is that during the 1920s all parties were willing to enact legislation for women; for example, the 1919 Sex Disqualification (Removal) Act which opened up the professions to women, the 1923 Matrimonial Causes Act which gave equal grounds for divorce, the Widows Pensions scheme of 1925, and the introduction of equal guardianship of infants in the same year. On the other hand, the politicians refused to tackle the more radical feminist causes such as equal pay, birth control, or family allowances, and they continued to pass protective legislation excluding women from certain occupations in order to reserve the jobs for men. Thus, by the late 1920s legislative progress for women had virtually come to a stop, in spite of their status as the majority of the electorate.

Finally, 1918 also involved a drastic redrawing of the constituency boundaries which, as in earlier reform bills, proved to be as important as the franchise. It considerably extended the existing trend towards equal-sized constituencies and to single-member seats. From this Labour gained a somewhat larger group of overwhelmingly working-class seats, especially in the coalfields. Both Liberals and Conservatives lost by the disappearance of many small boroughs. But on balance the creation of dozens of new suburban constituencies made the Conservatives the chief beneficiaries, for they had no effective competition in these areas. They also gained an advantage from the creation of the Irish Free State in 1921. In fact the decision of Sinn Fein not to attend Westminster in 1918 had already removed the 80 or so MPs whose votes had told against the Conservatives since the 1880s. As a result the Conservative Party was able to win a comfortable overall majority in the 1920s on the basis of less than 40 per cent of the popular vote.

Labour's Rise to Power

After his stunning victory in 1918, Lloyd George appeared likely to dominate British politics for the forseeable future: 'he can be prime minister for life if he likes,' declared the Tory leader, Bonar Law. In fact the golden aura of victory evaporated swiftly as the problems of peacetime loomed larger. Lloyd George's position was never as strong as it appeared: for he remained a Prime Minister without a party. His only chance in the long run was to create a new centre party before he lost office. In the event, however, he held the premiership for as long as the Conservatives, who comprised the bulk of his parliamentary majority, were prepared to tolerate him – just four years.

Up to a point Lloyd George fulfilled expectations by helping to keep the threat of Labour and class warfare at bay. Since the number of working days lost in strikes rose from 21 million in 1919 to 86 million in 1921, this mattered greatly to the Conservatives. Lloyd George steered a course through this phase by invariably granting big rises in wages; and he adroitly diverted the miners' pressure for coal nationalization by appointing the Sankey Commission in 1920. However, after this the Conservatives' grievances against Lloyd George steadily mounted. He made the mistake of continuing to govern like a wartime leader; his personal interference in foreign policy led him very near to another war with Turkey over the Chanak crisis in 1922. Lloyd George's willingness to reach a settlement with the Irish Republicans in 1921 also outraged some Tories, though the party leaders had by then reconciled themselves to the end of the Union. Others attacked the reckless sale of honours by the Prime Minister's henchmen, though this was in part because it deprived the Conservatives themselves of financial contributions.

Among the rank-and-file a reaction against high taxation had developed by 1920. This so-called 'anti-waste' campaign so alarmed Conservative leaders that they in turn insisted on policy changes. An official committee under Sir Eric Geddes in 1921 proposed economies in housing, education, and the armed forces, which were followed by a placatory reduction in income tax. However, such retreats only undermined the Liberal wing of the coalition. Dr Addison was led to resign after the abandonment of his housing policy, and Edwin Montagu, who was bitterly attacked by the Conservatives over his Indian reforms (see p.213), had been driven from office by 1921. The introduction of protective tariffs in 1921 and the brutal tactics used by the 'Black and Tans' in Ireland further embarrassed the coalition Liberals.

These grievances proved fatal when the electoral position of the government began to crumble. By-election gains by Labour and the independent Liberals led many Lloyd George Liberals to think about rejoining their old party. To the Conservatives Lloyd George no longer looked like an effective bulwark against the rise of Labour; instead he appeared increasingly a deadweight around the Conservative Party's neck.

In the autumn of 1922 these misgivings were brought to a head by the realization that Lloyd George was contemplating another term of office. As a result the Conservative leader, Austen Chamberlain, who had succeeded Bonar Law after his retirement for health reasons, agreed to hold a meeting of the parliamentary party in October. To the general surprise this produced a heavy vote against fighting another election in association with Lloyd George. The immediate result of this was the Prime Minister's resignation and Chamberlain's replacement as party leader by Bonar Law, whose health had improved. After nearly eight years of multi-party governments, normal party politics had been restored at a stroke. Bonar Law held an immediate general election, which he won comfortably though with only a minority of the votes (See Figure 12.2). The Conservatives had shrewdly severed their links with Lloyd George just when he had become a discredited figure. The party suffered some damage when a group including Chamberlain, Balfour, and Lord Birkenhead, who opposed the break-up of the coalition, refused to join the new government. But this at least created a chance of high office for some junior figures, notably Stanley Baldwin, who became Chancellor of the Exchequer. It was not long before ill-health again led to Bonar Law's retirement. With the coalitionists still detached, there was little material to choose from. Lord Curzon found himself ruled out, ostensibly because he sat in the House of Lords, but in reality because his colleagues found him so obnoxious. As a result the choice fell upon the apparently unremarkable Baldwin.

Within months of succeeding to the premiership in 1923 Baldwin astonished the political world by holding a fresh general election. Since he had a large majority and four years of his term still to run, this was an abuse of prime ministerial power for purely party purposes. He claimed that he wanted a mandate to introduce tariffs; but an election fought on this issue was expected both to reunite the Conservative Party and to drive Lloyd George back into the arms of the Liberals. Both objects were accomplished. However, in the election the Conservatives lost far more seats than expected, owing to fears that tariffs would lead to an increase in the cost of living, and the upshot was the first Labour government of 1924.

Few Labour politicians had expected to come to power quite so soon. As recently as 1918 they had only 63 MPs, and the leading figures, including MacDonald, Snowden, and Henderson, had all been defeated. Yet, Labour's new status as the effective alternative to the Conservatives was quickly underlined by the municipal elections of 1919. The party made

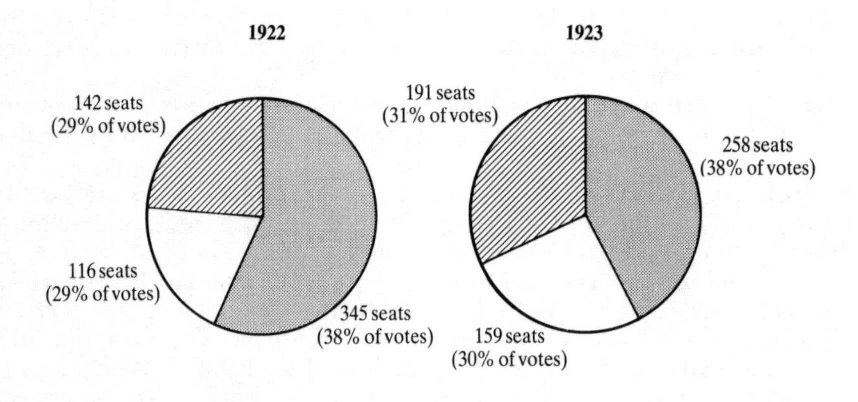

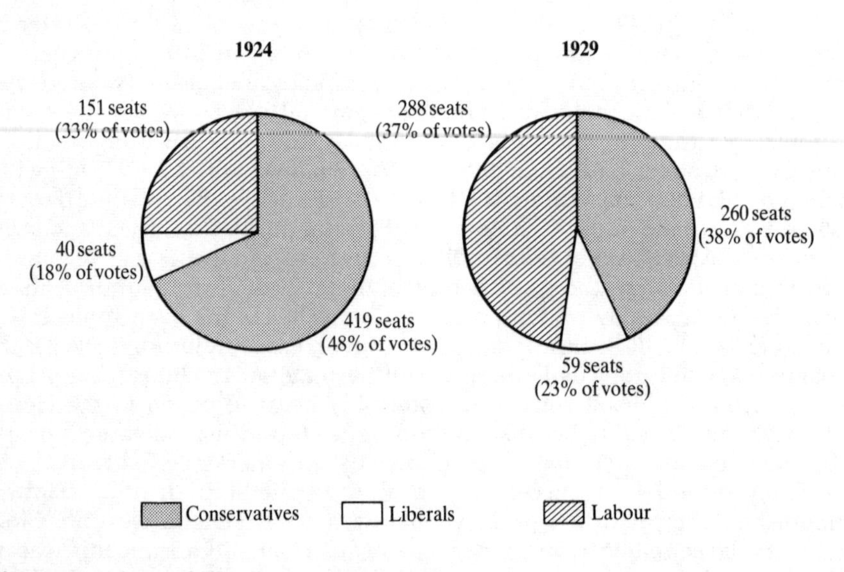

Figure 12.2 General Elections, 1922–1929

sweeping gains, especially in London, which proved to be the start of a trend that lasted throughout the 1920s and 1930s. Soon the Liberals and Conservatives began to be driven into an anti-Labour alliance in local government.

From the early 1920s onwards Labour enjoyed a powerful conviction that history was now on its side. By comparison with the discredited Lloyd George and the uninspired Tory leaders, the party seemed fresh and idealistic; Ramsay MacDonald, who returned to the leadership in 1922, was an able parliamentarian who appealed across the lines of class to a wide public. Many working-class families now felt aggrieved about the profiteering of wartime; with the collapse of the boom in 1920 and the sudden increase in unemployment the fruits of victory were seen to have withered already. This created fertile ground for interventionist policies. In particular, the proposal for a special levy on capital to pay off the National Debt and thereby free government resources for more constructive purposes proved appealing, and not just within the labour movement.

However, under MacDonald and Snowden the tactics were to tone down the party's socialism and play up its respectability. They distanced themselves from the capital levy and made no attempt to 'soak the rich' when in office in 1924 or 1929. Indeed, in many ways Labour's chief tactical triumph consisted in capitalizing on the continuity between its policies and traditional Liberalism. For example, the party staunchly upheld the free-trade cause, and sought to promote it by re-establishing commercial relations with Russia. By giving priority to housing, health, unemployment, and education the party effectively took over the Liberals' role as the champion of social reform. Above all, MacDonald's principled stand against British entry into the war placed him and his party in a good position to benefit from the gathering reaction against the prewar arms race. Labour carried more credibility as an advocate of disarmament and the League of Nations in the 1920s than either of its rivals. This goes a long way to explaining the attraction Labour exercised over many former Liberals, disillusioned by their party's prewar and wartime policies and divisions. In many ways Labour had become the effective heir to the radical tradition in both domestic and external affairs.

As a result the party's ranks were swollen by middle- and upper-class recruits such as Charles Trevelyan, Christopher Addison, Arthur Ponsonby, and Richard Haldane. But it also claimed several figures from Conservative backgrounds, including Susan Lawrence, Oswald Mosley, Cynthia Mosley (the daughter of Lord Curzon), Oliver Baldwin (the son of Stanley Baldwin), and Lord Chelmsford, the former Indian Viceroy. Thus, in spite of its reputation as the working-class party, Labour actually developed into a representative national party during the 1920s; arguably it was more national than the Conservatives, who still depended upon a very narrow social section for their candidates and MPs.

Yet there was no irresistible force carrying Labour to power in the 1920s, and one must take account of the large element of contingency in the party's rise. But for Baldwin's extraordinary decision to call an election over tariffs in 1923, for example, the first Labour government would have been much delayed. Similarly, errors committed by the Liberals played an important part in Labour's fortunes. Even after the disaster of 1918 there was nothing inevitable in the continued Liberal decline. During the 1920s they generated radical new policies to deal with the economic depression and effected a revival in their fortunes; even under a mass electorate they ran Labour neck-and-neck, with 29–30 per cent of the popular vote in 1922 and 1923. In addition, free trade remained a popular cry. Traditional Liberal causes such as Nonconformity, temperance, and Home Rule had clearly lost some of their prominence, though the real problem was that many of the advocates of these issues found it more electorally efficacious to vote Labour.

On the other hand, the Liberals were immensely handicapped because for five years after the 1918 election they continued to be riven by squabbles between the followers of Asquith and Lloyd George. Indeed, even after the formal reunion of 1923 the infighting was maintained by Sir John Simon, Walter Runciman, and others who detested Lloyd George. Asquith stubbornly held on to the leadership until 1926, though he was barren of ideas; in this way he deprived the party of abler leadership until it was almost too late. In addition, the Liberals suffered badly from the erratic and unrepresentative workings of the electoral system. This system now delivered big majorities to the Conservatives on the basis of only 38 per cent of the vote, and in 1929 nearly did the same for Labour. But the Liberals' support was not concentrated except in rural Wales, Highland Scotland, and parts of the West Country. Labour's growth in urban-industrial seats fatally undermined them, while in the Conservative areas the Liberal vote was split by the additional Labour candidates who came forward. As a result, the Liberals never obtained representation in proportion to their high vote. The failure to include proportional representation or the alternative vote in the 1918 reform bill proved to be a crucial error. As one election succeeded another, the non-Conservative voters increasingly concluded that Labour was better placed to defeat the Conservatives, and voted tactically. This process gradually left the Liberals as a party of the right looking to attract protest votes from middle-class Conservativism; this was somewhat at odds with their traditions and with many of the parliamentarians' instincts, which pointed to a restoration of Liberalism's radical, reformist strategy.

The Liberals had to make a crucial decision after the 1923 election when no party enjoyed a majority (see Figure 12.2). Since the Conservatives already held office and remained the largest party, they declined to resign at once. Yet as they had clearly lost the election, Asquith joined with Labour to vote them out of office. As the leader of the second largest party, MacDonald was invited to form a minority government. He made no arrangement with the Liberals to support an agreed programme of legislation. On the contrary, he preferred to keep them at arm's length so as to underline Labour's different status as a party of government. None the less the Liberals inevitably bore much of the blame, in Conservative eyes, for the MacDonald government.

The lack of a working majority scarcely worried the new Prime Minister. It gave him a good excuse for not attempting to introduce a socialist programme which he did not, in fact, possess. Instead, he filled his Cabinet with a mixture of former Liberals and Conservatives with experience of high office and worthy trade unionists unlikely to frighten anyone but the most extreme Tories. For the experiment was intended to be reassuring, to prove Labour's competence as a governing party. As Foreign Secretary, MacDonald played a constructive role in getting the French to withdraw from the Ruhr. At the Exchequer, Snowden produced a cautious budget which reduced some indirect taxes on food but involved nothing novel.

This was sound tactics, in that it denied the Conservatives the opportunity to claim that Labour government was a disaster. But the caution was probably overdone. MacDonald could have achieved something solid, but he missed the opportunity to introduce several widely expected measures, such as widows' pensions and equal votes for women, which would have had Liberal backing. He thus allowed Baldwin to gain the credit for these reforms after 1924. Also, after nine months in office Labour's own supporters grew critical about the failure to tackle unemployment. There were ominous signs here of MacDonald's incapacity as an executive leader. In the event he was relieved to be defeated in the Commons in September 1924, whereupon he resigned and held yet another general election. Although this produced a big Conservative victory, Labour increased its share of the vote, and the real losers were the Liberals, whose seats fell from 158 to 40.

From 1924 to 1929 Baldwin attempted to pursue what he himself called a Disraelian policy. By this he meant adopting a conciliatory attitude towards the working class and enacting social reforms. Up to a point he succeeded in implementing this strategy. The weakness lay in the failure of economic policies to tackle high unemployment and improve industrial growth. Handicapped by the over valuation of the pound in 1925, industry proved unable to recover its export markets. By 1929 the government was pinning its hopes on stimulating industrial development with an offer of relief from rates.

But Baldwin had already lost the initiative as a result of the General Strike in 1926. This originated in the loss of the export market for coal and the owners' insistence on reducing wages on a pit-by-pit basis according to profitability. The TUC had agreed to lead a sympathetic strike because of the prevalent fear that cuts in miners' wages would be the prelude to general cuts; this was a justifiable assumption, because if the new value of the pound were to be maintained industry would have to reduce its costs. Anxious to avoid a general strike, Baldwin delayed it from 1925 to 1926 by offering a subsidy to maintain existing wage levels pending the report of an official inquiry under Sir Herbert Samuel. But by May 1926 no agreement had been reached, and so the strike went ahead.

After only nine days the TUC called the strike off without securing any of the miners' demands and without seeking any safeguards for the men on their return to work. On the face of it, this was a great triumph for Baldwin's government. But the political consequences favoured Labour in the long run. For Baldwin felt obliged to give way to pressure which his right-wingers had been exerting for several years to tackle the legal rights of the unions. The result was the 1927 Trade Union Act, which outlawed general strikes and changed the system whereby union members had to contract out of the political levy to one of contracting in. For several years fewer members paid the levy and the Labour Party's income fell by a third. But the practical effect was marginal. The punitive legislation gave the industrial wing of the Labour movement a solid motive for rallying around the politicians with a view to returning another Labour government. Moreover, in the regions the strike had been widely supported by working-class families; the effect was to undermine the existing political allegiance of Liberal and Conservative working men and draw more of them into Labour's camp.

Labour also benefited from the increasing public preoccupation with the government's handling of the economy and in particular unemployment, which stood at 1.1 million by 1929. In 1928 Lloyd George, now Liberal leader, published a report of the Liberal Industrial Inquiry entitled *Britain's Industrial Future*; the fruit of collaboration between politicians and distinguished economists including J. M. Keynes, Walter Layton, and William Beveridge, this involved an unusually detailed programme of state investment in capital projects with a view to stimulating the sluggish economy and reducing unemployment by half a million. This was the kind of approach most Labour supporters wished to see from their own party, and at the election of 1929 it seems that voters expected a new Labour government to tackle unemployment in this way. Thus it was MacDonald rather than Lloyd George who reaped the benefit of the Liberal campaign claim 'We Can Conquer Unemployment'. Labour's vote rose to 37 per cent, and they won 288 seats to the Conservatives' 260. Though still short of an overall majority, MacDonald took office for a second time just as the international economic crisis was beginning to gather.

The National Governments

From 1.1 million in June 1929, unemployment rose inexorably to 2.5 million by December 1930. MacDonald's apparent inability to get to grips with the problem undermined his parliamentary support, leading to revolts by the ILP members from Clydeside, and also weakened Labour morale in the country, as was indicated in poor by-election results. The absence of a parliamentary majority was no real obstacle to bringing forward new policies; the problem was lack of ideas. Lloyd George repeatedly engaged in private talks designed to produce an agreed programme in return for regular Liberal support in the Commons. By February 1930 one frustrated junior minister, Oswald Mosley, drew up his own programme including tariffs, control of the banks, rationalization of basic industries, and a development plan for agriculture, rather like those of the ILP and Lloyd George. But it was rejected by Cabinet ministers still overawed by the Chancellor, Snowden, who adhered rigidly to the Treasury's views. The Cabinet appears to have been handicapped by a mixture of ignorance and naivety; when, after the fall of the government, the gold standard was abandoned and the pound devalued, one ex-minister commented: 'No one told us we could do that.'

As the economic and political situation deteriorated, MacDonald gradually succumbed to the logic of the situation, which was to reach an agreement with Lloyd George. By the summer of 1931 they were close to it; the government had introduced a bill for the Alternative Vote, to please the Liberals, while the latter were propping up the government's majority in Parliament. But the Prime Minister had allowed matters to drift

ANOTHER LAST STRAW.

CAMEL. "DON'T WORRY ABOUT WHERE YOU'RE GOING TO PUT THAT STRAW; I'VE GOT A BROKEN BACK ALREADY."

MR. SNOWDEN (*shifting the responsibility*). "AS I KEEP ON TELLING YOU, THAT'S NOT MY FAULT: YOU BROUGHT IT WITH YOU FROM THE LAST PLACE."

Philip Snowden as Chancellor juggling with precarious national finances for too long. By 1931 the mounting unemployment had exacerbated the budgetary deficit; deteriorating world trade and the overvalued pound had resulted in a serious balance-of-payments problem. This inevitably created fears that the pound would be devalued, which provoked withdrawals of money from London. There was consequently strong pressure to restore confidence in the pound amongst investors, which could best be done by balancing the budget. In August 1931 a committee under Sir George May estimated the deficit at £120 million and recommended expenditure cuts of £97 million, including a 10 per cent reduction in unemployment benefit. Eventually the Cabinet split 12–9 in favour of these proposals; but in view of the opposition of the TUC it was clear that the government could not go on in this divided state. Yet there was nothing inevitable in this breakdown. The financial interests had been anxious, not to destroy MacDonald's Cabinet, but to persuade it to take the necessary measures.

In the circumstances the normal response would have been for Baldwin, as the leader of the opposition, to form an alternative government. But against expectations the Labour government was succeeded by a National government under MacDonald's premiership, including the Conservatives, the Liberals, and a few Labour members. Initially Baldwin had not been enthusiastic about this. He had been a major critic of the last coalition in 1922 and retained vivid memories of its effects upon the party. Moreover, he had already been subjected to sustained attack on his leadership from fellow Conservatives since 1929, and feared that joining a coalition would only provoke more trouble. But as the King was keen and Samuel, the acting Liberal leader, willing, he reluctantly agreed to serve in a National government; but it was to be a purely temporary arrangement designed to implement certain necessary economic measures.

However, it was not long before these assumptions were overturned. The Conservative leaders rapidly began to see the advantages of keeping the National government together to fight an early general election *before* the economic crisis had been dealt with. They feared that any delay in holding an election would allow Labour the chance to recover and capitalize upon the unpopular decisions taken by the government. It was clearly going to take longer than anticipated to resolve the economic problem, and the measures would be more complicated than a mere balancing of the budget. In particular, the Conservatives saw the opportunity to introduce the protectionist policy they had so long advocated; but in view of earlier defeats over tariffs it seemed wiser to get an election out of the way first. Hence the decision to appeal for a vague 'Doctor's Mandate' to put the economy right. MacDonald had no objections; once his former colleagues had repudiated him there was no way back to the Labour Party. It was the Liberals, who had hoped to postpone an election, who were the most reluctant, especially as Lloyd George had stayed outside the National government. But they hesitated to withdraw so soon, particularly as their participation in the government would at least guarantee short-term protection at the election.

As a result, an election took place in October 1931 in which the National government won an overwhelming mandate with 67 per cent of the votes and 554 MPs. Labour's impossible position was not improved by incompetent leadership by Arthur Henderson. At least half of the ex-ministers supported measures of retrenchment, and Henderson failed to offer any clear opposition to the new government's policy. Consequently the party found itself confused and vulnerable to the accusation of Snowden and MacDonald that it had run away from the economic crisis and was not fit to govern.

A myth has developed that, in spite of the party's difficulties, Labour's support held up at the election. In fact, while the turnout was as high as in 1929, Labour's share of the vote fell sharply, to 30.5 per cent, and only 52 MPs held their seats. To some extent this reflected the loss of the advantage Labour had enjoyed in 1929, when it had won many seats on a minority vote in three-cornered contests. Now the non-Labour vote was concentrated on a single National candidate in most constituencies, and three out of five former Liberals seem to have switched to the government. Women voters were particularly ready to desert the Labour Party; since far fewer of them enjoyed the institutional links, such as union membership, which helped keep the men loyal, they were always a more volatile element.

The year 1931 also proved important as a turning-point for free trade. Though both Labour and the Liberals believed it was still a popular cry, they lost votes equally heavily. After their success the Conservatives naturally wished to press ahead by introducing tariffs, and as a result the Samuelite Liberals withdrew from the government in 1932, leaving behind the National Liberals under Sir John Simon, who no longer supported free trade. Consequently by 1932 the Liberals had lost their foremost unifying cause, and they had also discredited themselves by remaining for so long in the ranks of a government with whose policy they disagreed. Thereafter the Liberals were too closely associated with the right wing to be credible as a radical force. At the 1935 election they ran only 161 candidates, won 6.4 per cent of the poll, and returned twenty MPs. The events of 1931 had entirely destroyed the revival of 1926–9.

The election gave the National government a five-year lease in office, but in the event it remained in power until 1940. Why was this? Although ostensibly formed in order to defend the pound, the new government soon abandoned the gold standard and accepted a substantial devaluation. Bank rate had fallen to 2 per cent by June 1932, and in due course a slow economic recovery began. This was the key to political survival. Cheap money proved especially helpful in stimulating the building industry; a million houses were constructed between 1931 and 1935. By 1934, when unemployment at last began to fall, there was a sense that Britain had emerged from the depression. The cuts in unemployment benefit were restored, and new income tax relief was offered in the 1935 budget. The combination of deflation and lower taxes meant rising real incomes for middle- and working-class people if they had a job. This paved the way for another general election in 1935.

The Labour Party, bereft of its leaders after the holocaust of 1931, chose the elderly George Lansbury as its leader until 1935, when he resigned and was replaced by Clement Attlee. It was, however, still difficult to rebut the accusation that Labour's return to office would bring with it an economic crisis. In spite of that, Labour pushed up its vote to 38 per cent, a little higher than in 1929. But this produced only 154 MPs, by comparison with 288 in 1929. The electoral system was now working to the advantage of the National government, which, with 432 seats, enjoyed a majority of 249 over all other parties. Baldwin, now at the height of his popularity, took over the premiership from the failing MacDonald.

It suited Baldwin to project the National government as the epitome of moderation at a time when other countries were succumbing to the extremes of left and right. In many ways it was his own ostensible supporters in the Conservative Party who gave him the roughest ride. Indeed, since the election defeat in 1929 much of his party had been in revolt against his leadership. Winston Churchill left the shadow Cabinet in order to be free to attack his Indian policy at a time when Lord Irwin, a Conservative Viceroy, was making concessions to the Indian nationalists. At the same time Lords Rothermere and Beaverbrook used their control of the *Daily Mail* and *Daily Express* to wage a campaign for Empire Free Trade. This led them to sponsor their own candidates against official Conservative ones and thus split the party's vote. Their real target was clearly Baldwin, whom they hoped to drive from the party leadership.

Understandably, then, for Baldwin the National government assumed the role of an effective air-raid shelter. Many of the 470 Conservative MPs elected in 1931 preferred Baldwin's conciliatory, middle-of-the-road line, for they were not likely to retain their working-class vote if the party lurched to the right. With its huge parliamentary majority the government could afford to shrug off revolts by the diehard elements on the Tory back-benches. Churchill, for example, led a bitter onslaught on the Indian reforms during 1931 to 1935 (see p.214). But even though around 50 MPs supported him, this was insufficient to make the government back down. Churchill was, in any case, widely regarded as disloyal and a careerist by orthodox Conservatives. Having been brought out of the wilderness by Baldwin, who made him Chancellor of the Exchequer in 1924, he was now apparently bent on overthrowing his leader.

This view of Churchill's Indian crusade coloured reactions towards his next cause – rearmament and appeasement. Although his complaints about the poor state of British defences were well-informed and often well-received, he was seen as an incorrigible warmonger, and never attracted more than a handful of supporters in Parliament. Also, he suffered from the charge of inconsistency, for as Chancellor after 1924 he had been partly responsible for reducing military expenditure. It also seemed odd that one who had been so vitriolic in his attacks on the Bolsheviks in Russia should now be willing to cooperate with them in order to check the Nazi regime in Germany. However, by 1938 opinion had clearly begun to move against the appeasement policy. The resignation of Sir Anthony Eden as Foreign Secretary in that year provided a more respectable figurehead for the critics. Yet Eden and his allies continued to keep aloof from Churchill for fear of damaging their cause. In any case, Eden himself failed to keep up a sustained attack on government policy, partly because he had been so closely associated with it, and he never mobilized more than a small group of followers. Setbacks for National government candidates in by-elections in 1938 and 1939 heralded a collapse of public support for appeasement, though voters were rallying around independents rather than Labour candidates. Until the outbreak of war it still seemed highly probable that the government would win another general election in 1939 or 1940.

Although Labour had suffered a humiliating reverse in 1931, the party comfortably retained its position as the alternative government, and effected a considerable recovery. In 1932 it gained 458 seats in the local elections, and in 1934 it won control of the London County Council. Party membership increased by 100,000 to 380,000 in 1933, and reached 419,000 in 1934. The 94 gains at the 1935 general election, though disappointing, at least made Labour a powerful parliamentary force again. But the traumatic events of 1931 had affected the party in more profound ways. MacDonald was almost universally condemned as a traitor by Labour activists, who remained suspicious of anyone who was merely using the movement for personal advancement. Inevitably the question was raised as to how the party had reached the dilemma of 1931. It had preached socialism but neglected to produce a socialist policy, thereby leaving itself with the task of ameliorating the effects of a failing capitalist system rather than changing it. Some argued that socialism would never be attainable because, even if Labour won a majority of seats, the Establishment would manage to frustrate its policies. However, parliamentary traditions ran so deep that the bulk of the movement resisted the temptation to opt for an extraparliamentary strategy. In 1932 the ILP disaffiliated from the Labour Party in disgust, but thereafter it rapidly lost membership and became a peripheral organization. Increasingly the Labour Party's policy was influenced by such pragamatic socialists as Ernest Bevin, Hugh Dalton, and Clement Attlee. In foreign affairs this meant convergence on a programme of rearmament and resistance to the fascist dictators. Domestically, it involved a more realistic consideration of the methods by which nationalization of industry could be implemented; in the late 1930s the party committed itself to imposing state control over a fifth of the economy. In addition, union leaders like Bevin were increasingly impressed by the application of Keynesian methods for stimulating growth and boosting employment in the USA. As a result, Labour settled upon a policy of socialist economic planning which was in fact a limited mixture of nationalization, regional economic policies, and Keynesian management of demand.

The Stability of the British System

By comparison with many other European states, Britain weathered the economic and social strains of the interwar period fairly successfully. While parliamentary regimes collapsed in the face of both left- and right-wing autocracies, the British soldiered on under their parliamentary system and constitutional monarchy. Political divisions were, if anything, less acute than they had been in the Edwardian era. There are several contributory reasons for this surprising stability. The most obvious explanation is that the tradition of parliamentary politics and respect for political institutions was more deeply rooted in Britain because it was older. It is interesting that even an influx of new voters in 1929 did not diminish the turnout in the general election; indeed, it rose to 76.1 per cent; it was again at 76.3 per cent in 1931 and at 71.2 per cent in 1935. This suggests that the public was not on the whole alienated by the economic failures of successive governments.

However, this is only valid up to a point. Turnout at elections remained consistently lower than before 1914. This is partly because the reforms of 1918 and 1928 brought into the political system large numbers of new young voters who were much less committed to the conventional political parties and rather more open to alternative movements. This was compounded by the presence of many disillusioned ex-servicemen who felt somewhat let down as a result of postwar unemployment. In the 1930s some of them were attracted by the British Union of Fascists. The politicians were so worried about demobilized troops that several attempts at organizing them had been made in the immediate aftermath of war. The Comrades of the Great War was formed by extreme right-wing figures, the National Federation of Discharged and Demobilized Soldiers had links with Liberal politicians, while the National Union of Ex-Servicemen was attached to the Labour Movement. However, in each case these organizations dwindled after the initial surge of enthusiasm, and even the British Legion, which was established with a view to keeping former soldiers out of the clutches of either the right or the left, never managed to recruit more than 10 per cent of the eligible men.

Thus, in spite of their grievances, the ex-servicemen appeared to be no more likely than the population in general to be attracted into radical or revolutionary movements that were fundamentally opposed to the British system of government. Part of the explanation may be that even when the conventional parliamentary politicians became discredited – as many of them clearly did between the wars – the public continued to place its faith and loyalty in the figure of the monarch. King George V, who had reigned throughout the war years, attained his Silver Jubilee in 1935 amid universal acclaim. For many people the King provided a focus for national pride and loyalty above and beyond the political parties and untainted by their failings.

On the other hand, much the most important explanation for the relative stability of the interwar period lies in the social impact of economic developments. In Britain the depression was neither as severe nor as protracted as it was in much of Europe and North America. Deflation meant that many middle and working-class people enjoyed rising real incomes, had access to a growing range of consumer goods, and could make savings without risk. Of course, improving material conditions do not necessarily guarantee political stability; rising expectations often constitute the greatest danger to governments. There is, therefore, an element of contingency to be taken into account here.

Although some contemporaries claimed to see a threat to parliamentarianism in the rise of Labour, it seems that the labour movement never lost its confidence in constitutional methods. In part this reflects the experience of thousands of the party's activists in local government.

It was also of great importance that after the 1923 election the King had been quick to invite Labour to form a government, even though the party had won only 191 seats; this helped to cement the loyalties of the Labour MPs, who might easily have been alienated by any suspicion that they were being excluded from office.

No doubt some of the evidence suggests a different pattern. For example, in the early 1920s 'direct action' had been fashionable amongst some trade unionists. The dockers had offered a famous challenge to the government's policy towards Russia by refusing to load the *Jolly George*, a ship carrying arms to help the anti-Bolshevik forces. But Ramsay MacDonald and the party leaders were hostile to such initiatives, and it is a sign of their control that the party conference banned Communists from holding Labour Party membership. The General Strike represented a danger-point for this strategy because it provided many opportunities for left-wing extremists to work within the mainstream of the labour movement. Yet there is little evidence that syndicalism gained any lasting foothold. The strike involved remarkably little violence and no loss of life. But there was an element of luck in this, for if the strike had lasted more than a brief nine days the government would have made more use of troops, with an inevitable upsurge of violence as the result. It now seems clear that the mood in the labour movement was much more angry and militant around 1918–21. Subsequently militancy declined steadily, so that by the 1930s the response to mass unemployment was surprisingly subdued and small scale. The classic form of protest adopted in this decade was the 1936 Jarrow March, a deliberately small, well-behaved, and carefully organized affair which delivered a petition at Westminster and then returned home; it represented a dignified political protest rather than a portent of revolution.

This restraint on the Labour side was echoed by a certain moderation amongst the Conservatives. In the Edwardian period many Tories had shown a contempt for constitutional politics by promoting violent opposition to government policies in Ireland. After the war some in the party professed to see a subversive threat in the rising Labour Party, or claimed that Bolsheviks had penetrated the government and civil service. However, under the conciliatory leadership of Stanley Baldwin this view was confined to the peripheries of Conservatism. There is no reason to doubt, still less to disparage, his privately voiced words to a colleague: 'The main ambition of my life is to prevent the class war becoming a reality.' Of course, Baldwin did not always succeed in his intentions; but he promoted a good deal of bipartisan policy in terms of domestic reforms and imperial readjustment, and for nine months he averted the General Strike. A Conservative Party led by Austen Chamberlain, Curzon, Birkenhead, or Churchill would have adopted a much harsher approach to the labour movement and, at the least, polarized politics more sharply.

It was not until 1931 that the ingredients for a major crisis in liberal democracy seem to have been present in Britain: a deepening economic depression, long-term unemployment, a declining trade union movement, discredited political leadership, and millions of young voters. In these circumstances it is scarcely surprising that extremist movements of the right and left attempted to make capital at the expense of conventional politics. Critics of the TUC alleged that it had presided over a fall in membership from over 8 million in 1920 to 4.3 million in 1933, and was doing very little to organize the men. For a time the National Unemployed Workers Movement, led by Wal Hannington, stepped into the vacuum. It mobilized 50,000 members in 1930–1, organized demonstrations and hunger-marches, and presented petitions. Its especial objects were the abolition of the hated Means Test and the restoration of the cuts in unemployment benefit. Yet this was scarcely subversive. The NUWM reached a peak in 1932 with a 25,000 strong rally in Hyde Park, but then petered out.

Although excitable right-wing politicians regularly warned about the infiltration of Communists, the reality was rather sobering. For example, the British Communist Party, marginalized by its exclusion from the Labour Party, experienced an increase in membership from just 6,000 in 1930–1 to 18,000 by 1939. This represented the recruitment of middle-class intellectuals, stirred by the rise of fascism and the Spanish Civil War, rather than working-class members. The British Union of Fascists did its best to fan the embers of Communism by engaging in violent clashes with the Communist Party, notably at the 'Battle of Cable Street' in London's East End in 1936; but the publicity thus generated merely flattered the stage armies involved.

Undoubtedly the BUF posed the most dramatic and well-articulated challenge to parliamentary politics, at least for a time. Although tiny fascist organizations such as the Imperial Fascist League had existed in the 1920s, it was not until Sir Oswald Mosley formed the BUF in 1932 that the movement took off in Britain. The timing proved to be of great importance. During 1932–3, as unemployment continued to rise, it seemed quite possible that the National government would fail to deal with the depression, just as its predecessor had failed. In such a situation there would have been no alternative government, and the parliamentary system would have been severely discredited. This was the burden of Mosley's case; he and his movement would step in to save the country from chaos. Having served as a junior officer in the war, Mosley was well placed to appeal to the disillusioned ex-servicemen who felt badly let down by the politicians. Young people flocked to join the new movement in the early 1930s, much to the dismay of conventional politicians. The growing anti-semitism in BUF propaganda was also no disadvantage, for dislike of the Jews was common at all social levels in interwar Britain; on the other hand, anti-semitism was by no means strong enough to constitute the basis for a really powerful political movement.

The most significant source of strength for the BUF appears to have been disillusioned grass-roots Conservatism. Many of the MPs appreciated Mosley's claim that he could hold political meetings without suffering intimidation by the left; they also recognized that his proposals for imperial economic development represented the kind of programme they had failed to get from the official Conservative leadership for many years. Consequently, many of the Tory and upper-middle class elements were very tempted to make common cause with the BUF, or at least to use it to ditch Baldwin. It was not until 1934 that they felt able to dispense with this unpalatable necessity, largely as a result of the onset of an economic recovery. The 1935 election ensured the survival of the National government and by this time the BUF was in sharp decline. Even so, the government showed itself rather slow to intervene against the BUF's methods; the Public Order Act did not come into force until 1937, when it was clearly safe to use the full force of the law to check fascist activities. The final factor contributing to the BUF's decline at this time was the growing public fear about Nazi Germany, which led many people to regard Mosley as a mere agent of the fascist dictators on the Continent.

13

The Era of Domesticity

The traditional view of the interwar period as one of depression, despair, and unemployment is not without a good deal of empirical support, especially if it is contrasted with the happier decades after 1945. However, it does represent an exaggerated picture of life between the wars; indeed, in some respects the 1930s saw the beginning of social changes more usually associated with postwar Britain.

The Rising Standard of Living

For most of the time most of the British people enjoyed paid employment; but what mattered to them was essentially how much their wages and salaries would buy. Here we have to distinguish contemporary rhetoric from fact. Unquestionably politicians, civil servants, and employers regarded substantial wage reductions as an economic necessity in the 1920s; and the government's whole strategy for restoring the prewar value of the pound assumed that this could be achieved. But one must remember that the strategy was frustrated.

Between 1920, when the boom collapsed, and 1923, money wages fell sharply, though at the same time prices began to fall. Union fears about a general attack on wages, especially after the return to gold in 1925, led to the General Strike of 1926. Yet by 1929 wage rates were still at the 1923 level, despite the fact that retail prices had fallen by 6 per cent. This may seem surprising. It is a reminder that the General Strike was not the failure it is traditionally thought to be. For employers it represented an impressive manifestation of workers' solidarity and also of their fundamental reasonableness. After 1926 there appears to have been some kind of mutual understanding – formalized in the Mond–Turner talks of 1928 – between unions and employers about maintaining stable wages. This meant that the unions did not resist employers' attempts to cut costs by reducing their labour force, while the owners abstained from cuts in wage rates. Even in the depression of 1929–32 wages fell by only 4 per cent on average. In the same period wholesale prices dropped by 25 per cent and the Ministry of Labour's cost-of-living index by 12 per cent. From 1934 wages were rising again, as was the cost of living. Thus, although workers in hard-hit industries like coal and cotton suffered badly, on the

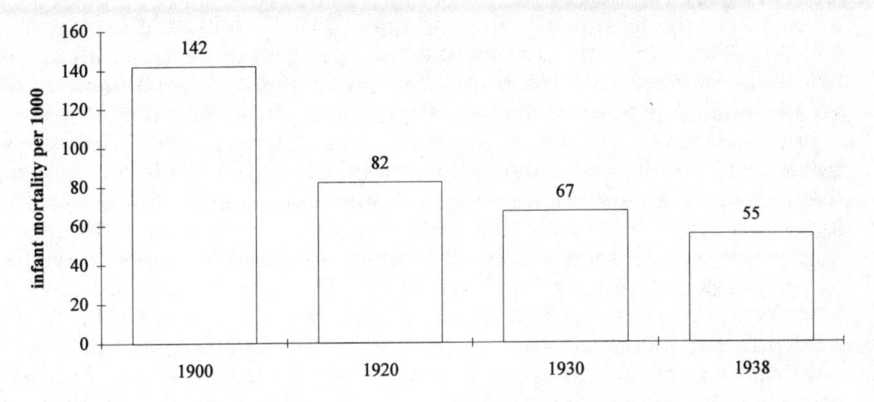

Figure 13.1 Infant Mortality Rates per 1,000 Live Births in Great Britain

whole employees enjoyed rising real wages in the 1920s and 1930s. One estimate suggests a 17 per cent improvement in real wages between 1924 and 1935.

This is obviously of considerable political and economic importance. Rather than being crushed by poverty and hopelessness, many people entertained gradually rising expectations. Improved family incomes opened the way to a wide range of inexpensive consumer goods and minor luxuries characteristic of this period: cinemas, dances, the radio, women's magazines, football pools, cosmetics, and mass-produced copies of fashionable clothes. The motor car, often taken as symbolic of the inter-war years, was still too expensive to be purchased by all but a minority; however, there were a million motor cars on the roads by 1930 and 2 million by 1939. Clearly, the expansion of the consumer-goods industries would not have been possible without the capacity of millions of middle- and working-class families to enjoy a small surplus left over from their essential expenditure.

The effects of gradual improvements in living standards showed up in improved health and longevity between the wars. Life expectancy for women increased from 55 to 66 years between 1910 and 1938, and from 52 to 61 years for men. The figures for infant mortality also continued to fall. However, the national statistics conceal much unevenness. In 1935, for example, infant mortality rates stood at 47 in south-east England but 68 in the north of England. Poverty meant that the working classes were much more likely to die young. Death rates per 1,000 in 1937 varied from 73 in Harrow and 80 in Oxford to 134 in the Rhondda and 138 in Wigan.

A major study, *Food, Health and Income* by Sir John Boyd Orr in 1936 suggested that 10 per cent of the population was badly fed, including one in five school children. But this evidence of poor diet for some should be seen in the context of notable improvement in society as a whole. By comparison with the pre-1914 period there had been major increases in the consumption of fruit, vegetable, and eggs. When B. S. Rowntree repeated his famous study of York in 1935–6, he found only 3.9 per cent of the population in absolute poverty. However, Rowntree revised his original poverty line and argued that a family of five required 43s. 6d. weekly income. On this basis 17.7 per cent of the population were below the poverty line. He found that low wages and casual labour were greater causes of poverty than either unemployment or old age.

Improved health and longevity reflected a mixture of trends and changes: better diet, a municipal water supply piped to virtually all homes, preventive measures against such diseases as tuberculosis and typhoid, and the long-term effects of Edwardian state welfare policies such as the school medical service. Contemporary fears about the declining birth rate helped to sustain social policies for mothers and children during and after the war. Typical of this thinking was the 1918 Maternity and Child Welfare Act, which required local authorities to set up infant welfare centres and antenatal clinics and to appoint health visitors; it also allowed them to provide home helps, day nurseries, and food for expectant mothers. By 1928 these services cost nearly £1 million per annum.

However, there were huge gaps in the system. The 1911 health insurance scheme covered 15 million people by 1921 and 20 million by 1938. But this left around 15 million, largely women and children under five, without assured help. The provision of hospitals was also patchy, involving a mixture of voluntary, municipal, and poor-law hospitals. Much concern focused on the failure to reduce the incidence of maternal mortality; the rate per 1,000 stood at 4.82 in 1923, rose to 5.94 in 1933, and fell to 3.25 by 1939. The Women's Co-operative Guild and several other women's organizations which campaigned for improvements in female health established a Women's Health Committee Inquiry in 1933, which studied 1,250 working-class women and published the results as *Working-Class Wives*, edited by Margery Spring Rice, in 1939. In the sample no fewer than 46 per cent of the women suffered from 'bad' or 'grave' ill-health, while 22 per cent were in 'good' health. In general it was abundantly clear that women's health was neglected, partly because wives minimized their own ailments but also because they lacked access to professional advice and treatment.

Feminists advocated the extension of health insurance to all women regardless of whether they were in employment, the introduction of family allowances, and the provision of advice on birth control. None of these remedies was adopted, with the partial exception of the third. However, the politicians were susceptible to pressure on behalf of women and children. Under the 1929 Local Government Act, the local authorities were allowed to take over the poor-law hospitals; they increased the number of maternity beds, and by the 1930s it was becoming usual to give birth in hospital. From 1934 they were also permitted to distribute free or subsidized milk in schools; by 1937 3.2 million children were benefiting from this. In 1936 the Midwives Act compelled local authorities to train midwives. All these *ad hoc* measures took Britain half-way to a national health service by 1939. Those who were not covered by insurance could always pay for private treatment, but in practice those most in need, such as women, did not do so. Private medicine was no answer to the problems of ill-health, but it would take another war to create a consensus in favour of a comprehensive state system.

The Housing Revolution

Another important aspect of improving living standards between the wars was the higher standard of housing; but this also coincided with the start of a radical change in housing *tenure*. Before 1914 only 10 per cent of homes were owner-occupied, almost all the rest being privately rented. However, most landlords were small property-owners and lacked the resources to maintain or improve their houses. They were not a popular class, and during the war the government lost little time in imposing restrictions on rents. Thereafter some form of rent restriction remained in place, and the private rented sector entered a protracted period of decline.

One alternative was the construction of municipal council houses for rent. By the end of the war the government felt very concerned both about the inadequate supply of housing and about the poor quality of the existing stock. Not only did it enact the Addison Housing Act, under which 170,000 council houses were built between 1919 and 1921, it also set up the Tudor Walters Committee, whose 1918 report recommended the provision of indoor toilets and bathrooms, larger separate kitchens, proper light and ventilation, and extra bedrooms. The policy of state subsidies for council housing was reintroduced in 1924 by John Wheatley, which led to another 520,000 houses being built by 1933. Arthur Greenwood's 1930 Housing Act subsidized slum clearance and boosted the municipal housing stock further. Altogether 1,100,000 council houses were built between the wars, representing nealy a third of the total number of new houses. Municipal housing set a high standard which private builders often failed to match, but with which they had to compete.

Unfortunately, the benefits for working-class families were not as great as they might have been, partly because the rents charged were fairly high and also because the Conservatives disliked municipal housing on grounds of the cost and because of the competition it posed to private builders. They therefore cut off subsidies in 1933 and so checked the expansion of council housing. On the other hand the National government's cheap-money policy helped to reduce mortgage rates to only 4.5 per cent; in combination with rising real incomes, this generated a fast-growing demand for owner-occupation. As a result, by the end of the decade 1.4 million people held building-society mortgages. Two-and-a-half million houses were built for sale between the wars, and by 1939 31 per cent of all houses were owner-occupied. By the 1930s a semi-detached house could be purchased for £400 with a deposit as low as £25. This soon produced miles of suburban roads filled with three-bedroomed semi-detached houses adorned with mock-Tudor beams and identical front and back gardens. In retrospect, suburbia has attracted a good deal of disdain. It is true that the properties were often inferior in quality to municipal houses; inadequate planning controls led to 'ribbon development'; and the developments suffered from an absence of community facilities. On the other hand the houses represented a major improvement in the living conditions of millions of families now able to enjoy proper drainage, toilets, baths, hot water, and private gardens. The reduction of overcrowding, especially in bedroom accommodation, made for happier home life.

Above all, women were the beneficiaries of the housing revolution. They spent more time in the house than other family members, and traditionally suffered most from the inconvenience and drudgery involved in keeping decaying property clean. The Women's Co-operative Guild and the Labour Party Women's Organization urged the authorities to take notice of women's own needs in terms of separate kitchens and internal bathrooms, and this was one of the few questions on which the authorities were really prepared to listen to women. As a result, the new housing was easier to keep clean and warm; between the wars working-class homes enjoyed twice as much floor space as they had had in the mid-nineteenth century; above all, two-thirds of all homes had been wired for electricity by 1939. None of this made women's domestic lives easy. The new equipment – cookers, vacuum cleaners, washing machines – were comparatively expensive, and did not reduce the time devoted to housework. They were chiefly useful for middle-class wives, who now found it difficult to find and retain domestic servants.

However, if the home continued to involve hours of hard work, it was at least becoming a more attractive and comfortable place. Women particularly enjoyed the cheap wireless sets which were to be found in three-quarters of all houses by the 1930s; the number of licence-holders leapt from 36,000 in 1922 to 8 million by 1939. The reduction in family size also helped to make the home more a place of leisure for both husbands and wives; increasingly this took the form of interior decorating, gardening, and entertaining. This pattern was consolidated by the growing habit for couples to spend leisure time outside the home in one another's company rather than in sexually segregated groups; they would go together to public houses, cinemas, and dance halls. By 1934 there were no fewer than 93 million cinema admissions in Britain; many people went once or twice a week.

Women, Family, and Marriage

After the war the public image of young women changed abruptly. Where papers like the *Daily Mail*, *Daily Express*, and *Daily Sketch* had recently portrayed the patriotic munitionette, now they saw the 'flapper'. 'The social butterfly type has probably never been so prevalent as at present,' complained the *mail*. 'It comprises the frivolous, scantily-clad, "jazzing flapper", irresponsible and undisciplined.' Young women were widely accused of lowering moral standards in their pursuit of an exciting social life, while men, as usual, were regarded as innocent victims. Part of the problem was that the war had unbalanced the population even more than usual. 'Our Surplus Girls' in the *mail*'s words, outnumbered men by 1.9 million. It was feared that as a result of their wartime employment women would want to retain jobs now needed for the returning servicemen, though in the event they were almost all sacked. The refusal of many women to go back to poorly paid lives of drudgery as domestic servants was taken as a worrying sign in the servant-employing classes that women were getting ideas above their station. It seemed possible that their new political influence and their entry into the professions was attracting women away from marriage and motherhood. Indeed, according to Barbara Cartland, then a young journalist and self-appointed authority, the physical condition of the flappers was so changed by their lifestyle – dieting, dancing, masculine clothes, narrow hips, small breasts – that many would be unable to give birth to healthy babies anyway.

Not for the first time, nor the last, the popular press had things more or less completely wrong. There is some evidence that between the wars the younger generation was more likely to engage in sexual relations before marriage; but this should be seen in the context of a major decline of several of the great vices of Victorian and Edwardian England – prostitution and alcoholism. British society was actually becoming more moral between the wars, and its route lay through the spread of domesticity, not in a retreat from it. The participation of women in the labour force hardly changed between the Edwardian period, the 1920s, and the 1930s. Moreover, despite fears that the war had blighted the marriage prospects of younger women or distracted them from traditional roles, the fact is that during the 1920s the Edwardian decline in marriage rates was reversed; among women in their late teens and twenties a higher proportion now married. The proportion continued to rise through the 1930s and, indeed, in every decade up to the 1970s, with a brief interruption in the Second World War. It is an unavoidable conclusion that most of the rising generation of British women regarded marriage and motherhood as the major goals in life. Even amongst younger feminists like Vera Brittain, marriage was seen in a very positive light; to some extent they were able to take political rights and access to careers for granted; the new challenge was how to *combine* employment with marriage and motherhood, not choose between them. It is only fair to note that while marriage rates rose, so too did divorce rates. The reform of 1923 equalized divorce so that women need prove only adultery, and A. P. Herbert's 1937 bill added a number of additional grounds for divorce including desertion, which was especially important for women. However, in spite of the increase only 6 per cent of all marriages ended in divorce in the late 1930s. In view of the fact that marriages had to last longer as a result of greater longevity, it can be concluded that interwar marriage was a remarkably stable institution.

One key consideration affecting women's view of marriage between the wars was the belief that pregnancy and child-rearing need not be as great a burden as it had been in the past. The trend towards smaller families among certain middle-class groups had been clear since the late 1870s, but now it spread throughout society. By the late 1920s married women experienced on average 2.2 live births, by comparison with five or six in the Victorian period. One of the remarkable aspects of this profound social change is that it was accomplished in the teeth of much propaganda designed to deter couples from restricting family size. The women's magazines resolutely avoided the subject of birth control. The National Baby Week Council cooperated with the politicians in upholding the view that childbirth was a duty owed by women to the state. Governments offered few material inducements to have big families beyond increased tax allowances for second and subsequent children; but the *News of the World* continued to present free willow-pattern plates to all proud mothers of ten! In the

1920s the medical profession, the Church of England, and all the political parties condemned birth control, though many members of each of these institutions actually practised it. Indeed, those who printed birth-control literature were still liable to prosecution for obscenity.

In spite of all this, behaviour clearly changed markedly, though it is not easy to explain why. Over a long period of time the effect of rising standards of living and declining infant mortality invariably has the effect of encouraging married couples to reduce family size. In Britain many women wanted to avoid pregnancy, but were ignorant of the ways of doing it. Traditionally, couples either avoided sexual relations or attempted coitus interruptus. In addition large numbers of women obtained abortions or miscarriages by one means or another. As late as the 1930s it was estimated that 100,000 to 150,000 women died each year as a result of abortion. By the 1900s some improvements in the mechanical methods of birth control had been achieved. Condoms had become cheaper, more reliable, and were distributed to the troops during the war. However, they were not widely approved of, and reformers like Marie Stopes recommended caps, pessaries, and diaphragms. The drawback was that these required some cooperation from doctors, which was not generally forthcoming.

The key to change lay less in the adoption of any particular method of birth control than in a growing determination to make the attempt. The mood was well caught by Marie Stopes's famous little book, *Married Love,* which sold 400,000 copies between 1918 and 1923 and a million by 1939. Stopes's contribution was to make the idea of birth control respectable. She argued that frequent pregnancy was not in the national interest any more than the mother's; it simply ruined women's health and led to high infant mortality. By deliberately spacing births, a couple could ensure a healthy generation of mothers and children. But in addition Stopes addressed a very clearly felt need amongst married couples. Many women were anxious to be able to enjoy sexual relations with their husbands free from the perpetual fear of another pregnancy. Above all *Married Love* was a tract in praise of a modern marriage of equals, in which it was recognized that women enjoyed sex as much as men.

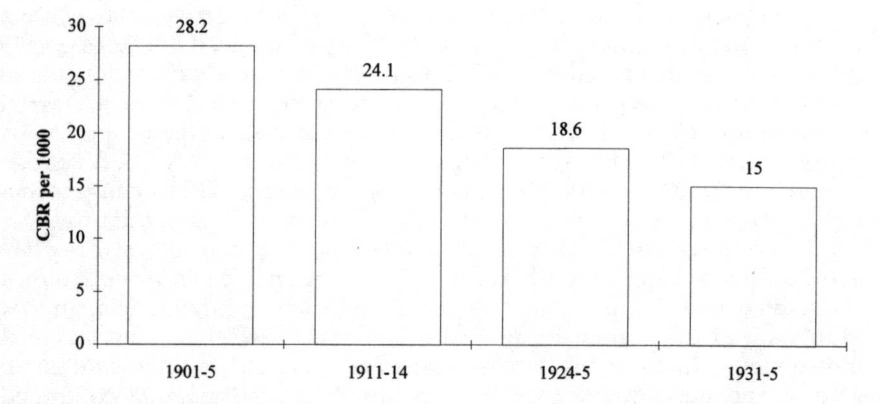

Figure 13.2 Crude Birth Rates (per 1,000 Population) in England and Wales

Marie Stopes followed up her literary success in 1921 by establishing a clinic dispensing advice on birth control in Holloway. Twenty or so voluntary clinics were operating by the late 1920s, but successive Ministers of Health claimed that it would be illegal for local authority clinics to give any such service. This view was not abandoned until 1930, the year in which both the Church of England and the British Medical Association backed down on the issue. The doctors were moved less by concern for women's health than by the fear that they were gradually losing business as women went elsewhere for medical treatment. However, as late as 1937 only 95 of 423 local authorities were actually providing birth-control advice, and then only to married women for whom pregnancy would be detrimental to health. There were around 70 voluntary clinics also in operation by this time. Thus the emergence of the two-child family continued to be largely a matter of private endeavour, achieved without professional assistance.

Social Welfare and Income Distribution

As a result of political pressure to reduce taxation and the attempts to balance the budget, government spending on social welfare fell between 1921 and 1924, though by 1929 it had regained its former level. In general pensions, housing, and unemployment benefit tended to expand their share and education and health to contract. As unemployment mounted, total expenditure rose, much to the government's dismay, with the result that by 1930 11.1 per cent of gross domestic product went on the social services; by 1932–3 the figure had risen to 13.1 per cent, though by 1935–6 it had slipped slightly to 12 per cent. But this compares with only 4 per cent in 1913. Actual expenditure rose from £100 million in 1913 to nearly £600 million by 1938.

As we have seen, governments showed some sympathy for measures designed to promote the health of women and children, though they tended to put the responsibility onto local authorities (see p.196). They resolutely avoided the one innovation that would have made a significant impact on mothers and children: family allowances. This idea was propagated by Eleanor Rathbone, among others, as a result of experience during the war which had shown that quite modest allowances, paid directly to mothers, were a very cost-effective way of keeping children out of poverty. However, the cost would have had to be borne by the national revenues, and none of the parties was yet willing to include this in its programme. The major initiative was Neville Chamberlain's 1925 Widows, Orphans and Old Age Contributory Pensions Act. This extended the 1908 pensions scheme to 65–70-year-olds, but only in return for contributions by employers and employees. The most important part of the legislation was the 10s. weekly pension for widows. This was a long-awaited and widely advocated measure which brought relief to thousands of hard-pressed women now able to save their children from the workhouse.

Education, by contrast, was not a priority for either Conservative or National governments between the wars. The school-leaving age had been raised to fourteen in Fisher's 1918 Education Act, but other innovations succumbed to the 'Geddes Axe' in 1921. After 1924 Lord Eustace Percy, the minister responsible, imposed further economies on staff which meant that classes of fifty children were common. At this time three children in every four received their entire education in the *elementary* schools. The Hadow Report of 1926 sought to address the problem of Britain's inadequate secondary education. It recommended raising the leaving age to fifteen and dividing schools into primary and secondary, the transition occurring at age eleven; at the secondary stage children would go either to 'modern' schools to receive a non-academic education or to grammar schools. By 1938 two-thirds of children were attending modern schools. Grammar and direct-grant schools charged fees, although a number of free places were available for pupils from poorer families who won scholarships. The National government dropped the promise to raise the leaving age to fifteen, cut teachers' salaries in 1931, and began to charge fees after a family means test even for formerly free places. As a result British education, already well behind that in Western Europe and North America, continued to be a serious impediment to national economic development.

Much the most controversial area of social policy concerned the provisions for the unemployed. Basically, interwar governments used Lloyd George's 1911 scheme in which workers, employers, and the state made contributions to the Unemployment Insurance Fund from which benefits were paid. However, this ceased to be financially viable because from 1920 onwards the scheme was greatly extended to cover 12 million workers and unemployment was consistently high. The fund had therefore to be subsidized by the Exchequer. There was also a growing problem concerning workers who had not paid enough contributions to qualify for normal benefits; they either received 'uncovenanted' benefits or simply applied to the poor-law guardians. One result of the pressure on the fund was a series of dismal expedients designed to reduce the costs. Conservative ministers imposed a means test in 1922–4 and 1925–8. Large numbers of women applicants were driven away by the requirement that they must accept work as domestic servants. Also, applicants were obliged to demonstrate to the local employment committees that they were 'genuinely seeking work'; this led to 3 million people being refused benefits during the 1921–30 period.

The other major controversy arose out of the treatment of those unemployed workers who, because of their ineligibility for benefits, were forced to apply to the poor-law guardians. Around 450,000 received relief from this source each year in the mid-1920s, and three times that number in the aftermath of the General Strike. As a result of the advance of the Labour Party in local government after the war, many of the guardians were now sympathetic to the plight of the unemployed, and were often willing to provide relief at higher rates than those approved by the Ministry of Health. This incensed Conservative ministers, and rekindled their long-standing hostility to the spread of elected local authorities during the later decades of the Victorian era. In that period Conservatives had particularly objected to the policies of the London County Council and many of the school boards, partly because of the increase in rates and partly because they felt that it was dangerous to provide too much education for the working classes.

The government went to law to stop school boards spending money on secondary education; but eventually Balfour decided the only solution was to abolish school boards, which he did in 1902.

Between the wars this suspicion of local government as a hotbed of radicals and socialists resurfaced in connection with the board of guardians in Poplar, where George Lansbury was mayor; the Poplar guardians paid above local minimum wage rates, applied no household means test, and supported men engaged in industrial disputes. In 1921 some of the Labour councillors there ended up in prison as a result of their financial policies. By 1926 Neville Chamberlain, the Health Minister, had taken powers to suspend and replace the guardians in West Ham, Chester-le-Street, and Bedwelty for paying excessive relief. Conservatives were not impressed by the fact that the boards had been elected to pursue the policy of which they disapproved. On the contrary, Conservatives believed it to be wrong that people who were dependent on welfare benefits should be able to vote in the elections for the administering bodies. By 1929 Chamberlain had concluded, like his predecessors, that the only course was to abolish the poor-law boards, which he did under the Local Government Act.

Their functions were transferred to committees of county and county borough councils, known as Public Assistance Committees. They were instructed not to exceed the payments that applicants would have received under unemployment insurance. But democracy was not so easily suppressed, and the PACs often paid more generously. The 10 per cent cut in unemployment benefit imposed in 1931 only exacerbated the tension between national policy and local practice. Moreover, the PACs were required to apply a stringent means test, which meant basing relief not on the individual applicant but on the family's income. As a result, fathers and husbands could be refused relief because of the earnings of sons, daughters, and wives. This inevitably caused friction within families and led some to leave the family home. So resented was the inquisition that it had the effect of deterring many from applying for relief; around a quarter of a million were removed from the registers, and another half a million had their payments reduced.

However, the payments were still insufficiently uniform for the government; they intervened again in 1934, with a new Unemployment Act which fixed new national scales and transferred the work of the PACs, which now dealt with 928,000 applicants, to Unemployment Assistance Boards. In the event the new rates turned out to be significantly lower than existing PAC payments, which led to massive popular demonstrations by the unemployed in January 1935. The National government, now contemplating a general election, suspended the new scales and introduced more generous ones in November 1936. The actual level of unemployment benefit stood at 23s. a week for a family of five in 1922, at 29s. in 1931–5, and at 36s. from 1937. This should be compared with the estimate made by Rowntree in the 1930s that 43s. was required to keep a family of five above the poverty line.

These shifts in official welfare policies, combined with the collapse of traditional industries and communities, made for a heightened sense of class consciousness and conflict between the wars. Critics of British society could truly say that income and wealth continued to be very unevenly distributed after the war, as it had been before 1914. On the other hand, the slight redistribution of income begun in the Edwardian period was sustained between the wars. There are broadly four explanations for this pattern. First, working-class wages tended to grow faster than middle-class salaries – five times faster between 1911 and 1938, though this was mostly concentrated into the war and immediate postwar years. Second, many of the largest incomes, especially those derived from land, agricultural rents, and incomes from abroad, fell quite markedly. Third, the transfer of income via government social welfare to poorer people continued, albeit at a modest pace. Fourth, the taxation of incomes continued to make a major impact. The period from 1916 to 1918 had forced the wealthy to get used to what were historically very high rates of taxation. Even after cuts in the early 1920s, the basic rate of tax was 4s. in the pound by comparison with 1s. 4d. just before 1914 and 8d. in the 1890s. In fact, the use of tax allowances in respect of children helped to keep many middle-class people with modest salaries out of the tax net, but the unmarried, the childless, and those on large incomes were comparatively heavily taxed. The effect of taxation alone was to reduce the share of income enjoyed by the top one per cent of the population from 29 per cent to 24.4 per cent between 1914 and 1938. While great inequalities clearly remained, the long-term trend was towards a more equal distribution of income in British society.

14

Imperial Climax and Decline

From the perspective of the Second World War it is very tempting to see interwar Britain as a great power already in headlong decline. No doubt she had come close to military defeat in 1918 and had been fortunate to receive American financial and military aid. In the postwar world relative power had clearly shifted in favour of the United States, while Britain continued to be hampered by a sluggish economy, a vulnerable empire, extensive international obligations, and a failure of political leadership.

Yet while this pessimistic view is not without foundation, it does exaggerate British weakness. Although Britain undoubted suffered from her old dilemma of inadequate resources and extensive obligations, her relative position, especially as far as the 1920s are concerned, was much stronger than the defenders of appeasement have admitted. This was largely because of the special circumstances affecting the other powers. The USA simply failed to play the influential role in economic and diplomatic affairs that her underlying strength would have allowed. Germany and the Soviet Union were largely consumed by internal turmoil. France and Japan were strong and showed a propensity to take initiatives, but essentially within limited regional spheres. Britain alone could claim to be a global power in the 1920s. Thus in the aftermath of the Treaty of Versailles the two allies, Britain and France, could view the world in general, and Europe in particular, from a position of great strength. This situation obviously changed over time. But it is a caution against the assumption that interwar Britain was obliged to adopt a policy of appeasement out of sheer physical weakness. In a twenty-year period, appeasement was more appropriate at some stages than at others; further, its value as a policy depended very much on the skill – or the incompetence – with which it was implemented.

Defence and Disarmament

From a purely British viewpoint the postwar peace settlements were largely satisfactory. The German navy had been sunk at Scapa Flow; the German military threat to France and the Low Countries had been eliminated both by enforced disarmament and by the demilitarization of the Rhineland;

the Empire had been preserved and even extended by the acquisition of mandates over former German colonies; and Britain's strategic position in the eastern Mediterranean had been strengthened by the dismantling of the Turkish Empire, the creation of buffer states, and the British occupation of Palestine, Transjordan, and Iraq.

On the other hand, the Tsarist regime in Russia had given way to a Bolshevik one preaching international revolution and the overthrow of imperialism. Japan and Italy were anxious to grab new territory in the pacific and North Africa respectively. In Europe a large number of new, weak states had been established, creating much potential for conflict and instability. France had to be restrained from further intervention against Germany. And finally, there was little confidence in any new methods for dealing with these problems; the League of Nations, set up to satisfy President Woodrow Wilson, was largely resented by British Conservatives, who thought it might undermine British sovereignty and the Empire.

After November 1918 the government embarked on a major and hasty demobilization. Inevitably the army suffered from a reaction against the massive commitment that had been made to a Continental strategy and from the wish to avoid a repetition of such a costly conflict. It was tempting to believe that Britain no longer needed an expeditionary force. In 1919 her defence policy was based on the 'Ten-Year Rule', an assumption that she would not be engaged in a great war for the next decade. To a large extent, all the governments of this period believed it desirable to avoid an arms race of the sort that had characterized the Edwardian period; and they felt that beyond a certain level expenditure on armaments would hinder the growth of the economy, a concern which increasingly preoccupied them. To this end conscription was abandoned in 1920, and the army, which had been 3.5 million strong in 1919, was reduced to 370,000 by the end of 1920. Its chief role now was to police the empire; some 60,000 troops remained in India, and others were kept busy in Ireland and the Middle East.

From 1919 onwards the Treasury began to press for severe cuts in the defence estimates, proposing a total annual expenditure of £110 million which was less than that in 1913–14. Yet the Treasury did not get its own way by any means. The views of the Foreign Office, the Admiralty, the War Office, the India Office, the Colonial Office, and the new Air Ministry had also to be considered. From 1923 the chiefs-of-staff subcommittee of the Committee of Imperial Defence also contributed to the debate. But by 1920 expenditure on the armed forces, which had been £600 million, was cut to under £300 million. The pressure of the Geddes Committee made for further reductions from 1922 onwards, which pushed costs down to £110 million, but this was partly because certain problems had now diminished and Britain was less committed in Ireland, Afghanistan, Persia, southern Russia, Egypt, Transjordan, and Mesopotamia.

Once down to £110 million, British defence spending remained fairly constant until 1935. Disarmament was a very real feature of interwar policy, but there were limits below which it could not be taken. Moreover, even the economies of the 1920s left some scope for rearmament of which the chief beneficiary was the Royal Air Force. The growing emphasis on air power is readily explicable. The raids by German planes and airships on

London and the east coast during the war had made a deep psychological impression because of their novelty. Sir Hugh Trenchard, the Chief of Air Staff, capitalized on this to establish the RAF's independence from the two senior forces; he got a separate ministry and, thus, some political influence for the RAF. It was widely believed that a future war would begin with a massive air attack which would deprive Britain of the advantage hitherto conferred by her island position; her population, concentrated in a few major conurbations, would be highly vulnerable, and massive loss of life and destruction of industry seemed inevitable. The demoralizing effects of such an attack might oblige Britain to sue for peace. 'The Bomber', in Baldwin's notorious words, 'will always get through.'

Thus Trenchard argued that the creation of a British bomber force capable of deterring hostile powers was essential to British defence. But he also used other reasons to justify an expansion of the RAF. By comparison with the army, an air force could be an economical means of meeting some of Britain's obligations. In the Middle East, where Britain had to police large areas but preferred not to become too deeply involved, it was possible to withdraw some of the expensive infantry and rely instead on small numbers of aircraft. In Iraq, for example, where Britain was trying to maintain an unpopular regime, she used air squadrons to bomb the villages of the rebellious Kurds into submission; and similar tactics were adopted in Transjordan and Aden.

Since the French air force was the only one capable of inflicting damage on Britain in the early 1920s, it was taken to be the national 'enemy'. French aggression in the Ruhr in 1923 prompted the government to examine carefully how large an air force Britain required. In 1922 she had 23 squadrons, of which 14 were bombers and 9 fighters. A committee under Lord Salisbury advised that Britain should have 52 squadrons by 1929, and as a result there was a 50 per cent increase in RAF estimates between 1923 and 1926. By 1935 the ratio of expenditure on the army, navy, and air force was 36:49:15.

To some extent the Royal Navy was being displaced as the politician's favourite peacetime force; it had lost its claim to be Britain's first line of defence to the RAF. However, as the navy still had a major role to play in keeping open Britain's worldwide trade routes and defending imperial communications, it maintained its share of resources more successfully than the army. For the navy, attention shifted from Europe, where there was no serious rival, to the Far East, where the rapid building by both Japan and the USA posed a challenge. In 1920 Lloyd George's government had decided to allow the Admiralty to build against the USA, but it was impossible for Britain to maintain a two-power standard against these formidable rivals in the Pacific. Several expedients were adopted to meet the problem. In 1921 it was decided to build a naval base at Singapore to which, in the event of war, a major fleet would be dispatched. Meanwhile the Anglo-Japanese Alliance of 1902 was abandoned. At the Washington Naval Conference in December 1921, the USA, Britain, Japan, France, and Italy agreed to restrict their capital ships to the ratio 5:5:3:1.75:1.75, and thus to cease building capital ships for a decade. Taken together, these policies were probably unwise for Britain. They damaged her shipbuilding industry, which was now starved of orders. The Admiralty was dismayed at the decision to abandon the policy of maintaining superiority over any one rival navy. Above all, Britain had given up her prewar policy of safeguarding her position in the Far East by cooperation with the Japanese. All the British government really had to put in its place was a vague conviction that it could rely on the Americans; but for many years that proved to be a baseless expectation.

The only result of these modifications in the 1920s was to leave British defence in an incoherent condition. The three services each planned for a different war; the RAF with France, the navy with Japan, and the army – now that a Continental role was ruled out – concentrated on the defence of India's north-west frontier. Yet none of these options was wholly realistic. Singapore was never adequately prepared, and in any case battleships on their own were far too vulnerable; the RAF was thinking about the wrong enemy altogether; and if India had ever faced a serious invasion Britain simply did not have the large number of troops that would have been required. Above all there was no equivalent to the coordination of planning and resources that Britain had undertaken before 1914. Nevertheless, as the 1920s wore on and tension was reduced by diplomatic means, it did appear that Britain was coping well despite the threats to her position.

Popular Opposition to War

Not surprisingly, the interwar period in Britain saw a strong reaction against the huge casualties suffered in the Great War; there was also much criticism of the prewar arms race and an unwillingness to pay high taxes to finance additional rearmament. This has often been interpreted as imposing a check on the government's defence policies and thereby forcing it to appease the fascist dictators. The origins of this view lie with some of the leading politicians closely associated with appeasement. Baldwin, for example, claimed that the public were wholly opposed to rearmament in the early 1930s and that the government could not obtain a mandate to change course until 1935. The Munich settlement was, and still is, defended in part on the grounds that it bought an extra year or two of peace in which Britain could rearm. The implication is that the politicians were invariably waiting for public opinion to turn around before they could be tough in their foreign policy.

However, this explanation originated as an excuse amongst discredited politicians, and must therefore be regarded with some suspicion; it has never been demonstrated that at the time governments were influenced primarily by public opinion in determining their policies. Since all governments, in spite of party-political disagreements, actually offered rather similar policies on disarmament and the League of Nations, there is no obvious means of estimating public reaction, and none of the general elections from 1922 onwards can be said to have turned on foreign or defence policy.

When assessing the manifestations of popular opposition to war it is important to take account of both the extent of their support and the depth of feeling. The outright pacifist organizations, for example, were quite small groups of Quakers and Socialists, often of very long standing. On the other hand, several much larger bodies of opinion adopted an anti-war stance without necessarily being pacifist. The Labour Party Conference regularly voted in favour of complete disarmament, and in Parliament the party opposed the military estimates until 1937. However, it cannot be assumed that these decisions, taken on the basis of the block votes cast by big trade unions, necessarily reflected the views of the ordinary members. Politicians showed some concern about the views of women – largely because they were a recently enfranchised group; certainly women were often considered to be naturally inclined to oppose war, though the experience of 1914–18 provided very little basis for such a generalization. Among the women's organizations the Women's Co-operative League was the most important one which actively propagated the cause of peace. On armistice day each year its members distributed white poppies in protest against the slaughter of the Great War. However, this has to be set against the larger numbers of people who marked armistice day in the conventional way. Perhaps the most significant and typical pressure group in this period was the League of Nations Union. Founded in 1918, the LNU boasted 225,000 members by 1925, rose to a peak of over 400,000 in 1931, and then declined to under 200,000 in the later 1930s. It was not a *pacifist* organization, however; rather, it represented middle-of-the-road opinion in all classes and all parties which wished to avoid the mistakes of the pre-1914 era and was anxious for the British government to cooperate in helping the League of Nations to resolve disputes peacefully.

It is probably significant that many of the best-known symptoms of anti-war sentiment were concentrated in a fairly short period in the late 1920s and the early 1930s. These were the years of the 'King and the Country' debate in the Oxford Union (1933), when the students voted in support of a proposal that 'this House will in no circumstances fight for its King and Country', the East Fulham by-election (1933), in which a Conservative majority of 14,000 was turned into a Labour one of 5,000 ostensibly because of the unpopularity of rearmament, and the so-called 'Peace Ballot' of 1934, in which 11 million people voted overwhelmingly in favour of further disarmament, even to the point of total disarmament. There was a flood of anti-war literature including Robert Graves's, *Goodbye To All That* (1929), Seigfried Sassoon's *Memoirs of an Infantry Officer* (1930), and Vera Brittain's *Testament of Youth* (1933) all of which, judging by their large sales, appeared to catch the mood of the British people at the time.

However, the meaning of these manifestations is complex and easily misunderstood. For example, the Fulham by-election probably turned on domestic issues rather than on foreign policy. In any case the seat was normally a marginal one, so that a government defeat was not very startling. This was only one of 40 seats defended by the National government, the majority of which were in fact retained even where candidates did support rearmament. In the case of the Peace Ballot, it is relevant to note that it was organized with the object of demonstrating to the government the extent of popular support for the League of Nations at a time when the withdrawal of Germany had placed a question mark over the future of both the League and the Disarmament Conference. Much the most significant question in the ballot was whether voters approved of the use of both economic and *military* sanctions against aggressor states; this was in fact supported by 6.7 million to 2.3 million. This was an indication that the public was not pacifist, and was in fact ready for an initiative, under the auspices of the League, to check the ambitions of the dictators. Thus the title 'Peace Ballot' is a misleading one.

What does seem clear is that public attitudes to war and peace fluctuated considerably during the interwar period. In 1918 voters were in a highly belligerent mood; and although the anti-war organizations gained fresh recruits, in the early 1920s they probably represented only politically active minorities. Later in the 1920s and in the early 1930s anti-war opinion evidently strengthened; this probably reflected the heightened hopes and fears aroused by the Disarmament Conference at Geneva and its subsequent failure. The rise of Hitler and the sudden appearance of the British Union of Fascists stimulated some rethinking amongst left-wing opponents of war, as well as confirming some right-wing critics of appeasement in their views. Then came a series of grave international crises, including Mussolini's invasion of Abyssinia in 1935, the remilitarization of the Rhineland, the outbreak of the Spanish Civil War in 1936, which severely weakened pacifism on the left, and the German advances into Czechoslovakia and Austria in 1938. The cumulative effect was to make war seem much more likely, and also a just and necessary risk to many people. One sign of the shift was the Labour Party's change of position during 1935–7; influenced by Ernest Bevin and Hugh Dalton, the party conference voted heavily in favour of 'collective security', which meant League of Nations sanctions against aggression, thereby precipitating the resignation of the pacifist party leader, George Lansbury. The moderate League of Nations Union lost membership at this time, some of which had concluded war was inevitable; the more thoroughgoing pacifists were attracted by a new body, the Peace Pledge Union, formed in 1936; by 1937 it had obtained pledges not to participate in a future war from 120,000 people.

How far interwar governments allowed themselves to be influenced by public attitudes is unclear. As we have seen, there was a limit to the extent to which they were prepared to disarm whatever the public may have wanted. Winston Churchill, who, as Chancellor of the Exchequer from 1924 to 1929, played a major role in keeping arms expenditure down, was not noticeably affected by popular views, any more than he was in the 1930s when advocating rearmament. It is abundantly clear that Lloyd George and the Conservatives paid lip-service to the League of Nations but had no intention of abandoning traditional methods in favour of open diplomacy. Baldwin's decision to appoint the popular Anthony Eden as Minister with responsibility for League affairs was a typically adroit move to keep on the right side of public opinion, but indicated no change of policy. All the key decisions were the result of 'expert' military and diplomatic advice.

A rather striking indication of the relationship between government and popular attitudes came in 1935. No sooner had Sir Samuel Hoare, the

Foreign Secretary, pledged Britain's support for the League than Mussolini decided to attack Abyssinia. The hollowness of Hoare's words was rapidly exposed when he offered to partition Abyssinia to satisfy Mussolini's demands. The public, having been led to expect a very different response, were outraged. Baldwin saved his government by sacking the Foreign Secretary; but he applied only half-hearted sanctions against Italy and dropped them as soon as the controversy had subsided. The truth was that the National government wished to appease Mussolini as a matter of policy, not because it thought the public opposed war. The whole episode suggested that the electorate was ready to take a tougher line with the dictators while the politicians were dragging their heels, not the other way around.

Later in the 1930s the government grew increasingly anxious to manipulate public attitudes, as voters began to give support to anti-appeasement candidates in by-elections. At Bridgwater in 1938, a Conservative appeaser was defeated by an Independent critic of Chamberlain's policy. Conservative Central Office and the Whips applied intense pressure to rebellious MPs, threatening to withdraw the whip and encouraging their constituency associations to select new candidates. One arch-opponent of appeasement, the Duchess of Atholl, resigned her seat under this pressure and fought a by-election in December 1938. She was only narrowly defeated by the official Conservative, despite being branded a warmonger. It was clear that by this time a large part of the British public were prepared to face the possibility of war, and that only Chamberlain's stubbornness stood in the way.

The Empire and Nationalism

In many ways British imperialists derived comfort from the experience of the First World War. India and the Dominions had contributed 2.5 million troops to the allied cause. The leaders of the white colonies had come to London to sit in the Imperial War Cabinet in 1917–18. Above all, the war not only strengthened imperial strategy by entrenching British control in the old Ottoman territory straddling the Mediterranean route to the East, it also extended imperial territory by the mandates over German East Africa, South-West Africa, and New Guinea. This, however, merely created an illusion of imperial strength and security. In reality the interwar Empire was weakened by economic depression, by nationalism, and by the loss of will and divided counsels at home.

Although the white colonies had accepted Britain's right to declare war on their behalf in 1914, their subordinate status was resented by much of the population of South Africa, Canada, and Australia. The effect of military participation was to strengthen their own nationalist pride as much as their loyalty to Britain. Australia, whose troops suffered heavy casualties in the Dardanelles, was inclined to be very critical of British military incompetence. Thus, in the aftermath of the war the dominions felt unwilling to pool their naval or military resources with Britain or to accept imperial federation. The most they would agree to was to meet at Imperial Conferences in 1921, 1923, and 1926. They insisted on being listed separately from Britain at the League of Nations; and their common refusal in 1922 to support Britain when she seemed about to plunge into a war with Turkey was a sign that they intended to pursue independent foreign policies in the future.

Consequently the British government accepted the necessity for a formal clarification and reassessment of the constitutional relationship between the dominions and the mother country. A new Dominions Office was established in recognition of the different status enjoyed by dominions by comparison with colonies. At the 1926 Imperial Conference Baldwin conceded self-government in external as well as in domestic affairs. The dominions were henceforth defined as 'autonomous communities within the British Empire, equal in status, in no way subordinate one to another in any aspect of their domestic or external affairs, though united by a common allegiance to the Crown'. In future the Crown would be represented by a Governor-General, and the dominion governments would deal with each other through high commissions. These changes were embodied in the Statute of Westminster in 1931. This redefinition of the meaning of 'dominion status' had important repercussions: while it pleased the white colonies, it greatly complicated the question of constitutional reform in India.

Elsewhere nationalism began to have disintegrating effects upon British control. In the Middle East, for example, where Britain had acquired responsibility over large tracts of territory, she had given undertakings to the Arabs of independence from the Turks, and in 1917 had made the famous Balfour declaration promising a 'national homeland' for the Jews. Consequently she found herself obliged to station troops to deal with the insurrections by nationalists whose ambitions had not yet been satisfied. In 1922 Britain granted Egypt a strictly limited independence which left her in control of Egyptian defence and the Suez Canal.

However, for British politicians much the most traumatic episode of the early 1920s was the renewed challenge of nationalism in Ireland. As a result of the Easter Rebellion of 1916, the Irish Home Rulers and their parliamentary strategy had been discredited, and in the 1918 elections the Sinn Fein swept the board outside Ulster. A Dáil or parliament for the 'Irish Republic' had been established under the presidency of Eamon de Valera, and from 1919 the Irish Republican Army waged a terrorist campaign against the British Army and the Royal Ulster Constabulary. Lloyd George felt obliged to employ an irregular force known as the 'Black and Tans' to counter the IRA's guerilla tactics, and while his own Tory diehards refused to abandon their opposition to Irish demands the Prime Minister became hopelessly enmeshed in a bloody conflict. In 1921 the government set up a new parliament in Belfast, based on the Westminster model, to run the internal affairs of Ulster. But it was not until 1921 that Lloyd George accepted that physical force was never going to settle the problem of republican Ireland. He had therefore to risk defying the reactionary Conservatives on whom his parliamentary majority depended. In December, agreement was finally reached on the establishment of the Irish Free State, which was granted the status of a Dominion under the Crown. Power was formally transferred in March 1922. In effect the bulk of Ireland now became an independent country, though there was still 'common citizenship' and Irish Free State citizens were allowed to vote in UK elections.

Although the ancient controversy over Home Rule now speedily subsided, the settlement had the immediate effect of undermining support for Lloyd George's coalition, which collapsed in October 1922. It also had wider repercussions in the sense that some Tories felt embittered and shocked at the final severance between Britain and Ireland. They could not help seeing this victory for nationalism as the start of a policy of 'scuttle', and were consequently all the more sensitive to the major challenge now being posed by the nationalists of India.

Although a nationalist organization in the form of the Indian National Congress had existed since 1885, most British officials regarded it, with some justification, as an anglicized elite with little following in the country. They comforted themselves with the thought that India comprised too many religious, social, and regional communities ever to be capable of generating a united movement for self-government. However, from the First World War onwards, this assumption was to be thoroughly undermined. During 1916 and 1917 Hindus and Muslims began to collaborate in order to pressurize the British, and fresh popular interest was aroused by the Home Rule Leagues. This development stimulated a major concession by the Liberal Secretary of State, Edwin Montagu. In 1917 he offered to involve Indians more closely in the administration 'with a view to the progressive realization of responsible government in India as an integral part of the British Empire'. This historic declaration was subsequently translated into the Montagu–Chelmsford reforms, which greatly extended elected Indian representation in provincial governments and gave them effective control over a limited number of departments. The new system was to be reviewed after ten years with a view to introducing further reforms.

This strategy was intended to encourage the moderate nationalists to continue their cooperation with the British government. However, it failed for two main reasons. First, Montagu's statesmanlike approach was undermined both by British officials in India and by reactionary politicians at home, who chose to believe that the British Raj would last forever if only a firm policy were pursued. During 1919 official overreaction to popular unrest in the Punjab led to the imposition of martial law and thus to the notorious 'Amritsar Massacre', when troops under the command of General Dyer shot and killed nearly 400 Indians who were not engaged in political activity. This was an unnecessary blunder, for which Dyer was quite properly sacked. Unfortunately a great deal of support was shown for him in Britain, which served to antagonize Indians and to weaken the credibility of those who wished to cooperate with the British.

The other key development was the remarkable transformation of Congress under Gandhi's inspired leadership in the early 1920s. He shrewdly focused Indian attention upon specific, material grievances, successfully engineered collaboration between Muslims and Hindus, and effectively created a mass movement for the first time. Nationwide campaigns such as

Non-cooperation in 1920–2 and Civil Disobedience in 1930–1 severely shook the morale of the British officials, and gave Congress a mass membership and an organization in every part of India. From 1920 onwards Congress was able both to organize massive campaigns of disruption and to win elections all over the country.

The crucial stage in the struggle came in 1928–31, under Lord Irwin's viceroyalty. Congress was by now pressing for Dominion status which, as a result of the 1926 Imperial Conference, effectively meant independence – rather more than the British were prepared to concede. Gandhi led a Civil Disobedience campaign which eventually came to an end when Irwin took the initiative by releasing Congress leaders from jail and negotiating directly with Gandhi. This led to a series of compromises, including an undertaking by Gandhi to attend the Round Table Conference in London. On balance the deal seemed to represent a gain for the British side. However, British official opinion was outraged and demoralized by the Gandhi–Irwin Pact because, by treating Gandhi as the effective spokesman for India, the Viceroy had raised his status greatly. Moreover, although nothing came of the Round Table Conference, Baldwin and the National government accepted the necessity to grant further constitutional reform. Eventually in 1935 Sir Samuel Hoare succeeded in enacting the Government of India Act, which gave Indians effective self-government at the provincial level and majority representation in the central government. This measure proved to be of major importance in two ways. In India it led to fresh elections in 1937, in which Congress made sweeping gains; thereafter, Congress governments were formed in the majority of provinces. In Britain, however, the act was the object of a bitter and prolonged campaign in which liberal Conservatives like Baldwin, Hoare, and Irwin were attacked by the diehards led by Winston Churchill. The Tory critics argued that Congress was unfit to govern and unrepresentative of India – a claim shortly to be exploded. They also contended that once Britain ceded control in India she would rapidly fall from the ranks of the great powers. Yet while Churchill and his allies enjoyed considerable support within the rank-and-file of the Conservative Party, they failed to deflect the National government from its objective. This was a decisive defeat for the defenders of the British Empire. There was never to be so stern a struggle again.

The interwar period also witnessed more subtle challenges to Britain's position in India. One symptom of the declining will to rule was the publication of E. M. Forster's famous novel *A Passage to India* in 1924. At the time this was regarded as a thoroughly subversive work because of its depiction of friendship between British and Indian people. In 1934 George Orwell's *Burmese Days*, coming as it did from a British official, suggested that many of the British themselves now detested their system of rule. Indeed, beneath the public controversy over reform important changes were taking place in the administration of India. During the war a high proportion of the members of the Indian Civil Service had left the country. Thereafter it proved difficult to recruit young men into what had once been a most prestigious career. Thus, during the 1920s and 1930s Indians joined the service in growing numbers, such that by 1939 they comprised almost half of the ICS. One result was that new British recruits often worked under Indian officers or alongside them. In this situation they inevitably adjusted their attitude, if they had not already done so. The old confidence in Britain's role steadily crumbled under the pressure of this experience. Moreover, taken in conjunction with the political role of Congress under the 1935 Act, the changes in the ICS meant that Indians were to a considerable extent actually governing themselves before the Second World War. The ground for full independence was thus well prepared.

This period also brought important changes in the economic relationship between Britain and India. For example, the costs of both British and Indian troops in India had traditionally been met from the Indian reserves even if the troops were used outside India. By 1918 the British government was already paying for Indian troops when involved outside their own country; but from 1933 it also subsidized the cost of British soldiers stationed in India. The country was ceasing to be an asset to Britain, and after 1940 it became a huge military liability.

More importantly, during the 1930s the advantages Britain had traditionally derived from trade with India dwindled. The war itself had greatly stimulated Indian manufacturing industries; then, as a result of the interwar depression she became increasingly self-sufficient, introducing protective tariffs and gaining control over her exchange rates. Consequently, by the 1930s India was exporting more to Britain than she was importing from her. In the economic sphere she had already moved towards a position of independence.

Faced with these developments in India, British imperialists like L. S. Amery cast around for a fresh strategy. Amery always felt that Winston Churchill and his followers were too sentimental and narrowly political in their approach; they failed to appreciate the underlying *economic* potential of the colonies. Africa not only held extensive mineral and agricultural resources, it offered opportunities for emigration and was not beset by troublesome nationalist movements. As Colonial Secretary under Baldwin from 1924 Amery contributed to a number of initiatives designed to strengthen the economic basis of the empire. An Empire Marketing Board was created to encourage the public to buy more produce from the colonies; there was an Empire Settlement Act in 1928 and, in 1929, a Colonial Development Act whose object was to promote the construction of the railways and harbours needed if Africa's economic potential was to be effectively exploited.

What effect did all this have? The popularity of the great British Empire Exhibition at Wembley in 1924 and 1925 suggested that the British were still a vigorously imperial people. But this was only superficially true. In spite of the depression at home and the official encouragement, only 130,000 people per annum emigrated on average during the 1920s, which was well below the pre-1914 level. By the 1930s more British people were returning than were emigrating. Those that did leave went largely to the white colonies, and only a few thousand to Kenya, Rhodesia, and South Africa, where Amery hoped to build up a large British population so as to ensure the long-term future of the Empire. Falling prices made it very difficult for those who emigrated to establish profitable plantations in Africa. Nor was the British government willing to devote resources on the required scale to these territories. Even in the Conservative Party, Amery was increasingly a peripheral figure leading a failing cause. Like the country as a whole, the politicians turned inwards to concentrate on domestic problems rather than striking out boldly for colonial development.

This is borne out by other Conservative policies affecting the colonies. For example, in pursuit of the appeasement of Hitler, Chamberlain, Baldwin, and Lord Halifax (the former Irwin) made it clear that they were willing, indeed anxious, to restore Germany's colonies to her. Clearly the mandates were not regarded as particularly valuable by Britain; but extensive territories were involved and their cession would have been another blow to the imperial cause. Though the policy was not carried out, it provides a telling indication of government priorities; colonies were assets to be disposed of if they could serve the more important object of improving Britain's relations with the Great Powers in Europe.

Appeasement and Rearmament

Although the appeasement of Germany is indelibly associated with Neville Chamberlain's premiership after 1937, this was only the climax of a policy pursued throughout the interwar period. It originated in the conviction, shared by Lloyd George himself, that the Treaty of Versailles had been too harsh on Germany. During the 1920s all governments engaged in acts of appeasement. For example, German reparation payments were steadily reduced before being abandoned, she was assisted with her currency problems, and the British helped to reverse the French occupation of the Ruhr in 1923. In 1925 Germany became a signatory to the Locarno Agreements which, by guaranteeing the boundaries of Western Europe while conspicuously ignoring those in the East, virtually invited a further revision of Versailles. Then in 1926 the allied powers made an early withdrawal from the Rhineland and Germany was admitted to the League of Nations.

In this period, appeasement represented an enlightened policy pursued from a position of strength. It was founded on the expectation that if Germany's humiliation was expunged and her economy restored, the new parliamentary regime under the Weimar Republic would survive; consequently the causes of war before 1914 would not arise. However, it is by no means certain that the limited forms of appeasement undertaken at this time could really have satisfied even the respectable politicians of Weimar Germany. Appeasement stopped well short of restoring Germany to her former military strength or of permitting the racial unity of the German people. German grievances continued to fester and the Weimar regime was undermined. Thus the question arose whether appeasement should be taken further.

Ramsay MacDonald, who became Prime Minister for the second time in 1929, recognized that the growing discontent in Germany over the low level of her armed forces imposed on her by the Treaty of Versailles called for some response. The hope was that, if the Western powers took their own disarmament further, the situation would stabilize. At the 1930 London

Naval Conference, it was agreed to extend the moratorium on capital ships to 1936 so that by the mid-1930s the ratio of British to Japanese capital ships would be only 15:9, a further deterioration from Britain's point of view. For fear that the Germans might renounce the military clauses of the Treaty of Versailles, a fresh initiative was taken in the form of a Disarmament Conference which met at Geneva from 1932 to 1934. This crucial period coincided with the Foreign Secretaryship of Sir John Simon, widely considered the least effective occupant of that office in the twentieth century. The congenitally indecisive Simon had the misfortune to be in office when events began to move fast. In autumn 1931 Japan flouted the League of Nations by invading Manchuria and then withdrawing from both the League and the Disarmament Conference. Britain did nothing, but the chiefs of staff advised the government to restart work on the base at Singapore and to abandon the Ten-Year Rule. Concern increased when Hitler came to power in 1933, and soon afterwards took Germany out of the League and the Disarmament Convention. It was now clear that Germany would try to accelerate the process of revising the Treaty of Versailles. At Locarno the Western powers had already signalled their readiness for her to do so in Eastern Europe. Thus British politicians in general took the view that the fall of Weimar required a more urgent application of appeasement, not that it had destroyed the logic of the policy.

However, their strategy was intended to be a combination of appeasement and rearmament. Britain now faced potential challenges from Germany, Italy, and Japan, each in a different part of the world; for the chiefs of staff it was now urgent that Britain's diplomacy be properly matched by military plans and resources. The result was the establishment of a Defence Requirements Committee under Sir Maurice Hankey in November 1933. It concluded, rather like the prewar Committee of Imperial Defence, that Germany, not Japan, must be regarded as Britain's most serious enemy. The scheme recommended to meet the danger involved recreating an expeditionary force of five regular divisions and expanding the RAF to 1,736 first-line aircraft, the majority of which would be bombers. The idea was to stop Germany seizing Belgium and Holland and using airfields there from which to bomb Britain; at the same time, Britain would be capable of threatening the Germans with a counter-offensive in the air.

If properly adopted, this might have enabled Britain to pusue a strategy of responding to Germany's grievances from a position of strength. The leading appeasers claimed that the purpose of their policies in the later 1930s was to buy time in which Britain could rearm effectively. This is not, however, consistent with their practice. For several years the government was most reluctant to finance rearmament, and Neville Chamberlain, as Chancellor of the Exchequer, strongly resisted the Defence Requirements Committee proposals on the grounds that extensive government borrowing would jeopardize the fragile economic recovery. He largely got his way; the extra £97 million of expenditure proposed was cut back to £59 million. British defence spending rose from 2.7 per cent of gross national product in 1933–5 to just 2.8 per cent in 1934–5 and 3.0 per cent in 1935–6. In the circumstances this was little more than a token effort; serious rearmament did not begin until 1936–7.

However, the important thing was not just the amount of money spent but the type of rearmament adopted. While willing to expand the RAF, the government remained adamantly opposed to a new expeditionary force. The result was a dangerously unbalanced form of rearmament which had the effect of undermining British diplomacy in the later 1930s. The French and the Russians were dismayed that Britain still lacked the capacity to put an army on the Continent; this inevitably made them much more cautious about resisting Hitler's demands. For his part, Hitler was fortified by the knowledge that there was unlikely to be any effective intervention by Britain. The point was underlined in 1936 when he remilitarized the Rhineland. It was vital for the Western powers to maintain the strategic advantage of a demilitarized zone on France's vulnerable eastern frontier. Its loss weakened Britain's capacity to negotiate still further, and seriously demoralized the French.

During 1935–6 a second element of the government's appeasement policy also disintegrated. Britain and France had hoped to capitalize on Italian fears about German designs on Austria by drawing closer to Mussolini, thereby keeping Hitler isolated. Unfortunately, they were so friendly towards Mussolini that he concluded it would be safe to pursue his long-standing ambition to seize Abyssinia. His judgement proved correct in the sense that the initial reaction of the British and French was to offer the Hoare–Laval Pact, a scheme to partition the country in Italy's favour. Only a popular outcry against this forced the governments into an embarrassing retreat, and into the application of very lame sanctions against Italy. This simply had the effect of antagonizing Mussolini and further discrediting the League of Nations. In addition, the Abyssinian crisis had major repercussions for British diplomacy. The joint front of Italy, Britain, and France collapsed, and Mussolini was drawn into the Rome–Berlin Axis; as a result Hitler calculated that he could now take the risk of moving his troops into the Rhineland. He also embarked upon a major rearmament in defiance of the Treaty of Versailles. The effect of all this was to shift the military-strategic balance in Germany's favour. France, recently so strong, saw her position threatened by two fascist powers on her frontiers; the outbreak of civil war in Spain added a third.

This was the rapidly deteriorating situation facing Neville Chamberlain on becoming Prime Minister in 1937. By comparison with the lackadaisical Baldwin and the indecisive MacDonald, he seemed to represent a marked improvement. Chamberlain took the view that the German leadership had a strictly limited list of outstanding grievances; the sooner these were resolved the sooner the tension in Europe would be lowered. If this involved an extension of German control in central-eastern Europe this was a price well worth paying, for it would restore stability to a region of undesirable little states now vulnerable to Soviet expansion. Contrary to the traditional view, Chamberlain was by no means ignorant of foreign affairs; but he was arrogant, and found it difficult to abstain from interference in the work of his colleagues. Consequently he came to rely upon a narrow group of colleagues who shared his opinions – Hoare, Halifax, Simon, and the Ambassador in Berlin, Sir Neville Henderson. Those who differed, such as Sir Robert Vansittart, the Permanent Secretary at the Foreign Office, were pushed out. Chamberlain began to pursue his own foreign policy because he considered that Sir Anthony Eden and the Foreign Office were too hostile to Mussolini; unable to cope with being undermined in this way, Eden resigned as Foreign Secretary in 1938.

For several years Hitler was systematically led to believe by both political and diplomatic communications that the British government would never ultimately resist him, and would effectively help him towards further territorial gains. Thus, when he occupied Austria in March 1938 Britain adopted the view that the German people there were fully entitled to join their fellow nationals, and the sooner the *Anschluss* was accomplished the better. Hitler next demanded the Sudetenland districts of Czechoslovakia, populated by 3.5 million Germans. Now Chamberlain took the initiative in bringing pressure to bear on the Czechs themselves and on the French, who had treaty obligations to defend Czechoslovakia, by warning that they could expect no support from Britain if they chose to resist. Secondly, he made his famous visits to negotiate directly with Hitler on an agreed partition of Czechoslovakia, culminating in the Munich Settlement on 30 September. In fact Britain came close to war at this time. But Chamberlain was determined to avoid it: 'How horrible, fantastic, incredible it is that we should be digging trenches and trying on gas-masks here because of a quarrel in a far-away country between people of whom we know nothing.' At the time, the case for another act of appeasement looked strong. First, it was argued that Britain had no vital national interest at stake in a war over the boundaries of Czechoslovakia or, indeed, any of the other East European states which were so disliked by British statesmen. Second, Britain was said to lack the effective military means to intervene even if she wanted to; nor were the Czechs, the French, or the Russians able and willing. Both arguments had been used on every occasion when the question of resisting the dictators came up. It was all too easy to minimize the importance for Britain of any individual act of aggression and fail to see the cumulative effect. Similarly the military strength of Britain's opponents was systematically exaggerated and that of her potential allies minimized. In 1938 the Air Ministry claimed that Germany had 2,909 front-line aircraft to Britain's 1,550; in fact they were nearly equal in number. On his return from Munich claiming 'peace with honour', Chamberlain enjoyed a fleeting popular triumph. But this collapsed in March 1939, when Hitler moved his troops into Prague and dismembered even the non-German parts of Czechoslovakia. By now the futility of Chamberlain's policy was widely recognized; he had been attempting to appease an opponent who was not really appeasable. Yet although the Prime Minister made a radical departure by proceeding to give guarantees to Poland and Romania, the next victims of Hitler's demands, he evidently continued to believe that further deals would obviate the need for war. When Hitler began to put pressure on the Poles the old argument was wheeled out: was it worth shedding British blood for Danzig? However, by this time Lord Halifax, for one, had recognized the folly of further appeasement and Chamberlain could not afford to lose a second Foreign Secretary. Even so, he prevaricated. When he addressed the House of Commons on 2

September 1939, by which time German troops had spent several days on Polish soil, he amazed MPs by failing to announce a declaration of war. Only the combination of dissension in the Cabinet and rebellion among the MPs forced him to do so the next day.

Part IV

Consensus: The Age of the Benign State, 1940–1970

15

The People's War

When Neville Chamberlain addressed the House of Commons on 2 September 1939, it was widely anticipated that he would announce a British declaration of war upon Germany. Commitments recently given to Poland and Hitler's fresh aggression against that country appeared to leave the Prime Minister with no choice. His failure to declare war shocked Parliament, and was taken as a sign that his heart was not in the cause. As a result, the remaining months of Chamberlain's premiership became known as the 'Phoney War'. It seemed possible that, if Britain avoided drawing German fire in the west, Hitler's success in Eastern Europe might bring the whole affair to an end without the necessity for major conflict on her part. The leading appeasers continued to entertain such hopes for some time, but the effect was to undermine confidence in Chamberlain's capacity as a war leader. While the country and the political parties were remarkably united about the justness of the war against fascism, the Labour Party remained bitter towards Chamberlain as a result of both his earlier policies and his arrogance. The Prime Minister was thus patently unable to act as a unifying force in the war; his lack of determination only exacerbated the problem.

Breaking the Mould

The first indication of a breakdown of the interwar pattern of politics came with the refusal of the Labour and Liberal Parties to join Chamberlain in a coalition government, although, following the precedent of the First World War, they agreed to a by-election truce. However, some steps were taken to put the government onto a war footing. New ministries appeared for Home Security, Economic Warfare, and Food and Shipping, and Winston Churchill accepted the prominent – but vulnerable – post of First Lord of the Admiralty. Yet the leading appeasers, Simon, Hoare, and Halifax, remained in office, while the nine-member War Cabinet soon became bogged down in arguments. The situation was not unlike that faced by Asquith's 1914 administration, in which a peacetime system was marginally adapted for war.

Before long, Chamberlain's plans for damage limitation were disrupted by two unpredictable protagonists: Hitler and Churchill. Within a matter of weeks the Poles had succumbed to German invasion from the west and Russian advances from the east. Meanwhile neither the French army nor the British Bomber Command took the opportunity of German absorption in the east to seize the initiative. Four infantry divisions were dispatched to France, the Royal Navy began to apply a blockade, and the RAF scattered propaganda over the German mainland. But it was all an anticlimax. By the beginning of 1940 the situation at home and abroad was beginning to deteriorate. Chamberlain made the mistake of removing a popular Minister, Leslie Hore-Belisha, which provoked the criticism of the press, already restless for lack of good copy. The Gallup polls began to record a slide in the Prime Minister's popularity. Churchill alone generated an air of purposefulness and excitement, but as in the previous war he showed a propensity for inventing wild military schemes. Chamberlain may well have calculated that he would discredit himself in this way. Such a prospect rapidly materialized as Churchill grew enthusiastic about sending troops to the Norwegian port of Narvik and mining Norwegian waters, with a view to checking German access to iron-ore supplies. This was both politically and militarily unrealistic, but eventually Churchill overcame his colleagues' opposition to the idea. In the event the campaign was incompetently managed, and Hitler moved in quickly to overrun the country. Although large amounts of German shipping were destroyed, allied forces suffered heavy casualties and had been driven back from Norway by April.

Although Churchill deserved to be blamed for this, it was Chamberlain who suffered. His record over rearmament and appeasement in the 1930s made him vulnerable, and the Norwegian campaign gave his critics an opportunity to attack him. In Parliament Conservatives like L. S. Amery and Lord Salisbury, and the Liberal Clement Davies, joined forces, backed from outside by the *Daily Mail*. The Labour opposition tabled a motion on the Norwegian campaign for debate on 7–8 May. Several weighty figures from the past, including Lloyd George and Admiral Sir Roger Keyes, joined the condemnation. Amery resurrected some famous words of Oliver Cromwell: 'You have been sitting here too long for all the good that you have been doing. In the name of God, go!' In the vote, the government's majority sank from well over 200 to 81. Even so, Chamberlain might have saved his premiership if the opposition had agreed to enter a coalition; but they refused, and he had to resign.

In this situation Lord Halifax was the man most generally acceptable to the Conservative Party, the Labour Party, and the King. But his reputation as an appeaser left him with less standing with the public, and Halifax chose to use the excuse of his membership of the House of Lords to back down in favour of Churchill. Thus at the age of sixty-six Churchill suddenly saw his fading career given a new lease of life. His elevation proved to be a turning-point, both because it destroyed the hegemony of the National government and restored Labour to power, and because it coincided with disastrous events abroad which were to leave a deep impression on the public mind.

In a way, Churchill's political position resembled that of Lloyd George in 1916; for as Prime Minister he was not the leader of a party, and had come to power only over the ruin of his leader's reputation. Like Lloyd

George, he formed a three-party coalition and established a small War Cabinet. But at first the break was not too sharp; Chamberlain, Halifax, and Simon retained high office. Several anti-Chamberlain Tories, including Eden, Kingsley Wood, Lord Lloyd, L. S. Amery, and Duff Cooper, joined the government, as did non-party figures such as Sir John Anderson (Home Office) and Lord Woolton (Food). From the Labour ranks Attlee and Arthur Greenwood became ministers without portfolio, Ernest Bevin became Minister for Labour, Herbert Morrison Minister for Supply, and Hugh Dalton Minister for Economic Warfare. The effect, by 1945, was to re-establish Labour as competent and patriotic in government after the damage done to the party's reputation in the 1930s.

Churchill enjoyed an advantage over Lloyd George in his relationship with the military authorities. He did not suffer sustained opposition from the generals, though this largely reflected the fact that Germany's sweeping success on the Continent left the British with far less room for manœuvre than in the 1914–18 war. Moreover, Churchill made himself Minister for Defence, thereby subordinating the three service chiefs. And although the chiefs of staff found him exhausting and unpredictable, they recognized the value of his own experience at both the War Office and the Admiralty, and contented themselves with restraining his wilder notions. For his part, Churchill was somewhat sobered by the reflection that with his previous record of military setbacks he could easily become the victim of the next fiasco.

Churchill's entry into office certainly coincided with a rapid deterioration of the military situation. Although the 400,000 troops of the British Expeditionary Force, combined with the French, Belgian, and Dutch, were roughly equal in numbers to the German armies facing them, their quality and staff work were inferior. The Allies misjudged German intentions. Following a main attack through the Ardennes, a breakthrough in the Allied line caused them to retreat towards the coast; by the end of May Britain was planning an evacuation from Dunkirk. This was well chosen, because it was close enough to be within range of fighter planes based in England. Some 200,000 troops were evacuated on 2 and 3 June alone, and eventually 338,000 made their escape. But by the end of the month the French had been forced to capitulate, leaving much of the north and west of the country in enemy hands and thereby exposing Britain to an acute danger of invasion. 'Personally', noted the King, 'I feel happier now that we have no allies to be polite to and pamper.'

However, the government now prepared for the worst. Male enemy aliens were interned, Sir Oswald Mosley and BUF members were imprisoned, and men aged between seventeen and sixty-five were recruited into a Local Defence Volunteer Force. But in spite of the military crisis, there was little sign of demoralization except among some leading politicians. The conventional view has been that public morale was boosted by firm official action and by Churchill's inspiring broadcasts. 'I expect the Battle of Britain is about to begin,' he declared on 18 June, 'Let us brace ourselves to our duty, and so bear ourselves that if the British Empire and its Commonwealth last for a thousand years, men will still say, "This was their finest hour".' In fact, public opinion was much more mixed. The surveys conducted by Mass Observation suggested that Churchillian rhetoric had a limited impact, particularly on women. As a group women were more resigned, less involved, and less optimistic about the war than men. This may have been partly because they bore the brunt of the conflict in terms of its disruptive effects on normal family life; they resented the evacuation of children, food queues, the blackout, the deterioration in the housing stock, and the loss of a social life. Certainly much of the population became indifferent or cynical towards the massive propaganda effort of the Ministries of Information, Food, and Labour. Official advice about the danger of giving away vital information by gossiping was regarded as ridiculous, and material on the economical use of food was either unread or ignored. The propagandists clearly had difficulty in judging the response to their message. The Ministry of Information's poster bearing the slogan 'Your Courage, Your Cheerfulness, Your Resolution Will Bring Us Victory' attracted so much hostility that it had to be withdrawn.

In a sense, the authorities were suffering not simply from their current policies but from accumulated resentment over their prewar incompetence. The most famous manifestation of this was the book, *Guilty Men*, published in 1940 and reprinted ten times in that year alone. A satirical attack on the leading appeasers, Chamberlain, Hoare, and Simon, it struck a chord with the public and continued to be discussed up to 1945. In the short term *Guilty Men* contributed to the feeling that the country was still in the grip of the kind of men who had called for sacrifices in 1914–18 and let the country down subsequently. The view that in the current war sacrifices ought to be equally shared amongst the population was echoed in J. B. Priestley's radio broadcasts and in the pages of the *Daily Mirror*, a newspaper then coming into its heyday on a timely blend of patriotism and social radicalism. One immediate effect of this critical mood was to facilitate interventionist policies, such as food-rationing, which would introduce an element of equality into wartime experience. The feeling was reflected in an observation by the Queen in the aftermath of a bombing raid which damaged Buckingham Palace: 'Now we can look the East End in the face.' But in a more profound way the crisis was helping to dissipate the political resignation typical of the 1930s and restore a spirit of irreverence which eventually ushered in a more left-wing agenda.

Mass War and Social Change

The Second World War focused attention upon the civilian and his living standards rather more effectively than the Great War had done. This was partly the result of military events. By the summer of 1940 Allied forces had been chased out of Europe, and for some time Britain's direct military participation was very limited. Conversely, the defence of the British Isles and the morale of the population at home moved to centre stage. In any case, the prewar propaganda had prepared people for a massive bombardment from the air – hence the early decision to evacuate 1.5 million children from vulnerable urban areas to the countryside. Although these fears proved to be exaggerated, in that the population became inured to regular bombing raids, some 60,000 civilians were killed and 4 million houses, representing about a third of the total housing stock, suffered damage.

The war also impinged upon civilians by stimulating the demand for labour, and succeeded at last in pushing unemployment below 10 per cent. However, even this proved insufficient; by the beginning of 1941 the government's Manpower Requirements Committee advised that an extra 2 million workers were required. Clearly, they would not materialize simply on the basis of the exhortation and propaganda used so far. Both politicians and trade unions showed as much reluctance as in 1914–18 for the obvious solution – the recruitment of more women. But by 1941 Bevin could see no alternative. Women were required to register in successive age groups at their local labour exchanges, where they would be allocated employment. By 1943, 46 per cent of all women aged between 14 and 59 were doing paid work for the war effort. The conscription of women was a striking achievement, not emulated by either Nazi Germany or Stalinist Russia, and made a material impact on the war by enabling Britain to mobilize a higher proportion of her economic resources. It underlined the superiority of a liberal parliamentary system over autocracy in rallying the people around the national interest.

The success of this policy is underlined by the evidence that on the whole women, outside the younger age groups, were not enthusiastic about being conscripted. To some extent, however, the authorities responded to women's problems by providing day nurseries and adjusting shifts so that housewives could fit in their shopping. Some 470,000 young women joined the three armed forces during the war, while thousands more served in the Auxiliary Territorial Service, the Land Army, Air Raid Precaution, and Auxiliary Fire Service. As in the First World War, the uniform was widely seen as a form of emancipation in itself.

For many men this sudden rise in status for women proved to be disorientating. There was a reaction against the appearance of large numbers of young, independent women, with money in their pockets, in public houses and other public places. This led inexorably to claims about deteriorating moral standards during wartime for which women were blamed. The government maintained its own double standard by distributing free condoms to soldiers to protect them against venereal disease, while leaving uniformed women unaided to risk pregnancy. Of course some married women, who faced lonely separation from their husbands, did pursue affairs, while many young women enjoyed brief relationships with American or Canadian troops. The result was a number of hasty marriages and a doubling of the illegitimacy rate to 9 per cent of all births by 1945. Five times as many people sought divorces after the war as before, and 58 per cent of divorce petitions were now lodged by men.

However, middle-aged married males were alarmist in interpreting all this as a collapse of traditional moral values. Extramarital affairs resulted from the boredom and loneliness of wartime life; they did not signify a revolt against marriage. Indeed, the Mass Observation surveys at the time underlined that for the majority of women marriage and motherhood remained their overriding goals. Though the marriage rate fell during 1941–5, thereafter it swiftly rose above prewar levels and continued to rise until the early 1970s. This helped to keep the birth rate buoyant and led to the 'baby boom' of 1946–8. By the end of the war men and women were anxious to return to a settled family life. In particular, Mass Observation reported that three-quarters of women workers wanted to drop their employment as soon as possible.

The coalition government recognized that the central part played by civilians virtually dictated not just promises about future social improvement but considerable innovation in social policy during the war. This manifested itself in four main ways. First, a system of rationing was applied to most food items except bread and potatoes, and free or subsidized milk, orange juice, and cod-liver oil was supplied to young children and expectant mothers, thereby helping to improve the health of the most vulnerable groups. Second, the government paid pensions and allowances to families whose income had been lost through war service by the head of household, and in 1941 they abolished the unpopular family means test. By 1945 a bill to introduce family allowances was going through Parliament. Third, they created the Emergency Medical Service from the existing patchwork of hospitals – a step towards the National Health Service. Fourth, as a result of pressure from Bevin, wage levels were raised for certain low-paid groups, including agricultural and railway workers. The general effect of labour shortages and the availablity of overtime was to drive average weekly wages up from around 53s. in 1938 to 96s. by 1945. An 80 per cent increase in wages, in the context of a 31 per cent rise in the cost of living, left most families better off.

Much the most striking indication of the mood engendered by the external crisis came in the response to the famous report of Sir William Beveridge in December 1942. By the time he was appointed to chair the government's Committee on Social Insurance in June 1941, Beveridge knew that the idea of a general reconstruction of social-welfare provisions enjoyed wide support amongst political, professional, and academic circles. In many ways his report was a consolidation of 'middle opinion' of the 1930s. For example, Political and Economic Planning pre-empted him with proposals for a national minimum wage, family allowances, and a national health service. Beveridge enunciated a sweeping plan to conquer the five giants – Want, Ignorance, Squalor, Idleness, and Disease. Government would be responsible not only for social insurance 'from the cradle to the grave', a health service, and family allowances, but also for maintaining the economy such that unemployment would not exceed 8 per cent.

If the scope of this was sweeping, its character was not revolutionary. Beveridge's ideas, after all, were moulded by Edwardian and Victorian National Efficiency ideology. His plan was based on the principle of *insurance*; contributions would bring entitlement to a national minimum income, but for those who fell outside the range of guaranteed benefits there was to be a scheme of National Assistance. And though Beveridge's proposals were to be of great advantage to women, his scheme recognized women essentially as wives and mothers. He had taken over the idea of family allowances, in spite of the original feminist thinking behind it, because it was now an urgent national interest, in view of falling birth rates, to promote larger families.

Essentially, the report crystallized opinion half-way through the war; it sold 100,000 copies in the first month and 630,000 in all. In this way it helped politicians to recognize the force of public opinion and thereby ensured that the Second World War would have lasting effects rather than the largely ephemeral impact of the 1914–18 war. Churchill and Kingsley Wood, the Chancellor, became worried about the popularity of the Beveridge Report, because it seemed likely to raise excessive expectations and distract the public from the tedious necessity for continued privations and discomforts. However, though cautious over the cost of the programme, they judged it expedient to show their commitment to social improvement. One sign of this was R. A. Butler's 1944 Education Act. In the same year Beveridge published his *Full Employment in a Free Society*, and government white papers appeared on Social Insurance, a National Health Service, and the famous one on Employment Policy which committed the government to Keynesian techniques for adjusting the level of public spending so as to maintain a high rate of employment. The following year brought legislation for family allowances which offered 5s. a week for second and subsequent children. By excluding the first child, this scheme departed from the feminist idea of the payments as a recognition of the value of every mother's work; moreover, the government's scheme proposed to make the payments to fathers, though Eleanor Rathbone forced them to back down on that point. Taken as a whole, the reform achieved during the war, and the momentum building up for further reform by 1945, ensured that social change would be underpinned by political change and thus sustained into peacetime in a way that had not been true after 1918.

Coalition and the Origins of Consensus

The experience of the First World War made the adoption of state controls designed to mobilize national resources easier and less fraught during 1939–45. The civil service expanded from 387,000 to 704,000; the Ministry of Labour took draconian powers to direct workers to needy sectors; and the government used the radio and, in effect, the press as vehicles for official propaganda, though stopping just short of formal censorship. Inevitably, government expenditure rose enormously, from £1.4 million in 1939–40 to £6.1 million by 1944–5, of which £5.1 million was for defence. But as early as 1941 Britain was exhausting her resources. Lord Lothian, the ambassador in Washington, had cheerfully told American reporters in 1940: 'Well boys, Britain's broke; it's your money we want.' The US government cooperated by agreeing to a series of 'Lend-lease' arrangements to enable Britain to maintain her war effort until 1945, though a high price was extracted for this assistance. By the end Britain had practically exhausted her reserves of gold, dollars, and overseas investments, and her debt had grown from nearly £500 million in 1939 to £3,500 million. However, though Lend-lease was an essential expedient, its importance should not be exaggerated; Britain largely financed the war effort from her own resources. The first war budget had raised income tax from 5s. in the pound to 7s. 6d., and it subsequently rose to 10s. In 1943 the Pay As You Earn system was introduced.

The war crisis helped to destroy the retrenchment philosophy in British financial policy. The war budget of 1941 is usually regarded as marking the adoption of a Keynesian approach to public finance. Even before 1939, Keynesianism had been establishing its influence in the civil service and Parliament, and the necessity to mobilize the country's economic resources completed the defeat of orthodox *laissez-faire* attitudes. During 1939–43 some 3 million additional men and women entered paid employment, thereby removing at last the stubborn unemployment that had dogged the interwar period. As a result Britain's rearmament advanced very rapidly, so much so that she was soon outproducing Germany in tanks and aircraft.

In the 1980s, however, this remarkable success came under attack by some right-wing historians, hostile to all forms of state involvement in the economy and society. They criticized the wartime coalition on the grounds that industrial production was inefficient and thus shored up sectors that should have been radically reorganized; they also felt that the high-minded liberal establishment imposed an expensive welfare commitment upon postwar Britain. This is largely invalid and irrelevant. It had been the case in the First World War that the extra output desperately needed in munitions had been obtained at a heavy cost and with lower productivity as new workers were hastily enrolled. In the Second World War much of industry operated with the advantage of firm orders; it could charge the actual costs of production plus a guaranteed rate of profit. However, in view of the situation faced by the government, the idea that it could devote itself to squeezing down the manufacturers' margins and generally driving a hard bargain is unrealistic and could easily have been counter-productive.

As far as new social policies were concerned, their costs were seen from the start as being linked firmly to the performance of the economy; Beveridge argued that a high level of employment must be maintained specifically so that the burden of state welfare would not become too great. Moreover, the assumption that governments entered into new social policies during the war *unnecessarily* seems very dubious. After the experiences of 1914–18 and the interwar period, much of the British population was unwilling to make great sacrifices except on the basis of current and future social improvement. During the war trade-union membership rose from 6.25 million to 8 million and, as in the Great War, only the first two to three years saw a reduction in militancy. By 1942 strike activity was growing again, and continued to do so to the end of the war, largely because men feared that peace would once again bring higher unemployment and lower real wages. In these circumstances, concessions were no more than a realistic means of achieving the measure of cooperation in the war effort that was so crucially necessary; it represented a hard-headed strategy, not simply an optional luxury for the liberal establishment.

Finally, any competent 'audit' for the Second World War would have to take account of the positive economic effects, not merely the negative ones. No doubt the conflict gave an artificial stimulus to traditional sectors like coal and steel by driving resources into them which might have been better employed elsewhere in peacetime. But this was scarcely avoidable. In any case the war gave a valuable boost to innovation in agriculture, chemicals, electronics, aircraft, and the motor industry, due to developments in radar, jet propulsion, antibiotics, and atomic power. This greatly strengthened science-based industry after 1945, and contributed to the ususually high rate of economic growth achieved in the aftermath of war.

In retrospect the aura of victory tends to make the Churchill coalition appear much more secure than it was; certainly in the summer of 1940 many observers thought it unlikely to last the duration of the war. However, by the autumn, with Britain winning the Battle of Britain, the prospect of a German invasion dwindled and the standing of the government naturally improved. Chamberlain's retirement for health reasons necessitated a ministerial reorganization and the selection of a new Tory Leader. The first task was accomplished by expanding the War Cabinet to eight members and appointing Herbert Morrison Home Secretary and Minister for Home Security. Churchill could have avoided the party leadership, but that would have left him in a vulnerable position comparable to Lloyd George's in 1918. He not only became leader but also removed his main rival, Halifax, by making him ambassador to the United States shortly afterwards. This allowed Eden to return to the Foreign Office.

In the House of Commons the government's position looked, on the face of it, impregnable. The absence of many MPs on wartime work weakened Parliament, while the absorption of the bulk of the Labour leadership into the government inevitably inhibited the usual role of the opposition. In the event, two chief centres of criticism developed. The first clearly took the government by surprise. War had an energizing effect on the back-bench women members, and drew Labour and Conservatives together in defence of women's interests. The leading figures in this were Edith Summerskill (Labour), Mavis Tate and Irene Ward (Conservatives), Eleanor Rathbone (Independent), and Megan Lloyd George (Liberal). In the early months of war they felt the government was slow to find employment for women, and deputations urged a greater use of women in the civil service and the abandonment of the marriage bar. The government made a concession by setting up a Womanpower Committee, which led to a debate in March 1941 in which Bevin came in for much criticism. The women MPs also demanded equal compensation for injuries suffered by women in the war, which resulted in a heavy defeat for the government by 229–95 votes in November 1942. They also surprised ministers during the passage of the 1944 Education Bill by obtaining a narrow majority – 117 to 116 – for an amendment to give equal pay to women teachers. The Cabinet overthrew this only by threatening to resign and seeking a vote of confidence.

The official Labour opposition enjoyed leadership from a few independent-minded figures such as Emmanuel Shinwell, but it only occasionally made a challenge. In May 1941, after the failure of the campaign in Greece, the House debated a confidence motion, but the coalition defeated it overwhelmingly despite being attacked by Lloyd George. In July 1942 there was another back-bench protest, over the level of pensions. Although Churchill's personal popularity remained consistently high, opinion polls showed sagging confidence in the government; by January 1941 support had fallen to 58 per cent, and by July 1942 to 41 per cent. Ironically, the entry of the USA and the Soviet Union into the war, and the British success in North Africa during 1942, stimulated criticism because it raised the prospects of an end to the war. In this situation the Labour Party found itself in a dilemma, being an integral part of the government, but also in a position to benefit from its unpopularity. Somehow it managed to be both patriotic and critical at the same time. By abstaining from by-election contests the party felt it was making a considerable sacrifice, because a number of safe Conservative seats were lost to independents and the new Commonwealth Party. However, the Conservatives complained that Labour was none the less maintaining normal party propaganda. There was some truth in this. The Conservatives disbanded their local associations and held no party conferences until 1945, whereas Labour continued to organize annual conferences and, up to a point, fulfilled its role as opposition in Parliament. However, this activity was clearly muted and sporadic. The greatest signs of revolt came in February 1943, when 121 MPs supported an opposition motion critical of the negative response of the Chancellor to the Beveridge Report. By that time the back-benchers were reflecting the mood of rising expectations in the country and the trade unions' apprehension about the economic effects of an end to the war. They therefore kept up the pressure on the party leaders. In October 1944, Labour's National Executive Committee decided that the party must abandon the coalition as soon as Germany had been defeated.

The Collapse of British Power

The Second World War made a reality of the nightmare long feared by the chiefs of the armed forces. Britain found herself simultaneously engaged in defending the British Isles from invasion, meeting the challenge of the Italian navy in the Mediterranean, and fending off Japanese aggression in the Far East. Her resources were hopelessly overstretched. This dilemma had been obvious since the 1890s, but before 1914 it had been dealt with by a mixture of diplomatic and military expedients; by 1939 Britain was relying more on bluff. However, the situation was never quite as bad as it appeared in the early stages of war. Fortunately, Hitler never regarded the military defeat of Britain as an urgent matter; he had not prepared for it, and in any case the Luftwaffe lacked the capacity for a major attack. Once Britain had begun to outbuild Germany it became increasingly unlikely that Hitler would attain the necessary air supremacy over southern England to justify risking an invasion. Britain's strength in fast fighter planes – Hurricanes and Spitfires – combined with the radar early-warning system enabled her to take a heavy toll of the relatively slow German bombers during 1940, and thus to remove the prospect of an enemy landing.

In the emergency, Churchill's instinct was to try to capitalize upon what he called 'the natural Anglo-American special relationship'. Half-American himself, Churchill devoted much of his time to cultivating President Roosevelt with a view to eventual entry by the USA into the war.

Meanwhile, his greatest contribution to the war effort was to secure the lend-lease deals and the use of American cruisers when Britain was at her most vulnerable. Even so, the relationship was not without its complications. The American government wanted Britain to announce withdrawal from India, and to give up Imperial preference and the sterling area, which Churchill saw as essential props to Britain's position in the world. There was also some disagreement over strategy. Britain wished to delay the opening of a western front against Germany, and preferred to tackle Italy which closely affected her Mediterranean communications. It suited the USA to concentrate on the Pacific, and thus postpone the landings in France until 1944. This delay crucially affected relations with the Soviet Union, whom the Allies failed to relieve during her time of greatest need. By July 1943 the Russians had turned the tide of war against Hitler's forces by themselves, which meant that they were able to begin an advance into Eastern Europe well before the British and Americans were able to bring effective power to bear on that area. In this way, Allied strategy was to lead to major problems in the postwar era.

Even in the worst days of the war Britain had not stood quite alone against Hitler. Altogether 5 million colonial troops, including 2.5 million from India, fought on her side. The Empire was retained – but only just. The entry of Japan into the conflict in 1941 exposed the crippling weakenesses in Britain's world position. For neglecting the proper defence of Singapore against a long-expected attack, and for underestimating the capability of the Japanese, Churchill was as culpable and as misguided as all the other politicians of his generation. In 1941 two battleships, HMS *Repulse* and *Prince of Wales*, were dispatched to Singapore without the necessary air cover, to be promptly sunk by Japanese attack. Then, during 1942, Singapore, Hong Kong, Malaya, and Burma rapidly fell to the advancing Japanese forces; soon the Raj itself was under threat.

These events helped to seal the fate of British India. British prestige was fatally undermined in the eyes of many colonial peoples in Asia and Africa. Moreover, the war finally turned India from what had once been a military asset into a huge liability. Nineteenth-century statesmen had spent years worrying about a possible attack on India via the North-West Frontier which had never materialized. Now the threat had come across the north-east border, and scarce military resources had to be dispatched to the south Asian theatre to meet it. Inevitable this military crisis impinged upon the internal stability of India. In September 1939 Lord Linlithgow, a tactless and mediocre Viceroy, had declared the country at war with Germany with no pretence of consultation. Since at this time Congress was ruling most of the provinces of India, and since Nehru fully accepted the necessity for defeating the menace of fascism, this was a foolish error on Linlithgow's part. He gave the radicals within Congress a good opportunity to withdraw their cooperation from the government and return to confrontation.

Back in Britain, the new Prime Minister ringingly declared, 'I have not become the King's First Minister in order to preside over the liquidation of the British Empire.' But for all his bluster, Churchill had been obliged by the need for American support to accept the Atlantic Charter which, among other things, encompassed the break-up of empire. Concessions did, in fact, begin to come thick and fast in several parts of the Empire during the war. Jamaica achieved adult suffrage, an elected majority was introduced into the Gold Coast legislature, and promises of self-government were given to Malta and Ceylon. In India itself Britain ceased to be in effective control. In view of the need to maintain some internal cooperation, the government decided to send Sir Stafford Cripps to India in March 1942 with a new offer to the effect that a postwar assembly should draw up a constitution for a self-governing India. But the Indian National Congress was not unduly impressed. Some thousands of captured Indian soldiers, under Subhas Chandra Bose, were already undergoing training with the Japanese with a view to joining in the expected invasion of India. Congress could not lightly agree to support the British in resisting an enemy whose forces included their own Indian nationals. The Cripps offer, therefore, had to be an attractive and watertight one. Unfortunately, the British had not grasped how important it was to Congress to avoid a Balkanization of the subcontinent, and Cripps's proposals seemed to allow for the splitting up of India into its provincial territories. In the end Gandhi dismissed the offer as 'a postdated cheque on a failing bank', fully conscious that the proximity of the Japanese would maintain the pressure on the British authorities. As a result, Congress went on to organize the 'Quit India' movement, its last great campaign against the British Raj, and the remainder of the war was a frustrating period of deadlock between Congress, the Muslim League, and the new Viceroy, Lord Wavell. Although the British took comfort from the gradual retreat of the Japanese, they lost control internally in the face of Hindu–Muslim rioting, widespread industrial strikes, mutinies in the armed forces, demoralization amongst the police force, and a major famine in Bengal. By 1945 it was unrealistic to expect British power to survive more than a couple of years.

1945: The Labour Landslide

Before 1939 British politics had apparently been firmly fixed in the mould established by the National government in 1931. At the last election, in 1935, its followers had won 53.7 per cent of the vote and returned 432 MPs out of 615. In spite of a handful of gains at by-elections in the late 1930s, there was no obvious indication by the outbreak of war that the opposition was within striking distance of winning a majority. Labour's best performance in terms of the popular vote was 38 per cent in 1935, but this had brought only 154 seats. In 1929 the party had won 288 seats, but only in conditions of an even, three-party competition; the subsequent demise of the Liberals made this unlikely to recur. Even after the traumatic events of wartime the political world was thus shocked by the scale of Labour's victory in 1945: almost 48 per cent of the poll and 393 seats, compared to barely 40 per cent and 213 seats for the Conservatives. In fact, the opinion polls had pointed to a big Labour lead since 1943, but such polls were still a novelty. Politicians preferred to rely on intuition and experience, which told them that polls and by-elections represented an ephemeral protest. There was a reliable precedent in Lloyd George's sweeping victory at the head of a victorious coalition in 1918 which pointed inexorably to the return of Churchill to power.

Clearly, in spite of the superficial similarities, the circumstances of 1918 and 1945 were not fully comparable. An obvious point for comparison lay in the structural changes in the electorate which affected both of these elections. In 1945 there had been no legal reforms, but since ten years had elapsed since the last election, a considerable number of voters had died and many new ones had joined the registers. One-fifth of electors were voting for the first time in 1945, and subsequent studies suggest that around six out of ten of them supported Labour. This reflected the making of a distinct political generation amongst voters who had been dismayed by the poverty and unemployment of the 1930s and were aroused by the experience of wartime to repudiate those responsible. The turnout in 1945 was much higher than in 1918, which meant that the anti-Conservatism among the troops made itself felt in the ballot box. This was sometimes attributed to the activities of the Army Education Corps and the Army Bureau of Current Affairs, whose bulletins helped to promote debate on topics which tended to reflect badly on prewar governments. However, the leftward drift in the army would probably have occurred without such activities.

This highlights a second part of the explanation for 1945. As the party in power before the war, the Conservatives were bound to suffer some blame for the country's unpreparedness, as Asquith's Liberals had done after 1914. However, the effect was compounded by the close association of most leading Conservatives with the appeasement of the dictators; this sharpened the distinction between Churchill himself and his party. In contrast, the prewar record of the Labour leaders had been blotted out by their prominent, patriotic service as war ministers. Whereas in 1918 many Labour and Liberal candidates had been attacked for being soft on Germany, by 1945 Labour simply reflected the general feeling that the Nazi regime must be eradicated and that Nazi leaders should be brought to trial for their crimes. As it was impossible to accuse the opposition of lacking patriotism, therefore, Churchill's political position could never be as strong as that of Lloyd George in 1918.

The third and most important explanation for the Labour landslide is that the election was fought over different kinds of issues from the 1918 election. In the Great War the British had not been expecting victory in 1918, and the sudden turn of events leading to the armistice in November 1918 meant that the December election had been held while the emotions of wartime remained at their height; voters had by no means readjusted to peacetime politics. In the Second World War, however, the public had begun to take victory for granted by 1942. Over a period of several years thoughts had time to focus on the postwar situation; hence the centrality of the Beveridge Report, for example. By the summer of 1945 the polls suggested that the electorate was chiefly concerned about housing, unemployment, and the implementation of Beveridge's proposals – issues on which Labour carried far more credibility than the Conservatives. The party's programme, 'Let Us Face the Future', and its propaganda, 'Ask Your Father', chimed in with the public mood and reminded voters they had been let down before in the aftermath of a great war. Labour's policies had not been significantly altered since 1935; what had changed was the public perception of them. In particular the proposals for nationalization had lost their association with extremism. This did not indicate a wholesale conversion to socialism; but it reflected wartime experience in which government was seen to have involved itself in the economy to good effect. As a result, there was relatively little objection to extending into peacetime interventionist policies that had worked well during the war.

The final element in the situation was the timing and tactics of Churchill himself. After Germany's surrender in May 1945, the Labour leadership came under strong pressure to withdraw from the coalition government. Churchill thereupon offered Attlee either a continuation of the coalition until the defeat of Japan or a general election in July. Attlee had no confidence of winning an election because Churchill's popularity was obviously high, the register was out of date, and it was possible many of the soldiers would not participate. But as he did not wish to split the party he accepted the early election. In the event, Churchill handled the election as if the country was still at war; though it was, the voters showed little interest in the fact. He also relapsed into the tactics of the 1930s, by labelling Labour as a party of extremists unfit to govern; he used the old argument that socialism could never be introduced without some form of autocratic government. In view of the record of Attlee and his colleagues in office, such attacks carried no credibility. None the less, Churchill's tours around the country became a triumphal progress in many areas. But voters evidently made a clear distinction between him and his party. Some revealed to the pollsters that they thought it possible to vote for Labour without losing Churchill as Prime Minister. This is credible only in the immediate context, in which he had been closely associated with Labour ministers, and in the longer-term context, in which he had been the foremost critic of Conservative ministers in the National government.

In spite of this, Churchill's personal standing almost certainly helped his party up to a point. Labour's lead in the opinion polls had been as wide as 16 per cent, but narrowed to 8 per cent in July 1945. The election had been decisively lost by the Conservatives two to three years before the actual poll, and the campaign only served to modify that a little. The result represented a watershed, for Labour had now extended its reach into many middle-class areas. For example, in London the party won 48 of the 62 constituencies. Elsewhere the remains of nineteenth-century loyalties were swept away; in Liverpool, 8 of 11 seats went to Labour, and in Birmingham, 10 out of 13. In this sense 1945 proved to be the culmination of the ambitions of the MacDonald–Henderson generation of leaders, who had aspired to unite a broad swathe of working and middle-class people around a programme of social and economic interventionism. This forced the Conservatives into a reconsideration of their past record, and made them more ready to accept some of the goals and priorities of their rivals. The centre of politics had shifted to the left, so that objectives such as full employment and state social welfare became common to the majority of politicians, not the distant expectations of a few idealists.

16

The Keynesian Era

'But this is terrible,' expostulated a diner at the Savoy Hotel in July 1945, 'they've elected a Labour Government, and the country will never stand for that.' Yet the Attlee Government, backed by 393 MPs and over 48 per cent of the vote, was unusually in tune with the popular mood in the aftermath of war. It has some claim to be the most successful government of the post-1945 period, and arguably of the twentieth century, both because of its record of concrete achievements and because it left the country stronger than it had been when it took over.

The new administration had an unusually clear idea of its objectives and how to accomplish them. No fewer than 75 Acts were passed during 1945–6 alone. The ministers also benefited from their experience of administration obtained during the war, and from the quiet efficiency with which Attlee managed his very talented Cabinet. The result was the enactment of a programme from which there was to be no major deviation for thirty years. 'Mr Attlee's consensus', as it has been described, commanded the respect of the Conservatives and the loyalty of Labour at least until the 1970s.

The Mixed Economy

In 1945 all parties accepted that it would be a mistake to abandon wartime controls rapidly as had been done in 1919–20. It was, of course, a matter for debate as to how many controls should be retained and for how long. But for Labour, the success of wartime policies in mobilizing the nation's resources underlined the efficiency of state control. In the country, food rationing and price controls commanded wide support down to the late 1940s. Import restrictions were also necessary in order to check the shortage of dollars in the immediate postwar years. Investment was another aspect of state interventionism; the Labour government used its powers to promote a regional economic policy to counter the high unemployment in the 'Development Areas'. As a result, 51 per cent of all new factories during 1945–51 were sited in these regions. The control over labour proved to be more problematical in peacetime. Here the government resorted more to moral and political influence to mobilize more workers where needed and to restrain wage claims; as a result, wages rose on average by only 2.8 per cent per annum up to 1949.

However, by 1948–9 a number of wartime controls had been abandoned as unnecessary; the young President of the Board of Trade, Harold Wilson, tried to pre-empt Conservative criticism by announcing a 'bonfire' of controls in 1949. Against this, however, the government had embarked upon a major extension of state intervention in the form of the nationalization of key industries. This programme began in 1946 with the Bank of England, civil aviation, and cable and wireless, moved on to coal and railways in 1947, electricity, gas, and long-distance transport in 1948, and ended in 1951 with the steel industry. With the exception of steel, about which the Cabinet itself was very divided, the takeover by the state engendered little controversy. This was partly because several utilities like gas and electricity were already municipalized. Further, there existed an acceptable model for nationalization in the form of Herbert Morrison's 1934 London Passenger Transport Board. This involved running the nationalized industries through small boards of experts appointed by the relevant minister. The takeover could be justified in terms of national efficiency as much as by socialist ideology. In any case, industries like coal and railways had become seriously run down under private ownership; the free-enterprise system was clearly not going to provide the investment required, and the former owners were lucky to receive generous compensation for the loss of their wasting assets.

The politician who best grasped the role of the nationalized industries in the context of overall economic planning was Sir Stafford Cripps, who served successively as President of the Board of Trade, Minister for Economic Affairs, and Chancellor of the Exchequer. It could not simply be left to the market, especially in the aftermath of a war, to see that scarce resources were sensibly allocated. Unless essential sectors like coal, the railways, and steel, upon which the rest of the economy depended, were put on a sound footing, industry as a whole would be held back. In many ways, this limited approach to socialist economic planning proved to be a success: it enabled the government to boost exports, restrain inflation, and maintain full employment – achievements which eluded all subsequent governments down to 1992. In the years up to 1948 retail prices, for example, rose by only 3.3 per cent per annum, and by only 2 per cent from 1949. Nationalization left about one-fifth of the economy under state control – enough for planning purposes but not enough to undermine free enterprise where it was more appropriate.

There were, however, several flaws in the nationalization programme. By concentrating on obviously failing industries and public utilities, it seemed to set natural limits upon the extension of the policy. There was a strong case for nationalizing consumer-goods industries, like sugar, which operated as virtual monopolies against the public interest. But the government clearly shrank from the political consequences of such a move. In this connection it was probably a mistake to exclude any workers' co-partnership in the state industries. This reflected the arrogance of the bureaucrat-intellectuals of Fabian socialism, who regarded the workers as incompetent; by creating an unresponsive bureaucratic structure, they denied nationalization a broader base of support in the country.

The other major factor which weakened the government's capacity to plan the economy was simply the force of events beyond their control. The first Labour Chancellor, Hugh Dalton, inherited horrendous economic problems in 1945. With the end of the war, the lend-lease arrangements came to an abrupt stop. Britain's exports in 1945 stood at only 46 per cent of their 1938 level because of the diversion of resources to war production. The sale of £1,000 million pounds of investments during the war further weakened the balance of payments; altogether, 28 per cent of Britain's wealth had been wiped out. Moreover, the country's liabilities had risen to £3,500 million by 1945.

The first step taken to deal with this situation was to raise a loan of $3,750 million from the USA and $1,250 million from Canada for a three-year transition period while the British economy adjusted to peace. Though a necessary expedient, the loan came with stringent conditions attached. The Americans insisted on making sterling convertible into other currencies, and on the maintenance of stable exchange rates – a practical impossibility. Moreover, the value of the loan was reduced by US inflation and by a movement of the terms of trade against British goods. As a result, Britain suffered a severe dollar shortage by 1947.

The government managed to concentrate resources into the manufacturing industries most capable of selling goods abroad. As a result, it achieved something rarely accomplished by governments between 1945 and 1992 – an export-led boom. This notable success would have been greater but for the diversion of vital labour resources into the armed forces; Labour's defence policy was clearly detrimental to its economic strategy. The other factor that hampered growth was the coal shortage, especially in the winter of 1946–7. The Minister for Fuel and Power, Emmanuel Shinwell, bore the blame for not recognizing the shortage quickly enough, and for failing to divert more coal from domestic use to industry.

However, the recovery of industrial output by 1948 made it appropriate to take another beneficial step – a devaluation of the pound; the growth industries could now take full advantage by expanding exports. In fact, the Labour government was extremely reluctant to do this. All parties regarded devaluation in emotional terms as an unpatriotic act, rather than as a technical adjustment. However, by 1948 the recession in the USA had reduced Britain's dollar earnings, undermined the balance of payments, and thus led to increasing speculation against sterling. Eventually the decision was made by Hugh Gaitskell, Douglas Jay, and Harold Wilson. In 1949 the pound was devalued from $4.03 to $82.90. This was a perfectly sensible readjustment which reflected American economic supremacy. It not only improved Britain's balance of payments, but in the longer term led to a beneficial trade balance between the dollar economy and the non-dollar world.

Unfortunately, the government accepted the conventional wisdom that regarded devaluation as a great failure. Admittedly, devaluation required a deflationary policy during the last two years of the government's life. Cripps – astonishingly high-principled – thought it improper to present a budget in the spring of 1950 just before an election. Moreover, Attlee mistimed the election by holding it in February 1950, while the effects of deflation were apparent, but before the beneficial consequences of devaluation had materialized. This error led to Labour's majority being cut to five. Two further mistakes followed. Gaitskell, now Chancellor, put excessive strain on the economy by financing the rearmament programme. Attlee, who had lost his grip on power, then called a second election in 1951. He should have been replaced by a more vigorous leader who would have held on longer; with the improvement of economic conditions by 1952 the government would almost certainly have retained power. In the event, Labour did remarkably well in the 1951 election by winning 48.8 per cent of the votes. This was the party's all-time best, and also exceeded the Conservatives' vote. However, the first-past-the-post electoral system failed to ensure that the votes cast were fairly reflected in the seats won: the Conservatives, with fewer votes, won an overall majority of seventeen.

The Welfare State

The creation of the welfare state was the greatest and the most enduringly popular of the achievements of the Labour governments. It involved a willingness by the community to accept responsibility for insuring its citizens against the perils of sickness, unemployment, injury, and old age, as well as for providing adequate housing and education. By comparison with earlier state systems, the post-1945 scheme was intended to offer both comprehensive benefits and ones to which the entire population would enjoy equal access; no stigma was to attach to the receipt of state benefits. How far the welfare state provisions actually met or fell short of this ideal is a matter of debate. Inevitably, the chief architect, Beveridge, left the mark of his Edwardian Liberal creed upon Labour's programme in the form of the insurance principle. There was obviously an element of continuity with interwar policies such as the 1925 pensions scheme and subsidized council housing. But there was also a definite socialist contribution in the work done by the Socialist Medical Association, for example, in promoting the idea of free treatment and a unified national hospital system. No doubt, wartime innovation by the coalition government made some impact too, but one can easily exaggerate the continuity. It is unlikely that the Conservatives would have been as radical as Aneurin Bevan in insisting on nationalizing the hospitals, or that they would have given priority to raising the school-leaving age.

The measures that compose the welfare state were enacted largely by two eloquent and effective Welsh politicians, Jim Griffiths and Aneurin Bevan, between 1945 and 1948. Strictly speaking, the Family Allowances Act was already on the statute book when the new government took office. It was, however, an important measure, which brought weekly payments of 5s. for all children, after the first, to nearly 3 million families; it was a cost-effective means of adding to family income where it could do most good – in the housewife's purse. 1946 brought the Industrial Injuries Act, the National Insurance Act, and the National Health Service Act. A payment of 26s. for a single person per week was adopted; this represented an increase over existing provisions of 2s. a week for unemployment, 8s. for sickness, and

16s. for the old-age pension. This last is a reminder of the major impact of the reforms upon the elderly, who had been chief victims of poverty in the past. In 1948, when the new National Health Service came into operation, the whole population enjoyed for the first time free medical treatment, medicines, spectacles, and false teeth, and a national system of hospital provision incorporating the various voluntary and local-authority hospitals. Access to medical services was now based upon need, not on ability to pay. Women, whose health had traditionally been neglected because many were not wage-earners and because they sacrificed themselves to their families' needs, gained more than any other section of the population. The NHS has justly been described as 'the most beneficial reform ever enacted in England'.

However, the welfare programme attracted some contemporary criticism for not going far enough towards the ideal of a welfare state; and conversely, in the 1980s it was regarded by some on the extreme right as having gone too far. First, let us consider the limitations and shortcomings of the new policies. The pensions were not linked to the cost of living. At first this did not matter, as the government held down price rises, but in time the payments began to lag behind inflation. As a result, many people had to apply for National Assistance, which involved benefits assessed after a personal means test. By 1955 over a million people were receiving National Assistance. For socialists it was a matter of regret that the welfare state failed to create a completely classless system of provision in education and health. Ellen Wilkinson suffered criticism for implementing the 1944 Education Act and leaving public schools intact. However, in 1945 no great political pressure had been generated to remove the private sector, and it was not until 1951 that Labour became committed to 'comprehensive' schools. In fact there was a good deal of Labour support for the traditional grammar schools, and the new system, based on the eleven-plus examination, did in fact facilitate the passage of many able working-class children into higher education. Wilkinson's other achievement was to raise the school-leaving age from fourteen to fifteen.

In health provision, the doctors were allowed to continue in private practice while working for the NHS, and a number of 'pay-beds' were retained in public hospitals. This compromise resulted from the resistance Bevan encountered from the vested interests of the medical profession. As in 1911, the British Medical Association played a shameful and obstructive role; as late as 1948 doctors were, by a margin of eight to one, opposed to working the NHS system. The maintenance of a private sector was not only socially unjust; but also added an element of inefficiency which became more serious as private medicine expanded, and fed off the resources of the NHS.

Housing has often been regarded as one of the least successful aspects of the new welfare state up to 1951. Bevan admitted that this enjoyed a lower priority while he was engaged in setting up the NHS. The government did make financial support available for council house-building, but gave lower priority to private housing for sale. However, the chief obstacle was not ideology but simply the shortage of resources for the construction industry in the late 1940s. In the circumstances, the government actually had an impressive record; in 1948 it was building over 200,000 houses, and the total for 1945–51 was 1.35 million.

From the opposite perspective, some critics have claimed that the welfare state was a hugely expensive burden which damaged the British economy in the long term. Obviously large sums of money were involved, notably in health. The total expenditure on health was put at £478 million by 1951, about five times the 1938 level. However, to see this simply as a *cost* would be simplistic, to say the least. The very fact that expenditure rose after 1945 indicated that a great deal of ill-health had gone untreated – with very damaging effects on the economy. Any genuine audit of welfare policies would have to include some assessment of the economic gain from a more fit and healthy labour force. Many of the welfare benefits, such as family allowances, were very cost-effective ways of relieving hardship. When the famous Rowntree study of poverty in York was repeated in 1950, the conclusion it reached was that only 2.77 per cent of the working class suffered from poverty. If the welfare state did not abolish poverty altogether, it represented the most effective single campaign against it.

Finally, it is important to remember that any suggestion that state welfare expenditure got out of control has no basis in fact. Beveridge himself had prepared his programme on the basis that it would be sustainable provided that unemployment did not exceed 8.5 per cent; beyond this the costs of welfare would be excessive. The government aimed to meet this requirement by using Keynesian techniques to maintain a high and stable level of employment. It succeeded in this; in 1947 unemployment was only 1.6 per cent, and indeed, it rarely exceeded 2 per cent in the postwar period. Nor did the government neglect economic considerations in order to meet its social objectives. It deliberately ensured that resources went into the construction of new factories at the expense of hospitals, schools, and house-building, for example. In any case, the level of welfare payments in the 1940s was far from generous – hence the resort to National Assistance. It was not long before other European states developed more costly welfare programmes than Britain, but their liberal provisions did not appear to damage the economic efficiency of countries like West Germany and Sweden. Finally, the Labour Chancellors of the Exchequer, Dalton, Cripps, and Gaitskell, were all rather severe about keeping expenditure within bounds – rather more so than their Conservative successors, in fact. The most obvious sign of this was the imposition of charges on false teeth and spectacles in 1951 by Gaitskell. In fact, the money actually saved in this way was negligible, and since the decision precipitated the resignation of Aneurin Bevan and Harold Wilson it was scarcely worth the trouble it caused. However, if this showed poor political judgement on Gaitskell's part, it indicated his determination not to spend more than could be afforded on welfare benefits.

The Politics of Consensus

The experience of a coalition government in mobilizing national resources for the war effort had helped to generate a broadly based consensus in

British politics by 1945. Something similar had appeared during the First World War, but that had been a superficial and ephemeral mood; the ideas of 1945 proved very enduring. This consensus can be seen in terms of broad agreement about the policies, the style, and the institutions of government. In this sense, consensus was far from new. Periods of bitter debate over the constitution are relatively rare in British history; what is characteristic about the post-1945 consensus is the agreement on the substance of policies. There was a very marked contrast with the Liberal government of 1905–14, which also used a landslide victory as the launch-pad for major innovations in policy. It encountered bitter opposition which had no parallel in 1945–51. The nearest Attlee's government came to a constitutional issue was the decision to trim slightly the House of Lords veto by allowing the Commons to overrule the peers by passing a disputed bill in two sessions rather than three.

There were five major areas in which Labour and the Conservatives adopted the same approach. First, the welfare state was consistently respected. Second, full employment was accepted as a legitimate and central aim. Third, the mixed economy, involving a much larger state sector than before 1939, became an established fact. Fourth, it was considered that the participation of the trade unions by consultation and conciliation was as necessary in peacetime as it had been in war. Fifth, it is often forgotten that the fundamentals of foreign, defence, and imperial policy were also common to both parties. This involved commitment to NATO, the nuclear deterrent, the gradual run-down of Empire, and enthusiasm for the Commonwealth. This last dimension is an important reminder that the consensus was not simply a matter of concessions by the Conservatives; for, if they moved to the left in social and economic policy, Labour moved to the right in external affairs.

The idea of consensus is, of course, somewhat embarrassing to partisans of both the main political parties; and those historians who have dissented from the interpretation tend to be associated with one or other party. This does not necessarily invalidate their views, however, and there are some qualifications to be made about consensus. It does not imply complete agreement; as we shall see, there were differences in emphasis and in priorities between successive governments. Nor does it mean that the *whole* of each party embraced consensus policies. In the Labour ranks there were those who campaigned against nuclear weapons; while on the Conservative side a right-wing section was to form the League of Empire Loyalists in protest against their leaders' policies. There was clearly more opposition to consensus policies among rank-and-file Tories than among the parliamentary elite; however, the right-wingers were largely excluded from office by Churchill, Eden, and Macmillan, and consequently enjoyed little effective leadership.

The obvious test for the consensus view of postwar politics consists in the response of the Churchill government after its return to power in 1951. It is easy to be misled by the party rhetoric designed to rally the activists by exaggerating the difference between the two parties. Labour freely predicted a return to extreme and divisive policies if Churchill regained office. Conservative propaganda tempted voters with 'good red meat' after years of austerity, proposed to 'set the people free', and described Britain as a 'socialist state monopolizing production'. Yet the reality after 1951 was more prosaic. The new government cut income tax a little and reduced subsidies on food; it continued the abolition of controls begun by Harold Wilson; it moved more resources into house-building, and it spent less on defence. But when all these are added together they made no more than a marginal difference, at most, to the Labour programme. In the 1970s, when the supporters of Mrs Thatcher investigated the record of the Churchill Cabinet, looking for a truer Conservatism, they were rapidly disillusioned by what they found. Churchill's party pledged itself to full employment. It made no attempt to dismantle the welfare state; in fact, by 1955 the real value of pensions and other benefits had increased since 1951. There was no denationalization except in steel in 1953 and road haulage in 1954, industries about which Labour had been rather uncertain. It was scarcely surprising that by 1955 Labour found some difficulty in attacking the Tory government. Both parties were beginning to find themselves in the position of claiming, not that they would make fundamental changes, but that they would manage the existing policies more competently than their rivals. By 1951 Labour had already moved the emphasis of its economic policy away from planning and physical controls towards Keynesian management.

Hence the famous characterization of the policies of Hugh Gaitskell and R. A. Butler as 'Butskellism' by *The Economist*. Butler cheerfully accepted the Keynesian label for himself, while Gaitskell was to be active, if unsuccessful, in attempting to remove the famous Clause 4, which committed Labour to state ownership of the means of production, from his party's constitution. The most important indication of the shift in Labour thinking was Anthony Crosland's *The Future of Socialism* in 1956. In this he argued that capitalism had been reformed, and that governments enjoyed sufficient means of control over the economy without resorting to further measures of nationalization.

Why did the Churchill administration sustain the consensus? One reason is simply fear of the electoral consequences of rejecting it. 1945 had been a shock, and the party appreciated that it would not have been returned to power if it had proposed to undo Attlee's reforms; Labour had won more votes than the Conservatives in 1951, and was thought likely to recover power at the next election. The second explanation is that many of the new ideas had already influenced Conservatives in the 1930s. After 1951, men of the liberal or 'one-nation' Tory school, like Butler and Macmillan, achieved high office, while another generation of progressives, including Reginald Maudling and Iain Macleod, was being advanced via the party's research department. Under its director, Butler, the department produced fresh statements of party thinking, notably *The Industrial Charter*, designed to educate the rank-and-file and bury the legacy of the 1930s. Third, Churchill himself was responsible for consolidating consensus politics within the party. As an old man still basking in the role of wartime leader of the nation, he doubtless wished to avoid spending his last years as Prime Minister beset by controversies; he wanted to promote peace and reconciliation. Moreover, Churchill, married to a good Liberal, had never entirely forgotten his Edwardian record as a Liberal social reformer. By marginalizing the right-wing critics and promoting the conciliatory figures to ministerial posts, he helped to ensure the continuation of one-nation Toryism for some years, under the premierships of Eden and Macmillan. The latter proved to be a determined interventionist, and Keynesian in the later 1950s and early 1960s when prices and incomes policies came into vogue. But the ultimate proof of the Conservative attitude was its use of state resources. In 1950, under Labour, government expenditure accounted for 39 per cent of gross national product; by 1960, after nearly a decade of Conservative rule, the figure stood at 41 per cent. The post-war consensus had been sustained, and was to survive into the 1970s.

"The Rt. Hon Mr. Cripp-skell-but-mac-croft-ory . . ."

The politics of consensus

The Affluent Society and the Stagnant Society

For most British people the 1950s and 1960s were decades of rising living standards and an expanding economy. To a considerable extent this reflected developments outside the control of British governments. From 1952, the winding down of the Korean War contributed to a resumption of economic growth and facilitated a reduction in British defence expenditure. R. A. Butler's 1953 budget set the tone by taking 6d. off the standard rate of income tax. Britain benefited greatly from the rapid expansion in economic activity in the Western world generally, particularly as a result of the reduction in trade barriers by the General Agreement on Tariffs and Trade and the development of the European Economic Community. The terms of trade also moved in Britain's favour, thereby enabling her to consume more of her own output at full employment. Finally, British industry was, by 1951, well placed to take advantage of the improved conditions, as a result of the diversification of manufacturing under the stimulus of the war and the concentration of resources in manufacturing under the Attlee government.

Attlee's successors, Churchill, Eden, and Macmillan, led administrations dominated by pragmatic, conciliatory Conservatives, anxious to keep on the right side of public opinion and re-establish their credentials as a party of government. This concatenation of domestic political motives and broader economic forces made the period one of unusual affluence. Throughout the years 1945–1969, unemployment hardly ever touched 3 per cent, and then only briefly, typically standing at 2 per cent or less. Inflation was on average 4 per cent per annum in the 1950s and 1960s. Average money wages rose from over £8 in 1951 to over £15 per week in 1961. By comparison with the high unemployment of the 1930s, and of the 1980s, this was unquestionably a golden age for the British people.

Naturally enough, the further the war receded the more the population demanded access to consumer goods and an end to sacrifice and austerity. The Conservatives were quick to exploit this mood, and consistently promoted consumerism by relaxing credit and lowering taxes during the 1950s. However, an equally important reason for high spending on consumer luxuries lay in the capacity of the economy to generate extra unskilled and part-time jobs for women, thus making possible the rise of the two-income family. As a result of this there was an increase in the number of motor cars from 1.5 million in 1945 to 5.5 million in 1960. Whereas only one household in seven enjoyed a television set in 1951, two out of three did so by 1960. These years also saw a huge increase in spending on other forms of leisure, notably foreign holidays in Spain and other European countries. But perhaps the most conspicuous form of private consumption was housing. This was not new; by 1939 home ownership had risen to 35 per cent, compared to around 10 per cent in 1914, but it continued to grow, reaching 59 per cent by 1981. The new Churchill government and the flamboyant Housing Minister, Harold Macmillan, deliberately made housing a priority. The 1950 Party Conference had pledged the party to build 300,000 new houses a year. In 1953 this target was reached; some 318,000 were built, though it is worth noting that at this stage four out of five were local-authority homes for rent, not sale. None the less, the 1950s saw a major expansion of building societies which offered cheap mortgages, so that home-ownership came well within the reach of many working-class families. This began to make a reality of the Conservatives' aim of creating a 'property-owning democracy'.

In spite of the mood of optimism among contemporaries, it has, however, become clear that the rising living standards of the 1950s and 1960s were by no means fully justified by the record of the British economy. Annual rates of economic growth averaged around 2.2 per cent, which were, in historical perspective, respectable, but well behind those of other Western countries – France with 4.6 per cent and West Germany with 4.9 per cent. The governments of this period enjoyed an advantage over their predecessors in possessing Keynesian techniques for moderating the pattern of booms and slumps in the economy. However, two things detracted from this advantage. One was the incompetence of the advice given by the Treasury over a long period; its data and forecasts were so unreliable that governments found themselves reflating the economy when unemployment was about to fall, or deflating when it was about to rise.

The second general problem lay not in the Keynesian methods as such but in the systematic misuse of them by a succession of Conservative Chancellors for political purposes. The first guilty politician in this series was R. A. Butler, who served as Chancellor under Churchill and his successor in 1955, Sir Anthony Eden. Butler actually pursued an erratic economic strategy, influenced heavily by the fact that the government had a small majority and a Prime Minister whose powers were obviously failing. When Eden took over in the spring of 1955, he capitalized upon his own novelty by holding a general election in May. Butler's April budget lowered taxes and thus helped the Conservatives to extend their majority to a comfortable 70 seats. But Butler's blatant election giveaway simply boosted consumer spending, stimulated inflation, and weakened the balance of payments; after the election the boomlet had to be dampened down – the start of what was to be a depressing stop/go pattern in the economy.

Eden's short premiership was chiefly notable for the Suez affair (see p.278), which abruptly ended the comfortable, soporific mood of the early 1950s by introducing deep and bitter divisions into politics. The government's policy has been widely credited with reviving the Labour opposition under Gaitskell and alienating much of the intellectual middle class from the Conservatives, to the benefit of the Liberal Party. However, the impact seemed rather ephemeral at the time. This was partly due to the skill of Eden's successor, Harold Macmillan; but it also seems likely that Suez affected only the politically aware minority. The public at large were little disturbed by the immorality of their government's policies, and responded more readily to economic pressures, as the outcome of the 1959 election suggests.

Suez proved to be a turning-point, in that it brought home to the politicians the extent of Britain's decline as a world power, and the fact that the decline reflected economic weakness. In 1957 Britain suffered a big fall in her gold and dollar reserves and a £500 million balance-of-payments deficit. By bringing Macmillan to power, Suez also helped to accelerate a readjustment in British policy. Macmillan initiated three shifts of considerable long-term importance: he began serious economies in defence spending in 1957, he used Ian Macleod to accelerate the policy of decolonization; and he made the first British application to join the EEC in 1961 (see p.280–1). In all this Macmillan played with great skill the classic role of the twentieth-century Tory leader, in gently but firmly trying to educate his party to accept the adjustments necessary now that Britain was no more than a second-rate and declining power. Macmillan's smooth Edwardian style, and his shrewd forays on the international stage, helped him to soften the blow and minimize the right-wing reaction. In a famous phrase, Lord Salisbury denounced Iain Macleod, the reformist Colonial Secretary, as 'too clever by half', which may not have helped his career, but amounted to a mere expression of frustration that the party was so firmly in the hands of liberal-minded Tories.

Unfortunately, in domestic affairs Macmillan showed himself stronger on style and presentation than on substance; he gave way too easily to short-term electoral pressures. By 1957 it was clear that incomes were rising much faster than output, and the Chancellor of the Exchequer, Peter Thorneycroft, wanted to check inflation by holding expenditure for 1958–9 down to the level of 1957–8. He warned the Prime Minister: 'With relatively few assets and large debts, we continue to live upon the scale of a great power.' However, ministers in charge of the spending departments unanimously opposed the proposed check on their programmes, and Macmillan conspicuously sided with them against his Chancellor. Eventually Thorneycroft resigned along with his junior ministers, Nigel Birch and Enoch Powell – the first shots in the battle for monetarist policy in the Conservative Party.

Macmillan dismissed the whole episode as 'a little local difficulty', and proceeded to put pressure on his new Chancellor to come up with more tax cuts with a view to preparing the scene for an election. Both the 1958 and 1959 budgets included some tax cuts, the latter to the tune of £370 million, the largest single give-away ever recorded. The British voters were once again lulled into a false sense of security by all this. With the memory of 1956 safely expunged, they returned Macmillan with an increased majority of over 100 at the 1959 general election. But by 1960 the precariously contrived boom had collapsed; excessive credit and spending produced inflation and increased imports which the Chancellor felt obliged to curb by raising bank rate to 6 per cent.

As a result of the erratic and opportunistic economic policies of the 1950s, something of a reaction seems to have set in around 1960. Critics began to dwell on the fact that economic growth was much faster in the countries of the European Economic Community than in Britain. But they also began to look more deeply into British politics for explanations. A typical critique of this period was Michael Shanks's *The Stagnant Society* (1961), which identified the rigidity of the social-class system as central to the problem; this resulted in poor industrial relations and insufficient opportunities for people of ability, but low social origins, to rise to positions of influence.

Friction between employers and workers certainly distinguished British society from that of the more successful economies, such as West Germany. The level of strikes, survival of restrictive practices, and wage-driven inflation seriously hampered the output and efficiency of British industry. The problem began in 1951, when Churchill deliberately appointed the conciliatory Walter Monckton as Minister for Labour with instructions to avoid antagonizing the unions. In major strikes in the public sector, such as the 1954 railway dispute, the government frequently intervened, or appointed an inquiry which invariably led to the concession of the unions' demands. Churchill firmly closed the door to any legislation designed to interfere with picketing, the closed shop, or the political levy.

On the other hand, it is easy to exaggerate the unions' responsibility. Wage demands were driven by inflation, often arising out of government policies; when union leaders tried to restrain the pressures, they could be undermined by shop stewards and unofficial strikes. Also, the employers as a group showed very little capacity to put industrial relations on a sound footing; they failed to follow the German example by instituting co-partnership schemes designed to break down the barriers between workers and management. Both the main political parties were bereft of constructive ideas in this area.

Industry in general suffered from poor productivity due to inadequate investment. This, of course, was already an established feature of the British economy. Governments did little to help; indeed, by diverting excessive resources into house-building, and by providing financial inducements to take on large building society mortgages, they exacerbated the drift of investment into housing at the expense of manufacturing industry; by the 1970s and 1980s, this was to become a major cause of economic weakness. In addition the banks and the City of London continued to play an unhelpful and inadequate role in industry. Those companies that invested heavily in long-term research and technological improvement tended to find their share value marked down by the stock market; but a short-term bid for profits at the expense of training and investment was rewarded with rising share prices. In this area, market forces could not be relied upon to promote the national interest.

The governments of the period clearly weakened the productive economy by their concentration on short-term expedients. They invariably found that full employment led to inflation, an inflow of imports, and a balance-of-payments crisis. This put pressure on the exchange rates, because investors rightly judged that in view of the weakness of the economy there was a danger of a devaluation, and sold sterling accordingly. Every British government, however, chose to regard devaluation as anathema. In order to avoid it, they repeatedly raised interest rates and imposed a dose of deflation to dampen consumers' demand for imports. These measures undermined manufacturing industry by adding to its costs, deterring investment, and making long-term planning difficult. The maintenance of the pound as a major international currency thus became a serious handicap as industry's interests were sacrificed to maintain it. By 1961 Macmillan had accepted that if Britain was to achieve the sustained economic growth and the exports that had eluded her, she would have to enter the European Economic Community and share in its buoyant economy; but this aim was not achieved until 1973, by which time it was too late to arrest Britain's decline as a manufacturing power.

In the mean time, the government was impressed by the success of economic planning in France which achieved rapid growth without the problems of inflation. As a result, Macmillan found himself attracted towards planning during the early 1960s. In 1961 he strongly backed the National Economic Development Council, whose object was to bring together employers, the TUC, and ministers to hammer out an agreed strategy; the NEDC set a target of 4 per cent annual growth for 1961–6. The unions, however, refused to cooperate in restraining wage demands. Therefore the Chancellor, Selwyn Lloyd, imposed an eight-month pay pause on the public sector in 1961, and asked employers in the private sector to cooperate. Subsequently he indicated a guideline of 2 or 2.5 per cent for wage settlements, on the understanding that they would be matched by productivity agreements. These efforts at imposing a policy for prices and incomes did have a measure of success. However, by 1962 the stop/go cycle had attracted a great deal of criticism of the government as a result of the increase in unemployment. In 1963 Lloyd was replaced as Chancellor by Reginald Maudling, whose preference for a Keynesian expansionist policy was reinforced by the need to prepare the ground for a general election in 1964. Yet with rapid inflation and a balance-of-payments deficit that was to reach £800 million, there was a strong case for dampening down consumer spending. In the event the election was postponed until the last possible moment, and Maudling made yet another irresponsible bid to expand consumers' spending power for fear of losing votes. The result was to exacerbate the balance-of-payments problems and to bequeath to his successor an unavoidable dose of deflation. It was a depressing conclusion to thirteen years of missed opportunities.

The Liberal Revival

One of the ironies of the 1945–55 period was that the fortunes of the Liberal Party reached their nadir when the policies of the British government were largely moulded by two great Liberals, J. M. Keynes and William Beveridge. Neither was a strict party man, though Beveridge did sit as the Liberal MP for Berwick-upon-Tweed in 1944–5, and the party never benefited fully from their reputation. The Liberals had suffered from a dilemma of their own making since 1931–2 when, by associating themselves with the National government, they had become a right-wing party. With Labour established as the heir to the radical reformist Liberalism of the Edwardian period, there was no future for the Liberal Party as a moderate version of Conservatism. In 1945 there was a slight revival in Liberal activity, but only 9 per cent of the vote and 12 MPs. Having been outflanked by the Labour Party, they found it difficult to carve out any distinctive role. In the elections of 1950 and 1951 their vote was squeezed by the Conservatives; in the latter election only 109 candidates stood, and 6 MPs were returned.

The turning-point in Liberal fortunes may be dated precisely at 1956. In that year the young MP for Orkney and Shetland, Jo Grimond, took over the leadership; he was both intellectually able and an attractive performer on television. Grimond stimulated a good deal of solid work on the reconstruction of Liberal policies; but he also gave the party a greater sense of purpose by mapping out a new strategy. 1956 was important also in that the Suez fiasco gave Grimond the opportunity to re-adopt a radical anti-Tory tone which had been missing for so long. The stop/go economic policies of the Macmillan government in the late 1950s and 1960s made disillusioned Conservatives susceptible to Liberal appeals. But while many of the new votes attracted by the party seemed to be coming from former Conservatives, the Liberal activists were often drawn from people who were, or who would otherwise have been, in the Labour Party. Labour's repeated electoral failure and increasingly union-bound image alienated many radicals. Grimond's rejection of class-based politics represented a fresh approach, and his case for a realignment of the left seemed a credible way of ending the years of Tory domination. By 1958 unmistakable signs of Liberal revival appeared, in the form of a sharply rising vote in by-elections culminating in a gain from the Conservatives at Torrington. Although the promise was not sustained at the 1959 general election, a second phase of revival was underway by 1961, fuelled by dissatisfaction over inflation, rising rates, and higher mortgage costs. In 1962 the Liberals won a sensational by-election at Orpington, a Kent commuter seat. This appeared to herald a breakthrough into suburban England by a party long confined to rural strongholds in the Celtic fringe. The Liberals were often to use by-election campaigns to restore momentum, and in particular to attract new funds and members. The opinion polls began to suggest that a high proportion of electors were prepared to vote Liberal if they believed they had a chance of winning. The 1960s also marked a major expansion in the Liberal foothold in local government.

In the event, the Liberal revival at general elections was modest. In 1964 the party achieved its best postwar result, with over 3 million votes or 11.2 per cent. Only 9 MPs were elected, though the total rose to 12 in 1966. Although this was disappointing, the Liberals were no longer a marginal factor in politics. The other parties realized that the growth of Liberal support could decisively affect their own fortunes. And although they often claimed the Liberals had no policy, they frequently borrowed the ideas being generated by the party. In fact, the Liberals were responsible for propagating a number of the major policies of the period from 1960 to 1980. On the domestic front they advocated the reform of the unpopular rating system in favour of site-value rating, the introduction of schemes of workers' co-partnership to facilitate better productivity and industrial relations, and the devolution of power to elected authorities in Scotland and Wales. As in the past, the party devoted considerable attention to reforms of the political system such as the House of Lords, televising Parliament, and proportional representation. Above all, the Liberals began to campaign for British entry into the EEC years before the other parties were prepared to think seriously about the question. In many ways Europe filled the role once occupied by free trade in Liberal politics. It was seen as offering the prospect of higher living standards, but also represented a great *moral* good in Liberal eyes, in that it brought a promise that the European powers would never again go to war with one another.

In spite of the influence exerted by the Liberals, their revival, in the 1960s at least, never fully realized the party's potential. There were a number of explanations for this. First, the party lacked the financial backing from big business and the trade unions on which their rivals relied. The electoral system denied them the parliamentary representation to which their vote – the largest of any Liberal Party in Europe – entitled them. Although the party fielded more candidates, it had, as yet, no real strategy for concentrating its resources in the limited number of winnable constituencies.

Secondly, the Liberals found it difficult to make much impact on the voters with their new policies. In Parliament they were marginalized, and most of the press neglected them except after by-election victories. In any case, proposals on Europe and constitutional reform attracted no great interest; on devolution they were to some extent outflanked by the re-emergence of Welsh and Scottish nationalism in the mid-1960s, while their distinctive stand against restrictions on Commonwealth immigration was a vote-loser.

Thirdly, the party's support was somewhat incoherent. Many of the new activists were urban-based radicals. They sat uncomfortably alongside the older rural Liberals, who were less interventionist in economics and less liberal on social questions. Beyond the central core of activists on whom the party depended lay the larger and rather ephemeral protest vote, which often comprised middle-class erstwhile Conservatives unhappy about their rates and mortgages but by no means converted to radical liberalism. They were attracted more by the style of Grimond's party – something slightly removed from sordid party politics – rather than by detailed policies. Finally, although the Liberals established strong roots especially in local government, they found that the realignment of the left was not yet taking root in the country. Once Labour recovered a belief that it could win an election, which it did after Harold Wilson became leader in 1963, it showed no interest at all in the Grimond strategy. The narrow victory in 1964 pushed this off the agenda for some years to come.

Adjusting to Decline

The three successive election defeats suffered by the Labour Party in the 1950s inevitably created a mood of pessimism and frustration on the left of politics. In fact the party's loss of votes was slight, from 48.8 in 1951 to 43.8 per cent in 1959, but magnified by the electoral system. But it began to appear that Labour was too old-fashioned in the eyes of the electorate, too closely associated with the unpopular trade unions, and insufficiently in touch with the aspirations of a property-owning democracy. For a time, a fashionable thesis held that much of the working class had been seduced by the consumerism of the 1950s and now aspired to middle-class aims and values, among which was Conservative voting. This did not survive serious investigation, which showed that the marginal shifts by affluent working-class voters were ephemeral. However, Labour devoted much effort to diagnosing its problems. After the 1955 election, an investigation into party organization pointed to the drop in individual membership since 1952 and the lack of professional agents, especially in the marginal constituencies. These were deep-seated failings of the party organization, and little was done to remedy them.

However, Labour's gradual decline was interrupted by short-term political factors in the early 1960s. First came the untimely death of Hugh Gaitskell and his replacement by Harold Wilson. Though admired and respected as a man of intellect and integrity, Gaitskell had proved to be a poor tactician, unduly rigid, and a divisive leader. He became embroiled in major internal controversies over the removal of the socialist commitment in the party's constitution and over unilateral nuclear disarmament. From 1963 Harold Wilson brought a very different style of leadership. He had an exaggerated reputation as a left-winger, largely because he had resigned along with Nye Bevan in 1951 over NHS charges; but this reassured the left, who were more cooperative in the 1960s than they had been, especially as Wilson promoted their leading figures. In fact Wilson showed himself a very pragmatic politician, whose chief formative influences were his Liberal background and his wartime civil-service experience. He succeeded in conciliating the various interest groups within the party and in maintaining a degree of unity throughout the 1960s.

Wilson became the most successful leader of the opposition in the twentieth century. He cultivated the press very successfully, took advantage of the economic difficulties of the Macmillan government, and deftly exploited the pervasive concerns about British national security during 1963–4. The atmosphere of suspicion about security in official circles began in October 1962 with the resignation of an Admiralty clerk, John Vassall, who was convicted of spying for the Soviet Union; this led to the resignation of the responsible minister. Then in 1963 it became known that another of the many traitors within the Foreign Office, Kim Philby, had fled to the USSR. In June came the resignation of the Secretary of State for War, John Profumo. He had had an affair with a call-girl, Christine Keeler, who had also been involved with a Russian diplomat, Captain Ivanov. The intrinsic importance of this affair by no means justified the sustained public debate that it generated, for there was no significant threat to British security. But Profumo had lied to the House of Commons and the Prime Minister had been made a fool of. The political effect of the Profumo affair was thus to keep the government on the defensive for a long time and to demoralize Macmillan, who appeared to be seriously out of touch.

Wilson undoubtedly presented a fresh and more attractive image than either Macmillan or his extraordinary successor, Sir Alec Douglas Home, who had to renounce his peerage in order to become Premier. Many observers, then and since, gave great emphasis to Wilson's proposals to exploit Britain's scientific and technological skills in order to rejuvenate the economy. But the British voters were little interested in science and much more impressed by Wilson as a middle-class provincial who reminded them that he, too, had a mortgage to pay. In the event Wilson won the much-delayed election in October 1964, but with a majority of just 3. This result has been described as 'one of the major upheavals of British democratic history'; but it is difficult to see the basis for this claim. The real significance of the 1964 election is that, in spite of all their problems since 1961 and in spite of the very favourable publicity won by Wilson himself, the

Conservatives almost *won*. Labour's share of the vote rose fractionally from 43.8 per cent in 1959 to 44.1 per cent, with the Tories on 43.4 per cent.

In economic affairs the Wilson premiership showed a striking continuity with that of his Conservative predecessors. Economic growth rates during 1964–70 averaged 2.7 per cent per annum, respectable but much inferior to the record of Britain's rivals. As before, spurts of growth were regularly interrupted by balance-of-payments deficits and speculation against the pound; since Britain lacked the reserves necessary to defend the currency, there was mounting pressure for devaluation and futile deflationary policies on the government's part. As in the 1950s, wages rose faster than output. So, too, did public expenditure which absorbed an extra 5 per cent of gross domestic product by 1970. This reflected the new government's determination to raise benefits for needy groups like pensioners and to offer rate rebates to the very low-paid. But the Exchequer also came under financial pressure because of the steady increase in the number of elderly people in the population and the need to provide higher education for the postwar generation now reaching their late teens.

The Labour government inherited a serious economic problem in the shape of a balance-of-payments deficit of £800 million. This presented the Cabinet members with an opportunity to cut themselves free from the mess into which their predecessors had fallen by a quick devaluation of the pound. Wilson, however, was highly conservative, and seems to have been influenced by the political effects of the 1949 devaluation; he refused to countenance the idea. Instead James Callaghan, the Chancellor, opted for surcharges on imports and increases in taxation. This only led to a run on the pound and the raising of the bank rate to 7 per cent. In this way the new government swiftly destroyed the hopes of economic growth, and condemned the country to a phase of deflation and stagnation in order to maintain the value of the pound and their own political prestige. This proved to be a crucial error from which the government never fully recovered. It was, moreover, a failure, since devaluation could not, ultimately, be avoided. The strategy made nonsense of the elaborate scheme that had been prepared for a major new ministry, the Department of Economic Affairs. It produced a grandiose National Plan which projected economic growth of 25 per cent during 1964–70. But this was window-dressing, for the plan had no teeth and was, anyway, torpedoed by the key decisions about defending the currency made in the Treasury. The new department served the purpose of setting the Prime Minister's chief rivals, Callaghan and George Brown, in competition with one another, but was of little practical relevance to the economy. It demonstrated that Wilson enjoyed the same skills of manipulation and presentation as Macmillan, but little appetite for getting to grips with underlying economic problems.

During 1965 and 1966 the government sought to defend the pound by invoking the support of the USA. In September 1965 the Prime Minister reached a secret agreement with President Johnson under which, in return for American backing, Britain would resist pressure to devalue and would maintain her costly role east of Suez. Johnson also demanded active British involvement in the Vietnam War. Though Wilson drew the line at this, he continued to support American policy in south-east Asia. This humiliating deal simply locked Britain into two doomed strategies, external and internal, and ultimately damaged the government's standing at home. It meant that when there was a modest improvement in the balance of payments in 1966 which created favourable circumstances for a planned devaluation, Wilson again missed his opportunity. His strategy was primarily led by political calculations. At the end of March he held the long-expected general election, which transformed his tiny majority into one of 98.

After this high point in the life of the Labour government, its central policies collapsed and it went into a prolonged and almost unrelieved decline up to 1970. The worsening of the balance of payments in 1967 led James Callaghan to raise bank rate to 7 per cent again and impose a massive dose of deflation on the economy. This failed to defend the weakened currency, and in November the pound was devalued from $2.80 to $2.40. Once again a stubborn refusal to make timely adjustments had resulted in a panic devaluation and, more seriously, had led government to inflict prolonged stagnation upon the economy. Economic weakness also forced the government to abandon its east-of-Suez role and cut defence generally (see p.282) at this point. Moreover, Britain's second application to join the EEC in May 1967 was summarily rejected in November. To add to the problems, Labour now began to lose by-elections, even in very safe seats, both to the Conservatives and to the Welsh and Scottish Nationalists.

The central failure of Harold Wilson to fulfil the high expectations on the economic front now led much of the Labour movement to rebel over other aspects of policy. His handling of the rebellion by white settlers in Rhodesia, who had illegally declared their independence in 1965, was a case in point. The rebels were fearful of direct British intervention, and would have backed down before a threat of military action. Wilson, however, disastrously undermined his position by letting the Rhodesian regime know that he would never resort to force. Instead, he imposed economic sanctions which he claimed would bring down the regime in weeks if not months. In fact the rebellion survived the sanctions comfortably. Wilson had been playing safe with the British public, who were not interested in the issue, but at the price of antagonizing many in his own party.

A similar and more divisive question was Wilson's pro-American line in Vietnam. Foreign policy became far more important than economic issues in left-wing politics in the later 1960s, and by 1968 a large anti-war movement had emerged in protest at the destruction of life by the massive American force in south-east Asia. Wilson alienated a whole generation of young radical activists by his stance. They were driven outside the Labour Party into extraparliamentary pressure groups, while the disillusion and dismay within the constituency organizations resulted in a collapse of voluntary work for the party.

Finally, the Prime Minister continued to unite many of his left-wing opponents with the trade unionists by taking on the question of reform of the unions in 1968. Wilson, who had come to the Labour Party via Oxford and the civil service, had no base in the organized working-class movement. In 1964 he had tried to build bridges between government and unions by making Frank Cousins, leader of the Transport and General Workers Union, a Cabinet minister. But in 1966 Cousins resigned in protest against the introduction of the Prices and Incomes Board. Like many union leaders, he refused to countenance any legal interference in wages and industrial relations. None the less, in 1968, when the report of Lord Donovan's Royal Commission on trade unions appeared, Wilson determined to act. He felt, correctly, that the public was beginning to blame union militancy for Britain's economic difficulties, and would probably support a bold move by the government. The new Secretary of State for Employment and Productivity, Barbara Castle, prepared an ironically titled White Paper, *In Place of Strife*, in January 1969. She proposed to empower herself to require unions to hold pre-strike ballots of their membership, to insist on a conciliation period of twenty-eight days before a strike took place, and to impose a settlement where an inter-union dispute led to an unofficial strike; an Industrial Board was to be set up to fine those who contravened the new rules.

The trade unions reacted with anger and shock to these proposals. Yet the Prime Minister could probably have ridden out the opposition if the critics had not drawn support from both left and right of the party, and from the Cabinet as well as the rank-and-file. Since Wilson had always regarded the parliamentary left wing as the one essential prop to his leadership, he was very reluctant to alienate it over the White Paper. A new danger loomed as Callaghan sought to use the issue to restore his own fading career by making himself the unions' champion in Cabinet; he cynically argued that trade-union legislation ought to be left to the Conservatives. Not surprisingly, the party's National Executive rejected the government's proposals by 16 votes to 5, and the Chief Whip advised that he would be unable to guarantee a majority in the Commons. In these circumstances there was a real prospect of a break-up of the government similar to that in 1931. The Prime Minister therefore accepted a humiliating defeat and dropped the proposed legislation. This was not the only such setback in 1969; the government also abandoned its bill to reform the House of Lords under the pressure of internal opposition led by Labour conservatives such as Michael Foot. It began to appear that in spite of its large majority the government was unable to govern.

At the Exchequer, meanwhile, Roy Jenkins imposed a stern deflationary strategy. His 1968 budget had taken £900 million out of the economy in indirect tax rises, and had imposed a 3.5 per cent maximum on wage increases. The 1969 budget continued the deflationary line, and by the autumn the balance of payments had greatly improved. However, all this was achieved at the expense of economic growth; Labour's original hopes of breaking out of the stop/go economy had come to naught. Wilson was misled by the improvement in Labour's standing in by-elections and opinion polls during the winter of 1969–70 into thinking that he could win an election in the spring. But when Jenkins failed to provide an electioneering budget the party was doomed. Consequently the Conservatives returned to office at an election in which the turnout dropped to 72 per cent. This was a sign of the apathy and disillusion amongst Labour supporters, who no longer saw the purpose in returning their party to power.

Although Harold Wilson managed to return as Prime Minister in 1974, he has continued to attract a high level of criticism as Prime Minister.

This is partly because he has been blamed by both right and left. The younger generation of 1960s radicals experienced the high hopes and then the disappointment that Wilson aroused. On the other hand the right looks back to his premiership as part of the decadent phase of Keynesianism. It must, however, be remembered that by comparison with, say, the Thatcher governments of the 1980s, Wilson achieved an impressive record in terms of higher economic growth, lower unemployment, and a stronger balance of payments. He also presided over major achievements in the fields of education and social reform (see p.263–9). However, to say that the economy performed better in the 1960s than in the 1980s, however true, is not to claim very much! Although Harold Wilson's defenders have suggested that his reputation is certain to recover in time, his biographers have been a little faint in their praise. This is because it is very difficult to exonerate Wilson on three charges. First, despite being technically better qualified to manage the economy than any other twentieth-century premier, he made major misjudgements and failed to arrest Britain's economic decline. Second, he gave way to the prevailing conservatism in his party and missed the opportunity to begin the modernization of the political system. Wilson largely failed to capitalize on the Fulton Commission's report on the civil service or on Richard Crossman's enthusiasm for reforming the House of Commons, and the attempt at updating the Lords was abandoned. Third, in spite of his considerable skill in managing the Labour Party and keeping it united, Wilson left it seriously demoralized, and consequently vulnerable to the lurch to the left which was to condemn the party to the wilderness from the end of the 1970s.

The Decline of the Parliamentary System

One of the chief differences between Victorian-Edwardian politics and the twentieth century was the decline of the great debates over constitutional and electoral reform. After the House of Lords controversy of 1910–11 and the resolution of the suffrage question in 1918 and 1928, the parties largely neglected such topics. When Labour won power in 1945, there was little they wished to change in the system of government; to some extent, wartime experience in office and the feeling that the civil service was on their side strengthened the party's natural conservatism. Even the fact that Labour lost office in 1951 despite winning more votes than the Conservatives failed to stir interest in reform of the electoral system. On its return to power in 1964 the party showed itself to have become absorbed into the Establishment; it had no ambitions to implement fundamental change, aspiring instead to make the existing system work. Only two reforms of any importance occurred in the post-1945 era. In 1958 Macmillan introduced life peerages, a shrewd way of propping up the existing hereditary chamber which was an extraordinary anachronism in the second half of the twentieth century. The other change came in 1969, when Labour lowered the voting age from twenty-one to eighteen, thereby enfranchising 3 million extra voters. But this occasioned nothing like the debate which had traditionally attended a reform of the franchise.

In these circumstances the parliamentary system in Britain suffered a prolonged decline during the entire period from the 1940s to the 1990s. In fact, some features of this decline, such as the tightening grip of the party whips, had been evident since the 1880s. But by 1945 back-bench MPs were treated as obedient lobby-fodder, and a government, even with a small majority, was very rarely in danger of being defeated. The extensive use of patronage in the form of ever-increasing ministerial posts was one way of emasculating potential rebels. But in any case MPs were poorly equipped to do their job. Salaries remained low, working facilities very poor, and few were able to carry out the research or gain access to the information that would have been necessary to criticize government policies effectively. Parliament had, in fact, largely ceased to perform its traditional role of checking on government expenditure; debates on the estimates had deteriorated into largely ritual occasions for general political discussions.

Ministers increasingly took advantage of delegated legislation which, in effect, empowered them to take decisions without the need for parliamentary approval. When the Attlee government created the nationalized industries, it practically excluded any detailed parliamentary criticism on the grounds that this would impede efficiency. Both Attlee and Churchill kept the decisions over the development and testing of atomic weapons secret. Later in the 1950s, when the nuclear energy plant at Sellafield (Windscale) in Cumbria suffered a major accident which resulted in the widespread release of radioactivity, it was kept secret by Macmillan, then Prime Minister. These were exceptional errors. More typically, governments presented Parliament with decisions too late for them to be effectively scrutinized. The worst area was defence, in which the government allowed itself to be regularly overcharged by the industries supplying its weapons and machinery. It rarely exercised any effective control over schemes to design and produce new aircraft. The development of the supersonic airliner, Concorde, was the classic case of a very costly piece of technology which never found a market sufficient to justify the huge investment of taxpayers' money.

But Parliament was by no means the only institution which became more decorative than functional. The Cabinet found itself increasingly reduced to rubber-stamping policies, as power concentrated around the Prime Minister. Although the actual power of the Prime Minister remained, as always, subject to the vagaries of events, especially on the economic front, and his influence seemed to fluctuate according to the temperament and aims of each incumbent, none the less there was a definite tendency for government to become more prime-ministerial. It became normal for each premier to retreat into a clique of personal advisers, and to introduce new policies through carefully controlled Cabinet committees. They also made increasing use of the press and television to strengthen their position. Attlee was, in this respect, the last of the old-style premiers who took little trouble with the media. But to later Prime Ministers such as Macmillan and Wilson, the manipulation of the media was a key factor in their retention of power. It became normal for them to leak information – for which others were liable to be prosecuted – in order to promote their policies or to undermine their rivals and opponents. Moreover, finding that Britain's economic problems were intractable, figures such as Macmillan, Wilson, and, later,

Mrs Thatcher, increasingly succumbed to the temptation to cut a figure on the international stage, where it proved easy to earn plaudits and bolster their domestic prestige even though very little was accomplished. On the whole, the press and television acted as willing tools of their self-promotion strategy.

As Parliament became increasingly a spectacle, many of the important issues of politics were debated outside its walls under the aegis of pressure groups and protest movements of various sorts. Some of these did aspire to gain a direct footing in Parliament but were frustrated by the unrepresentative pattern of representation. From 1958 the commitment of the main parties to a nuclear weapons policy came under attack by the Campaign for Nuclear Disarmament. This was triggered by the events of 1956 and 1957, notably the Suez crisis and the Soviet invasion of Hungary, but it really reflected the conviction that the conventional politicians were too negative to be able to save Europe from an impending conflict. Though a classic middle-class, intellectual pressure group in its origins, CND rapidly reached far beyond this to mobilize a large section of British society. Its annual marches from Aldermaston in Berkshire to London became a major feature of the political year. In due course CND became divided between those who wished to remain above party politics and those who saw Labour as the best vehicle for achieving its aims. The triumph of the supporters of unilateral nuclear disarmament at the 1960 party conference marked a high point, but the Gaitskellite leadership soon overthrew the disarmers by mobilizing the votes of the right-wing trade unions. Thus, by 1961–2 CND suffered a decline, though it showed a capacity for revival during the next twenty years.

To some extent the momentum behind the CND fell off as a result of the resolution of the Cuban missile crisis in 1962 and the signing of the test ban treaty in 1963; but its supporters were also diverted into the campaign against the American role in the Vietnam War in the later 1960s. This was one of the chief issues that drew many of the new university students into political campaigning in this period. The British students were simply part of a broader rebellion by young, middle-class idealists against the destructive impact of Western weaponry in a poor, third-world country. This movement also involved considerable numbers of young women, and thereby had an indirect effect in promoting a new political awareness amongst the well-educated, post-1945 generation of British females. The Women's Liberation Movement of the later 1960s and 1970s marked a sharp break with established patterns of feminist campaigning in Britain. It was markedly anti-parliamentary, avoiding not only the political parties but the formal structures associated with male-dominated organization. However, during the 1970s and 1980s feminists modified this stance to some extent. They increased internal pressure upon the parties to adopt more female candidates and established the '300 Group' to coordinate this effort. It did not begin to bear fruit until 1987 when, after a prolonged period of stagnation, women's representation in the House of Commons rose to 41 or 6.5 per cent.

By the 1960s Parliament had became out of touch with another element of British society – nationalism in Wales and Scotland. During the prolonged period of Labour domination in the Celtic fringe, the politicians had largely taken these countries for granted. Despite the foundation of Plaid Cymru in 1925 and the Scottish National Party in 1934, it appeared that the cause of self-government appealed to no more than a tiny minority. The conventional politicians were therefore somewhat taken by surprise by events in the mid-1960s. They should not have been, because the Liberals had already begun to capture Scottish highland seats by capitalizing on the sense of neglect felt by these far-flung districts. In both Wales and Scotland, the long-term decline in the regional economy and the tendency to suffer much worse unemployment than the south of England underlay the revival of nationalist support. Both Labour and the Conservatives could be portrayed as remote, Westminster-based parties whose economic policies had failed. The Nationalists effectively exploited the growing volatility of the electorate especially at by elections. In Wales, opinion had been stirred by the decline of the Welsh language and by the flooding of several valleys to provide water for English cities. Labour had created a Secretary of State for Wales with a seat in Cabinet, but this was not enough to satisfy the sense of grievance.

In 1966 Gwynfor Evans captured a Labour seat at the Carmarthen by-election; this was followed in 1967 by an even more dramatic upset in the Lanarkshire seat of Hamilton, when a large Labour majority fell to Mrs Winnie Ewing, the Scottish Nationalist. In local elections from 1968 onwards, the Scottish Nationalists made large inroads into Labour support in the central industrial belt. Although both Nationalist parties lost their by-election gains at the 1970 general election, their revival was a lasting achievement, for they used their new membership and organization as a base from which to mount fresh attacks on the Westminster parties in the 1970s. In the long run this forced Parliament to reconsider its opposition to the devolution of power to the regions.

17

The Permissive Society

During the late 1950s and 1960s, British society exhibited many symptoms of outright rebellion or divergence from conventional social behaviour. This was often explained in terms of a reaction amongst the younger generation, which had been cushioned by material prosperity and thus felt able to turn its back on adult society by developing alternative ideals and cultures. These expressions of revolt began with the rather cultivated aggression of the Teddy Boys of the late 1950s and rock-and-roll music, but during the 1960s this gave way to a less anti-Establishment youth culture. On the literary front the so-called 'Angry Young Men' seemed to react both against the materialism of the consumer society and against the political elite, somewhat discredited in the aftermath of Suez. Foremost among them were John Osborne, author of *Look Back in Anger* (1956), the theatre critic, Kenneth Tynan, and John Brain, who wrote *Room At the Top* (1957).

Yet it is easy to exaggerate the extent to which the arts world became alienated from the Establishment. In many ways its vitality became a cause for national pride, partly because of its sheer commercial success. During the 1960s Britain generated a wealth of new novels, films, and musical styles; her theatres flourished and her television was widely regarded as the world's best. Newspapers alone experienced a decline in the form of the squeezing of the middle-market press and the growth of the disreputable tabloid papers. The outstanding successes of the younger generation were the popular-music industry, led by groups like the Beatles and the Rolling Stones, and the design and fashion industry, which produced models like Jean Shrimpton and the clothes shops of Mary Quant. By and large they were happily absorbed into the British Establishment. The Beatles, for example, received MBEs from Harold Wilson. Further political recognition of cultural achievements came in 1965, with the creation of a new ministry for the arts under Jennie Lee. A widely acknowledged success, the minister managed to increase the subsidies dispersed by the Arts Council from £3 million pounds in 1965 to £9 million by 1971.

On the other hand, those contemporaries who characterized the 1960s as the Permissive Society believed that social and cultural changes brought with them a general lowering of moral standards. A characteristic spokeswoman for this reaction against liberalization was Mrs Mary Whitehouse, who directed her attacks particularly against television because of the prominence she felt it gave to violence, sex, and bad language. Other critics focused on the prevalence of sexual relationships outside marriage, and on the use of drugs. Indeed, for a time drugs like cannabis became fashionable among the young; it was estimated that at least half a million people used cannabis, which led to some pressure for its legalization.

Certainly the campaigns of the moral reformers appeared to have little, if any, impact on behaviour. Perhaps the most signal triumph for liberalism came in connection with the censorship of books and plays. Traditionally, productions could be prevented by the Lord Chamberlain on grounds of obscenity or by prosecution in the courts. In 1960 Penguin Books found itself up on such a charge for publishing a paperback version of D. H. Lawrence's novel, *Lady Chatterley's Lover*. The case exposed vividly the shifts in social attitudes that had taken place since the interwar period. At one point the prosecuting counsel demanded: 'Is it a book you would wish your wife or servants to read?' The jury lost little time in returning a verdict of not guilty. As a result, the role of the Lord Chamberlain was severely undermined, and by 1968 it had been phased out altogether.

However, the Permissive Society was nothing like as liberal or as radical as its critics believed. In many ways the battles fought over changing social behaviour were very old ones. The Edwardians had believed that their children were in revolt; the 1920s saw much alarm over the social life of the 'flappers'; and the Second World War was believed by some to have undermined the institution of marriage. This is a reminder that each society witnesses a resurgence of fears among the middle classes about the disrespect for authority or about the lifestyles of the working classes, or among the middle aged about the young, or among men about women. In any case, liberal attitudes were by no means as widespread in British society in the 1960s as might be suggested by the attention attracted by the youth culture of the time. For example, the reforms in the law relating to capital punishment, women's equality, or homosexuality represented the work of an enlightened, educated, middle-class minority which succeeded in bringing its influence to bear upon Parliament but was never backed by a majority in the country. To some extent the legal reforms affecting homosexuals, women, and divorced people probably did help to promote a more tolerant attitude in the long run; but it seems doubtful whether the same was true with regard to racial prejudice or capital punishment.

Finally, we must beware of following contemporaries in equating changes in behaviour with a deterioration in moral standards. In some ways twentieth-century society became more, not less, moral. For example, there had been a major decline in drunkenness since the Edwardian period. Very slowly drinking habits were moving in a Continental direction; that is, they increasingly involved respectable entertainment for whole families rather than for adult males alone. Even the growth of drugs was no more than a reversion to Victorian practice. The other great Victorian-Edwardian vice which had gone into headlong decline was prostitution. This was forgotten by those who criticized the young for premarital sex, and were forever warning of the decline of the institution of marriage in the face of all the evidence of its growing popularity.

Liberal Reform

The spirit of innovation that produced the characteristic economic and welfare reforms of the late 1940s did not extend to social and moral issues. The leading personnel of both main political parties were fairly conservative in their attitudes. During the later 1950s, however, perhaps because interest in economic questions had died down, there was vigorous debate over several social questions. A younger generation began to dilute the conservatism of Parliament, though the full impact was not seen until the return of a new Labour government in 1964 which presided over major reforms of the law on capital punishment, divorce, abortion, and homosexuality. There was, however, nothing inevitable about this; but for the appointment of a comparatively liberal Home Secretary, Roy Jenkins, in 1965, few of these measures would have been enacted.

Throughout the postwar period the Labour MP Sydney Silverman regularly introduced bills for the abolition of capital punishment. The case for reform was threefold: it was seen as morally wrong for the state to take life; there was a danger of convicting the innocent; and the death penalty was not an effective deterrent. There were 312 murders in Britain in 1900, and the number remained very close to that right up to the 1960s. Up to the hanging of James Hanratty in 1962 Home Secretaries had invariably sanctioned the death penalty, but they grew increasingly uneasy about it. Eventually Silverman's bill obtained a large majority in the Commons, and it became law in 1966 with some assistance from Roy Jenkins. In subsequent years the opponents of abolition, who were conscious that the public still supported the death penalty, promoted further debates on the issue; but the abolitionist majority in Parliament remained large. MPs became increasingly reluctant to reintroduce capital punishment because of a collapse of confidence in the judicial system. Defendants were repeatedly convicted on what had previously been capital offences as a result of false confessions extracted by the police and other discredited types of evidence; it became clear by the 1980s that many innocent people would have been executed but for the 1965 Act.

The equally emotive issue of abortion also came to prominence in the 1960s. In the past, politicians had been almost totally uninterested in the topic. Yet very large numbers of working-class women, faced with the ruination of their health by frequent pregnancies and lacking the means to avoid them, sought abortions; even in the interwar period many died as a result. Yet in the 1920s only a few organizations such as the Women's Co-operative Guild had been prepared to advocate the legalization of abortion. In the post-1945 period the problem became more widespread because of the increase in extramarital sex among young people. As with the birth-control issue in the 1920s, however, the political parties were reluctant to adopt a view on a question which would attract the wrath of the Catholic Church.

In 1967 a Liberal back-bencher, David Steel, used his place in the MP's ballot to introduce the Abortion Law Reform Bill. This allowed women to obtain an abortion under the NHS within 28 weeks of conception; however, it required the consent of two doctors, which in many parts of the country was not forthcoming. As a result some 22,000 women had legal abortions in 1968, the total rising to 75,000 by 1970 and 128,000 by 1980. Though alarmingly high, the figure stabilized at around this level for some years; moreover, it was appreciated that without the change thousands of women would have been exposed to the appalling risks entailed in illegal back-street abortion. For this reason the repeated efforts made by anti-abortionist MPs to modify the Act in the 1970s and 1980s were largely unavailing.

The question of homosexuality also came to the fore in this period. Since the Criminal Law Amendment Act of 1885, acts of 'gross indecency' between men had been punishable by up to two years' hard labour. This was the legislation under which Oscar Wilde had been prosecuted. It was in 1957 that the report of the Wolfenden Committee recommended changes in the law so that homosexual acts should not be illegal when committed in private between consenting individuals aged twenty-one or over. Subsequently the position of homosexuals in society began to be more freely discussed, partly as a result of the establishment of *Gay News* in 1962. In 1967 the Labour MP Leo Abse introduced a bill along the lines recommended by Wolfenden. However, this was an embarrassing issue for politicians. Very few members – Tom Driberg was an exception – were known to be active homosexuals. Conservatives usually adopted particularly hostile attitudes on the subject of homosexual law reform; but this was probably a defence mechanism, whether conscious or otherwise, to divert suspicion at a time when public views were felt to be unsympathetic. In view of the habit of Conservative selection committees of choosing candidates who were unmarried men from public-school backgrounds, there seems little doubt that the Conservatives in Parliament included a relatively high proportion of members with homosexual experience. Whether attitudes began to change as a result of the passage of Abse's bill is hard to say. The mid 1970s saw the ruin of a major political career, that of the Liberal leader, Jeremy Thorpe, largely as a result of allegations that he had been involved in a homosexual relationship. But by the 1980s it had become possible for a number of public figures, including politicians, to admit their homosexuality without destroying their careers.

The Family and Marriage

Postwar discussion about the undermining of conventional behaviour and social institutions has to be seen in the context of the remarkable popularity of marriage for a quarter of a century after 1945. The war itself inevitably disrupted the normal pattern of married life, and during 1941–5 the marriage rate among women fell to 67.6 per 1,000. But war had done nothing to diminsh the popularity of marriage. The rate rose to 75.7 per 1,000 during 1946–50, which was higher than the prewar level, and thereafter it rose progressively, to reach 94.2 per 1,000 in 1966–70. The year 1972 marked the peak, after which a slight decline set in. The pre-1939 trend towards younger marriages was sustained, so that by 1971 60 per cent of all women in their early twenties were married, compared with 25 per cent before 1939.

Pessimists pointed to the increase during this period in the rate of divorce, which had become easier as a result of legal reforms in 1923 and 1938 which had put men and women on an equal footing. But divorce was still rare and not respectable. It was usually achieved only when one partner agreed to be the guilty party so that the other could prove a charge of adultery. Thus as late as 1951 only 7 per cent of marriages ended in divorce, and the reformers felt that couples often remained trapped in unhappy marriages because of their reluctance to admit to adultery. During the 1950s Eirene White MP had attempted to introduce a bill to overcome this obstacle, but without success. It was not until 1969 that a new measure was enacted which introduced the principle that where a marriage had broken down it could be dissolved by mutual consent of both parties after three years. By 1974 19 per cent of marriages ended in divorce after ten years. However, this increase was not primarily due to the change in the law, for the rise in the rate of divorce preceded the 1969 Act. In some cases it was the result of the growth of marriage amongst the very young. But it also reflected the fact that as life expectancy increased, so the period through which marriages had to survive lengthened; before 1914 the typical marriage had been a briefer affair, terminated by the death of one partner.

The other long-term influence upon marriage was the prolonged decline in fertility which had made the two-child family typical by 1939. The Royal Commission on Population which reported in 1949 reflected interwar fears that the growing resort to birth-control techniques would result in a fall in population. This did not occur, but the increase levelled off in the 1970s and 1980s. By 1961 the total UK population had reached nearly 53 million. Low fertility and greater longevity combined to make for an ageing population; whereas in 1939 65-year-olds made up 9.2 per cent of the population, by 1981 they formed 15 per cent.

The key influence upon family size was the use of contraceptive methods by 90 per cent of all couples in the 1950s. But because the birth rate was now so much within parental control it became rather volatile and almost impossible for demographers to predict, though this did not stop them trying, often with unhappy results for government policies. During 1941–5 the birth rate per 1,000 population stood at 15.9, about the same as before the war. A dramatic 'baby boom' pushed it up to 18 per 1,000 during 1946–50. Thereafter the rate subsided to its former level, but demographers were surprised when it jumped again to 18.1 in the early 1960s before dropping back to 14.0 during 1971–5. Policies for school- and house-building were hastily modified in response to these fluctuations.

Limited Emancipation for Women

After the liberating experience of wartime, it might seem paradoxical that most women were keen to abandon employment to take up motherhood and domesticity once again. As a result the 1950s have often been seen as a missed opportunity for women. However, much depends upon what criteria are used to measure women's role and the changes affecting their lives. An interwar feminist like Vera Brittain saw the welfare state as proof of the emancipation of her sex. Moreover, the young women who cheerfully opted for a life of domesticity in the 1950s could expect a more attractive version of married life than their mothers and grandmothers had known. Birth control allowed them to enjoy sexual relations without the perpetual fear of pregnancy; their lives were much less dominated by childbirth; and they had a longer period in which to choose to return to paid employment or pursue other interests.

They were, of course, subject to a variety of pressures to conform to the conventional role of women. First, the commercial pressures manifested by women's magazines, driven by the need to attract advertising revenue, continued to encourage women to devote themselves to the pursuit of consumer goods. As *Woman*, *Woman's Own*, and the new *Woman's Realm* reached a peak in sales in the 1950s, five out of every six women read at least one such magazine each week. Second, the political and religious leaders urged women to stick to motherhood in the interests of the nation. 'The great majority of married women', declared Beveridge, 'must be regarded as occupied on work which is vital but unpaid, without which their husbands could not do their paid work and without which the nation could not continue.' For this reason the welfare system was strongly biased in favour of married women; it neglected or discriminated against the woman who was single or separated, and, indeed, the married woman who took paid employment. Governments simply ignored feminist demands that the married but non-working woman should be allowed to claim benefit in her own right independently of her husband.

Finally, during the late 1940s and 1950s women's role was the subject of a good deal of propaganda by assorted experts who claimed some scientific authority for their views. Typical were Dr Benjamin Spock's famous book on *Baby and Child Care* (1947) and Dr John Bowlby's report, *Maternal Care and Mental Health* (1951). Such work was used to make women feel guilty of neglecting their children if they took paid employment; this was believed to lead to psychological problems, juvenile delinquency, and crime. As a result of the evacuation of children and the general disruption of family life during the war, many parents were strongly influenced by such scientific propaganda during the 1950s and 1960s.

However, despite all the pressures, increasing numbers of women combined motherhood and employment. A typical postwar rationale for this strategy was expressed in *Women's Two Roles* (1956) by Alva Myrdal and Viola Klein; they saw it as a matter of *duty* for women to raise children and contribute directly to the nation's labour force. Both working- and middle-class women increasingly undertook jobs, often part-time, in order to enable their families to enjoy the higher standard of living now available. This was largely a by-product of the development of the economy rather than the result of pressure from feminists. By 1947 the government, desperate to boost output in the export industries, appealed to women, especially in the 35–50 age group, to return to work. As a result there were 800,000 more women in paid employment in 1947 than there had been in 1939. Subsequently they benefited from an expansion of the light engineering and electrical industries, which produced consumer goods using unskilled, production-line methods. Further, the steady growth of the clerical, administrative, and lower professional sectors continued to generate opportunities for women in shops, schools, hospitals, and welfare services. As a result, by 1981 45 per cent of women enjoyed paid employment compared to 34 per cent in 1931. The major change was the rise in *married* women's employment. For many decades only 10 per cent of married women had worked, but the figure rose to 22 per cent by 1951, 30 per cent by 1961, and 47 per cent by 1981.

Many interwar feminists regarded this as a considerable measure of emancipation. But the organized women's movement dwindled during the 1950s; the death of Eleanor Rathbone in 1946 symbolized the passing of Edwardian feminism. The last achievement of the earlier wave of feminist organizations came in 1954 and 1955, when R. A. Butler was at last persuaded to introduce equal pay for women civil servants and teachers. Significantly, the Equal Pay Campaign Committee dissolved itself after this success – making the assumption that there was insufficient popular support to extend the cause to industry. This is consistent with the minimal impact made by women in politics at this time. After the election of 24 women MPs in 1945 there was virtually no change; as late as October 1974 the total was only 27. Neither Labour nor the Conservatives were prepared to nominate more than a handful of women in winnable constituencies, nor did they regard feminist causes as of more than marginal importance. Those women MPs who were prepared to take up issues like divorce reform or equal pay were very much on their own. The rising generation of women politicians – Barbara Castle, Margaret Thatcher, and Shirley Williams – devoted themselves to the major party questions, on the assumption that to concentrate on women's causes would be to marginalize themselves. Their generation was surprised to find itself overtaken in the late 1960s by the emergence of what became known as the women's liberation movement, the product of a university-educated and prosperous group of women born in the postwar baby boom. Their successes owed a good deal to Harold Wilson, who was much more alive to women's issues and ambitions than any other Prime Minister. He included several women in his Cabinets – Barbara Castle, Judith Hart, Shirley Williams, Peggy Herbison – and enacted the 1970 Equal Pay Act and the 1975 Sex Discrimination Act.

The Rise of an Educated Society

The neglect of education had been a distinguishing feature of British society throughout the nineteenth century; and in this respect the interwar years had merely continued the tradition. After the war the implementation of R. A. Butler's 1944 Education Act went some way towards improving educational opportunity, but was soon recognized to be inadequate. Basically the Act proposed a system in which children would be transferred at the age of eleven to either grammar, secondary modern, or technical schools according to their 'age, aptitude, and ability'.

In practice there were many problems with this. The secondary modern schools, which actually catered for about three-quarters of all pupils, had no clear purpose at all. Moreover, very few technical schools were ever built. As a result Britain failed, as in the past, to generate the supply of skilled workers that industry needed, in radical contrast to practice in Germany, where most children received vocational training. But the problem was more than a matter of educational policy: in Britain, businessmen largely failed to provide the apprenticeships for school-leavers that were normal in the more successful economies. In addition, the effect of the eleven-plus examination system was stultifying; it led what were regarded as the best primary schools to concentrate narrowly on the mathematics and English needed to pass the examination and to neglect many other aspects of education. Moreover, the number of children who actually passed into the grammar schools was often determined not by the pupils' ability but by the availability of grammar-school places, which varied from one part of the country to another. Above all, the examination at eleven was too early in the pupils' lives; it simply had the effect of channelling middle-class children into grammar schools and excluding many intelligent working-class ones.

The overall result was that Britain continued to allow a high proportion of her pupils to drop out of school before they reached higher education. It is true that between the late 1940s and the late 1950s there was a 66 per cent rise in the numbers staying on into sixth forms, but Britain continued to lag behind other comparable countries. As late as 1969 only 13.7 per cent of seventeen-year-olds were in full-time education, compared to 16.9 per cent in Germany, 36.7 per cent in France, and 75.6 per cent in the USA.

The Labour government of 1964–70 tackled these problems in two ways. First, the Secretary of State, Anthony Crosland, announced in 1965 the replacement of the tripartite secondary-school system by multilateral or comprehensive schools, which all pupils would attend without the need for a special examination. Such schools had been pioneered in thinly populated areas such as rural Wales. By 1970 30 per cent of children attended comprehensive schools. This policy attracted controversy for many years. However, although the Conservatives posed as defenders of the grammar schools, when in office after 1970 they abolished most of them. The main flaw in the new system was that in some areas the retention of selective independent schools had the effect of depriving the local comprehensives of many able pupils. However, in spite of much fashionable criticism of the schools and teaching methods over the next twenty years the comprehensives were a considerable success. Far from neglecting examination performance, as traditionalists and some middle-class parents claimed, the comprehensives achieved a steady increase in the proportion of their pupils who passed the GCE 'O' and 'A' level examinations. They were held back largely by the failure of successive governments to put sufficient resources into the recruitment of teachers qualified in foreign languages, sciences, and mathematics. The lower pay and status of teachers in Britain by comparison with successful industrial states like Germany and Japan continued to be a crucial weakness in British society down to the 1990s.

The second major achievement of the Labour government was the expansion of opportunities in higher education. In 1939 only 50,000 students attended British universities. Many more possessed the academic qualifications for entry but were barred by lack of financial support. By 1961 the student population was still only 107,000. Two sections of the population were especially severely handicapped. Only 3 per cent of the children of the working-class majority reached university. And while 5.5 per cent of boys attended, only 2.8 per cent of girls did. The economy was being deprived of large resources of talent.

In 1963 the Robbins Report recommended that university places should be increased to 197,000 by 1967–8 to cater for the postwar 'baby-boomers' who would be ready for higher education by that time. The Wilson government responded by building nine new universities in the 1960s; in addition, from 1966 onwards twenty-one polytechnics were set up under local authority control. Finally, in 1969 a charter was granted to the Open University which was designed to offer higher education to older students who had missed the conventional route to university but could undertake study by correspondence courses. The other key reform was financial. From 1962 local authorities were required to provide a grant for every student who was offered a place in higher education. But for this subsidy, the expansion would hardly have taken place. As a result there were just over 200,000 university students by 1968, which represented 6.3 per cent of the age group. Britain still lagged well behind her competitors in Western Europe and North America. Nevertheless, taken with the other reforms in schools, the university expansion helped to make education the greatest achievement of the 1964–70 government. It was an irony that by 1970 that government had become the object of bitter attacks by many of the students who had benefited from its policies.

Race and Immigration

Immigration was far from being a novel feature in British society; in the nineteenth century large numbers of Irish people, as well as Jews and smaller numbers of Italians and Germans, had come to reside in Britain. But up to the 1920s Britain had been a net *exporter* of people, largely because she had a surplus in her labour force. After 1945 the surplus turned into a shortage; consequently the governments of this period were happy to boost the output of the economy by welcoming immigrant labour. There were three main sources of immigrants. The West Indian influx began in June 1948, with the arrival of the SS *Windrush* bearing 492 passengers from Jamaica. During the 1950s large numbers came from the Indian subcontinent, notably Punjabi Sikhs, Gujarati Hindus, and Muslims from Pakistan. Thirdly, during the later 1960s many Asians were driven out of Kenya by political pressure.

As Commonwealth citizens these immigrants enjoyed the right of entry into Britain under the 1948 Nationality Act. Although initially they were largely unmarried males who often contemplated a temporary stay in Britain, during the 1960s growing numbers of dependants joined them. In this way complete communities grew up, especially in London, Slough,

Bradford, Nottingham, Leicester, and other towns in which work was available. By 1971 there were 705,000 people of Carribean or Indian origin living in Britain. It was not long before a number of extremist organizations began to exploit feelings of hostility towards the new black- and brown-skinned immigrants. In the 1950s Sir Oswald Mosley attempted through his Union Movement to use the issue to relaunch his career, as he had used the Jews in the 1930s. Colin Jordan organized the White Defence League, and the Conservative MP Sir Cyril Osborne began a campaign against immigration on the grounds that the newcomers were diseased and criminal – echoes of Edwardian attacks upon Jewish communities in London. Such politicians helped to legitimate the resentment towards immigrants felt by less well-off English people. The result was a number of violent clashes between members of the two communities in 1958 in the Notting Hill area of London and in Nottingham.

Throughout the 1950s governments had been considering whether to place a limit upon the number of immigrants, but it was not until 1961 that the Conservatives introduced the Commonwealth Immigrants Bill. Their motives were largely political; in areas like the West Midlands the party was under strong pressure from its own supporters, and it soon became clear that three-quarters of the public approved of immigration controls. In addition, the sluggish stop/go economy no longer required unskilled immigrant labour.

However, the legislation, while eventually effective in reducing black immigration, proved to be unsuccessful in several respects. First, it provoked a rush of extra immigrants anxious to enter Britain before the Act came into force in 1962. Second, the reduction in numbers of entrants did nothing to eliminate racial prejudice. On the contrary, many Conservative candidates felt encouraged to exploit hostility to immigrants in both local and parliamentary elections. The most notorious case was Smethwick, a Labour seat in the West Midlands, where the Conservatives defeated the incumbent member against the national swing in 1964 by exploiting the race issue. The popularity of a television comedy programme 'Till Death Us Do Part', in which expressions of racial abuse were a staple element, testified to the spread of intolerance in the 1960s; though the programme was a comedy, it helped to make racism respectable in Britain. Consequently, politicians continued to exploit the issue. 1967 saw the foundation of the National Front which campaigned for the repatriation of immigrants. The low point came in 1968 when a senior Tory politician, Enoch Powell, made a number of highly provocative speeches prophesying violence as a result of the presence of Commonwealth immigrants. Such warnings were self-fulfilling, in that they gave legitimacy to the more extreme and violent elements in society.

However, there were some counter-moves against the rising tide of racism. Powell was sacked from the shadow Cabinet by Edward Heath, and no other leading Conservative adopted Powell's tactics until Mrs Thatcher in the mid-1970s. The Wilson government passed a Race Relations Act in 1965 which set up the Race Relations Board; and a further measure in 1968 replaced this with the Community Relations Commission. The idea was to give coloured people a legal means of seeking redress for discrimination over housing and employment. However, Labour had also been driven to abandon its original liberal attitude towards immigration for fear of losing any more votes. In 1968 James Callaghan was instrumental in restricting the entry of Kenyan Asians to those who could demonstrate a 'patrial' tie with a British resident. In effect, primary immigration ceased and Commonwealth citizens could not get into Britain except on the basis of a temporary work permit. Many families were split up as a result.

By the 1970s and 1980s a large proportion of the Indian and Caribbean community had been born in Britain. Consequently the idea of repatriating such people was not feasible. The younger generation grew up speaking in English accents and much better integrated into the broader community, through their education, than their parents had been. Many were highly successful businessmen, and by the 1980s a few had won election to the House of Commons. On the other hand, prejudice was very slow to disappear. In some districts of London Indian and Pakistani families suffered violent atacks upon their persons and on their homes. The West Indians, especially young males, experienced severe harassment from the police, and often sank into a subcommunity of crime and drug-dealing. Taken as a whole, the experience of Commonwealth immigration is a strong reminder that the liberalism of the Permissive Society remained confined to enlightened minorities.

18

The Loss of Great-Power Status

In foreign policy, 1945 brought no shift to the left. This was partly the result of the wartime experience of the Labour ministers and partly the accident of personality. Postwar policy was dominated by Ernest Bevin as Foreign Secretary and by Clement Attlee, both old-fashioned patriots who remained keen that Britain should retain her status as a Great Power. 'There'll be no messing about with the British Empire,' Bevin reportedly said on entering the Foreign Office.

However, these statesmen were also the victims of circumstances, in particular the collapse of the wartime alliance and the development of the Cold War. It rapidly became clear that, instead of honouring the commitments given to Churchill at Yalta to allow democracy in Eastern Europe, the Soviet Union proposed to treat the territory as a buffer zone and an economic resource. She was also interested in extending her influence via the Black Sea into the Mediterranean, as had the Tsarist regime in the past. As a result, during 1946 the British government began to regard the Soviet Union as a greater problem than Germany. This came all too naturally to Bevin, whose experience as a trade-union leader had left him with a rooted suspicion of Communists. His policy was founded on a belief that Russia intended to capitalize upon the weakness of Western Europe, upon the withdrawal of American forces, and upon dissension within the British Empire to spread the Communist revolution. It was thus alarming to see the reduction in American troops in Europe from 3.5 million to 200,000 between 1945 and 1947, the absorption of President Truman in domestic affairs, and his government's decision to renege on previous undertakings to cooperate with Britain over the development of atomic weapons.

Cold-War Defence

These worries led the Labour government into several critical decisions in 1947. It decided that Britain should develop her own atomic bomb. This was partly the product of fears that without it Britain would lose status among the great powers, and partly a reflection of her vulnerability to Soviet military might; the case for an atomic bomb as a deterrent seemed attractive. The government also extended military conscription – an unprecedent policy in peacetime – which provoked 72 Labour MPs to vote against the government and a further 76 to abstain. As a result Britain had an army 900,000 strong by the early 1950s. Though huge by British standards, it compared with 4.75 million Russian and 3.25 million American soldiers. Britain had embarked on a futile and debilitating policy of keeping up with the two superpowers, both of which were becoming active not only in Europe but in Africa, the Middle East, the Far East, and South America. It was unrealistic to suppose that in the long run Britain could sustain such a worldwide role. This is not to deny the sound and even compelling reasons for Britain to maintain sufficient forces to be able to offer the USA serious support; she could hardly be expected to be responsible for defending Europe alone. Indeed, Bevin has won praise from several historians for the vigour with which he maintained British influence in those postwar years. Unfortunately, Bevin adopted too narrow a view of foreign affairs and allowed British policy to be excessively cast in terms of American strategy, which in the longer term meant that Britain missed the opportunity to begin to readjust her ideas. The Cold War made it all too easy for Britain to be beguiled by the traditional role that she had once enjoyed playing in the world by hitching a ride on the American juggernaut. In the short term Bevin's approach imposed an unnecessarily heavy burden on British public finance and inhibited the economic recovery the government was attempting, with some success, to bring about. Attlee showed more awareness of the dangers of saddling the country with an outmoded worldwide defence role. But the counter-pressure of the Chiefs of Staff and the Foreign Secretary carried the day.

The year 1947 proved to be crucial. Impressed both by Soviet expansionism and by British decisions to withdraw from Greece, Turkey, Palestine, and India, the USA prepared to step into the vacuum. The following year brought the Marshall Plan, and 1949 the foundation of the North Atlantic Treaty Organization. As a result of the Russian blockade of Berlin and the Berlin airlift American aircraft returned to their former bases in East Anglia, and by 1950 the USA had committed four divisions to the defence of Europe. In supporting this initiative Britain steadily increased her own defence spending from £2.3 billion to £4.7 billion in 1951, which represented 14 per cent of gross national product.

This policy was seen by its practitioners in terms of a continuation of the 'special relationship' that had grown up during the war. Even at this stage, however, there were a number of causes of friction between Britain and the USA. The American government wished to accelerate the break-up of the Empire, at least in India, to destroy the sterling area and Britain's trading bloc, to deter Britain from the pursuit of atomic bomb, and to promote the integration of Europe into a single economic unit. Since each of these aims ran counter to British policy, it must be doubted how much scope was really left for a special relationship. Moreover, there was no question of equality in the two countries' relations. Although by 1950 Britain had become extremely vulnerable as a result of her role as the USA's chief base in Europe, and as the point from which American planes carried atomic bombs, there was no real consultation between the two over vital questions of war and peace: even Churchill's return to office in 1951 made no difference in this respect. In a sense Britain had little room for manœvre because until her own atomic bomb and her long-range bomber force became available she was entirely dependent on America for her deterrent. In reality, Britain was a client of the USA. The importance of the 'special relationship' was more a matter of domestic politics than anything else; over subsequent decades a succession of premiers, including Harold Macmillan, Harold Wilson, and Margaret Thatcher, were to exploit the relationship as a means of boosting their prestige and distracting attention from a record of failure on the domestic economic front. This had the unfortunate effect of inhibiting British governments from rethinking their relationship with their European neighbours and abandoning an essentially anachronistic role in world affairs.

Significantly, the return of the Conservatives to power in 1951 saw a reduction in defence spending – a reflection partly of a lowering in international tension and partly of the excessive policies Labour had pursued under Bevin's regime. But the search for independent nuclear weapons continued. In October 1952 Britain tested her own atomic bomb, and in 1954 the government decided to develop the H-bomb, which, according to one Tory MP, Julian Amery, 'will make us a world power again'. Yet the practical significance seemed dubious, for the long-range V-bombers were not expected to be ready until 1955–7. Britain's defence rested ultimately on American commitment to Europe, but it was politically difficult to recognize the fact. In spite of the economies made after 1951, Britain still devoted 8.2 per cent of her national income to defence in 1955, twice that of Germany. In 1956 Macmillan warned the Prime Minister, Sir Anthony Eden, that 'it is defence expenditure which has broken our backs'. The difficulty lay in the fact that Britain attempted *both* to become a leading nuclear power and to retain her conventional military role around the world, which eventually proved too costly. France, in contrast, was to demonstrate that it was feasible for a middle-rank power to achieve a genuinely independent nuclear capacity, but only if other defence commitments were reduced accordingly. In Britain the chiefs of the three armed forces enjoyed strong support in the Conservative Party, and lobbied successfully to slow down any cuts. In 1955 even the attempt to reduce national service from two years to eighteen months was defeated.

However, after the Suez incident in 1956 the new Premier, Macmillan, his Chancellor, Peter Thorneycroft, and the Defence Secretary, Duncan Sandys, managed to introduce economies in the 1957 defence White Paper. Conscription ended in 1960 and the army fell from 690,000 men to 375,000. Some troops and RAF forces were withdrawn from Germany. These reforms reflected the ministers' judgement that the economy could no longer sustain such a high defence burden. Increasingly the government justified its nuclear-deterrent policy on the grounds that it offered an economical alternative to conventional defence, and that it was, in any case, unlikely that future wars would involve the protracted use of ships, troops, and planes. Britain's needs were thus twofold: a credible deterrent to the Soviet Union, and a modest but efficient force capable of tackling subversion in the Third World.

Yet even this proved a complicated and costly option because of the problems involved in delivering nuclear warheads. Already by the late

1950s the V-bombers were obsolescent. A new supersonic bomber perished in the Sandys economies. This left Britain with a new missile, Blue Streak, which was unlikely to be available until 1962–5. By 1960 the Cabinet had decided to abandon this too, in view of the spiralling costs. In March of that year, therefore, Macmillan visited the USA where the new President, Kennedy, agreed to sell Britain the Skybolt ground-to-air missile which would prolong the life of the V-bombers. In return Macmillan offered the USA a base at Holy Loch on the Clyde for submarines carrying the Polaris missile.

Clearly, by the 1960s the attempt to maintain Britain as an independent nuclear power had become increasingly unrealistic, since she was wholly dependent upon the USA. The rationale for the policy was essentially political – to allay the dismay among Conservative MPs over the palpable loss of Britain's status as a great imperial power. Although Macmillan enjoyed a warm personal relationship with President Kennedy, Britain's true position was cruelly exposed by the Cuban missile crisis in 1962. If the USSR had not backed down and a war had followed, Britain would have been a leading target for Soviet attack. Yet her government was relegated to the sidelines, being denied influence and consultation about the use of American weapons based in Britain. British humiliation deepened in November 1962, when the American government abruptly cancelled the Skybolt programme on which she now depended. Again, the political pressures within Macmillan's party dictated that he should seek another lifeline from the USA, and a reluctant President agreed to provide a Polaris force on the grounds that it had become a 'political necessity' for Macmillan. Although the 1963 deal on Polaris was generous financially, it could be sustained only by squeezing the other armed forces, in particular the RAF, which now lost its strategic role to the navy, which did not want the Polaris submarines. However, while politics made it impossible for Macmillan to escape from the anachronistic nuclear policy, he had clearly become alive to the dangers of pursuing the goal of transatlantic interdependence with the USA; for that reason he initiated the first steps that were to lead Britain to Europe.

Decolonization

Neither Attlee nor Bevin had any wish to dismantle the British Empire. This was partly for reasons of national status and partly because they believed the standard of living in Britain benefited from colonial relationships. Empire provided markets for our goods and outlets for capital exports. In 1950 47.7 per cent of all British exports went to these territories. However, the new Labour ministers were also realistic enough to see that, as far as certain parts of the overseas territories were concerned, the liabilities had begun to outweigh the advantages. India had been been helped half-way to self-government by 1939, and it was impossible to backtrack on the concessions made during the war. In any case, by 1946 the subcontinent was descending into disorder on a vast scale, and the government could not contemplate shouldering the military costs of maintaining the Raj. In fact, no British government ever had been willing to pay; they had never maintained more than a small number of British troops there, and had generally expected Indians to bear the costs. What delayed a final solution was the difficulty of deciding to how many successor states power should be surrendered. Eventually Attlee broke the deadlock by dispatching Earl Mountbatten of Burma as last Viceroy in March 1947, with instructions to hand over control by June 1948 at latest. Similar pressures in early 1947 led the government to shake off other British obligations. In particular, the burden of keeping 100,000 troops in Palestine to be the targets of Zionist terrorism, while Britain vainly tried to balance Israeli demands for free immigration against the Arabs' fears about losing their homelands, was no longer bearable. The Palestine problem was simply referred to the United Nations.

Thus by 1947 India and Pakistan had received their independence, to be followed by Ceylon and Burma in 1948. Palestine was evacuated by 1949. 'Scuttle everywhere is the order of the day,' moaned Churchill. But how decisive were these initial steps towards the dismantling of the Empire? In one sense India's loss was surely crucial. In most of the African colonies British influence had only ever been superficial. But in India there had been, for all the controversies, far deeper ties of pride and emotion. Once the British had wrenched themselves away from India they were never likely to dig in very deeply elsewhere.

On the other hand, one cannot ignore the fact that, after the loss of the Indian subcontinent, there was no sudden collapse of the Empire. The chief phase of decolonization was not to come until the 1960s. Meanwhile, India's willingness to remain in the Commonwealth and retain Mountbatten as her first Governor-General conveyed a reassuring sense of continuity. There were several reasons for this situation. For one thing, although nationalist movements existed in other territories they were far less developed than that in India. Also, the Labour government tried to implement a positive policy, usually referred to as 'trusteeship', which implied eventual self-government when the circumstances in each territory were right. They were keen to promote economic development especially in Africa, to which end £120 million was allocated over a ten year period. It was hoped that after several decades of development a substantial middle class would emerge, capable of taking over the government. This would also enable Britain to entrust power to sympathetic, pro-Western regimes.

Finally, the loss of India itself still seemed to leave the strategic rationale for Britain's imperial role largely unimpaired. The elaborate system of overseas bases and communications was to remain until the late 1960s. In this respect, the anti-imperial influence of the USA proved rather short-lived. The onset of the Cold War made Britain's worldwide connections useful assets not to be given up lightly, and the fact that her colonial peoples seemed vulnerable to Communist subversion only made the American government more supportive of Britain's role abroad. The Middle East loomed especially large, partly because of the West's dependence upon its oil and partly because of its strategic role – the USSR could be threatened by aircraft stationed in the region. Whether these interests were sensibly pursued is another matter. When in 1951 Iran nationalized her oil refineries, including the British companies, Britain proved unable to intervene. Although she engineered a coup in concert with the USA, in 1953 Britain's subsequent share of Iranian oil fell from 53 per cent to 24 per cent. Egypt was similarly mishandled. In 1951 Britain still maintained 40,000 troops there. But the coup by Gamel Abdel Nasser in 1952 largely undermined what remaining influence Britain had. Even the Conservative government bowed to Nasser's determination to surrender the imperial role, and by 1954 Churchill had reluctantly agreed to withdraw British forces from the canal zone. The Western powers took comfort from the willingness of other states in the region, notably Turkey and Iran, to offer them bases now that Egypt had been effectively lost.

During the 1950s Britain conducted a number of rearguard actions with a view to slowing down the loss of imperial control or ensuring that when power was handed over it would be granted to sympathetic successor states. These initiatives enjoyed mixed success. For example, between 1948 and 1955 Britain fought a war against Communist guerillas in Malaya in the belief that the local rubber supplies made the area too valuable to lose. Britain felt confident enough of success to grant Malaya independence in 1957, but continued to take an interest in the security of the area. She was to retain control of the foreign and defence policy of Singapore, and in 1963 promoted a federation between Singapore and Malaya. Similarly, Britain went to some lengths to retain influence in the East African territory of Kenya because Mombasa provided an important base on the Indian Ocean. During 1952–4 she waged a war against the Mau Mau or Kikuyu tribesmen. Self-government was conceded by stages, and led to the eventual takeover by pro-Western forces.

Elsewhere, the rearguard actions enjoyed less success. In many African territories Britain attempted to use tribal chiefs as anti-nationalist bulwarks. As with the Indian princes, she found that this rarely worked. Thus elections in the Gold Coast and Nigeria in 1951–2 were won by the nationalists. The authorities allowed Kwame Nkrumah, the Gold Coast nationalist leader, out of prison in the belief that he would be a loyal collaborator. But after obtaining independence Nkrumah shocked the British by adopting a markedly anti-Western policy. An even worse mess was made by British initiatives in Northern Rhodesia, Southern Rhodesia, and Nyasaland. It was hoped that the white settler population there might serve as the basis for long-term British influence. To this end the three territories became the Central African Federation in 1953. However, it lasted for only seven years. Even in Southern Rhodesia there were 13 blacks to every one white, and 31–1 in Northern Rhodesia. Nationalist movements developed under Kenneth Kaunda in Northern Rhodesia and Dr Hastings Banda in Nyasaland, so that by 1962 Britain had conceded elective majorities in their legislatures. The Colonial Secretary, Iain Macleod, had recognized the impossibility of maintaining the Federation, though no government felt able to dissuade the beleaguered white community of Southern Rhodesia from holding out against African nationalism for some years to come.

One of the most unfortunate effects of the successful military initiatives was the misguided attempt to retain the island of Cyprus. It had never been of any practical use since the time Disraeli had acquired it from Turkey in 1878, but British governments chose to regard it as necessary to protect their oil interests. This led to the involvement of some 30,000 troops in a guerilla war against the forces of the majority Greek population during 1954–9.

It was during the course of the Cyprus conflict that Britain suffered her greatest disaster, following Nasser's decision to nationalize the Suez Canal in 1956. Britain had consistently failed to understand Egyptian pride and aspirations to be completely free from the imperial relationship. She also became antagonized by Egypt's readiness to turn to the Soviet Union for arms, and by the dismissal of a British General, Sir John Glubb, by the Jordanian government in March 1956. Britain and the USA compounded the situation by refusing to finance the Aswan High Dam in the hope of bringing President Nasser to heel. To the general feeling that Western influence was slipping in the region there was added a powerful element of personal involvement by the new Prime Minister, Sir Anthony Eden, who felt he had been fooled by Nasser. The British government claimed that its vital trade through the Suez Canal would not be safe if the Egyptian government took over control, an absurd proposition to which none of the other trading nations who also depended on the canal subscribed. However, Eden insisted on treating Nasser as a latter-day Mussolini and determined that he should be removed. The result was a clumsy and misguided invasion plan. By the time the British troops arrived in early November the public opposition of the USA had led to a run on sterling, thereby exposing British weakness most acutely. Eventually Eden felt compelled to withdraw the troops because there was no other way of securing American support for sterling.

The significance of the Suez fiasco has generated some debate amongst historians. On the face of it, the consequences were considerable. It exposed graphically the limitations of British power and aroused almost universal opposition to her in the UN; indeed, only Australia and New Zealand were prepared to support her. In the Middle East the affair greatly weakened Britain's influence while strengthening Nasser's position in Egypt. His successful defiance of imperialism made Nasser a hero in the third world, and helped to stimulate existing anti-imperial movements.

On the other hand, Suez was an isolated example of imperial overreach. Other forward moves were either successful or, at least, less costly and demoralizing. Nor is it clear that the measurable effects of Suez were very great. Relations with the USA were quickly patched up under Macmillan's premiership. And it has been argued that the basic policies leading to decolonization were already in place by 1956. This, however, is rather dubious. It was clearly assumed in the 1950s that it would be several decades before most of Britain's colonies would be ready for self-government; in fact only three won their independence – Sudan in 1956, Malaya and the Gold Coast in 1957. Suez had a vital medium-term impact on this leisurely process in that it destroyed the premiership of Eden. Ironically, it brought to power a man who was almost as responsible for the disaster. As Chancellor, Macmillan had failed to grasp the American view of Suez and the consequences for sterling, and had misled Eden after his talks with President Eisenhower.

However, once in office Macmillan accelerated the process of decolonization, subject only to the check imposed by his own party. After his 1959 election victory, Macmillan felt more sure of his own position and appointed Iain Macleod as Colonial Secretary with a view to relieving Britain of what appeared to be her increasingly dangerous liabilities. Ideas about trusteeship went out of the window as colonies were given up with little regard for the state of political or economic development they had reached. This was not simply the result of a loss of nerve after Suez. Macmillan had already been conscious of the incongruity between Britain's imperial role and her economic weakness. As Chancellor he had drawn up a profit-and-loss account for every colony in order to determine where the balance of advantage lay. The result of this hard-headed approach was that no fewer than 27 colonies received independence between 1960 and 1969. By the mid-1970s all that remained were a scattering of islands too small for effective independence, plus a collection of troublesome oddities from the days of imperial greatness – Gibraltar, Hong Kong, and the Falkland Islands. In the space of twenty-five years Britain had abandoned her imperial role; adopting a new one was to prove more uncomfortable.

For some years both Conservative and Labour politicians had placed high hopes on the Commonwealth as an alternative vehicle for British influence in the world. Although Burma had promptly left, most colonies followed India's example by remaining as members. Britain was keen to develop a common defence policy, to some extent in order to check Soviet influence over countries like India. But this failed to materialize, partly because the newly independent countries wished to be neutral in the struggle between Communism and capitalism. Moreover, Canada looked to the USA for its security, while in 1951 Australia and New Zealand reached a defence agreement with their American neighbour without consulting Britain. In addition, the Commonwealth gradually divided along racial lines. In South Africa the defeat of the pro-British General Smuts by Afrikaner nationalists in 1948 paved the way for the imposition of the apartheid policy which culminated in the expulsion of South Africa from the Commonwealth in 1961. After 1965 Britain's failure to bring to heel the white settlers who declared unilateral independence in Southern Rhodesia attracted widespread condemnation from the black countries. At home, relationships with the Commonwealth also began to be adversely affected by large-scale immigration into Britain from the West Indies, India, and Pakistan. The restrictions placed on immigration by the Conservatives and their exploitation of the issue attracted increasing criticism from Commonwealth leaders, with the result that by the mid 1960s the whole idea of the Commonwealth had largely lost its appeal for the right wing in Britain.

Reluctantly into Europe

As in the aftermath of 1918, so after 1945 Britain allowed her close relationship with France to lapse, and she showed herself slow to appreciate the significance of developments in Western Europe. This was partly because her own trade was, as yet, dominated by the sterling area, and because Europe, in its devastated condition, seemed likely to take years to recover fully. But British detachment reflected very basic assumptions that, in Bevin's words, 'Great Britain was not part of Europe' and had no wish to be. Thus, when Jean Monnet, one of the founders of the Common Market idea, visited Britain in 1949 to propose Anglo-French economic union he was rebuffed. But Monnet simply turned to Germany, and Britain missed the first of several opportunities to participate constructively and at an early stage in European cooperation.

The first concrete initiative was the proposal for a European Coal and Steel Community in 1950. 'The Durham miners won't wear it,' announced Herbert Morrison, as though nothing more need be said in the matter. But by 1952 Britain was surprised to find that six members had joined the ECSC. By this time a new British government had taken over, but the consensus on external policy continued unaffected. Churchill saw no great relevance for Britain in the closer cooperation between France and Germany; he remained chiefly interested in consolidating the Commonwealth, and in relations with the USA. Consequently he adopted a negative view of proposals for a European defence community in 1954, though eventually Eden pledged the permanent commitment of British troops in Europe.

Once again, however, the British were surprised to find that no rebuff by them ever finally killed off the momentum for greater unity. In 1955 six European states invited Britain to take part in talks on further European intergration. This centred round a 'Common Market' involving the removal of internal trade barriers, a common external tarrif, and the free mobility of labour and capital within the union. The Churchill–Eden governments took the view that such an arrangement might lead to political federalism and threaten the sovereignty of the member states. Moreover, it might damage Britain's advantageous commerce with the Commonwealth. In the mid-1950s it was easy for the government to convince itself that Britain had re-established herself as a Great Power on the basis of her nuclear capacity, her large army, and her largely intact imperial role. For a time, therefore, Britain actually hoped to kill off the Brussels negotiations; but by early 1956 this had clearly failed, and the six states went ahead in March 1957 to sign the Treaty of Rome which created the European Economic Community. Meanwhile, Britain inspired a countermove in the form of a European Free Trade Association involving Denmark, Norway, Sweden, Austria, Switzerland, and Portugal.

Once again Britain had missed an opportunity, as soon became apparent. Already by the late 1950s Europe accounted for as large a proportion of British trade as the Commonwealth. Unfortunately, EFTA was not an adequate alternative to the EEC either economically or politically; indeed, far from being a bridge to Europe it only complicated Britain's subsequent application for membership. Had Britain agreed to play a constructive role in the formative phase of the EEC, she would have had no great difficulty in negotiating concessions to reflect her special interests in Commonwealth trade and her access to cheaper food. But the opportunity passed, and Britain was to be in a much weaker position in the 1960s.

By 1961 it had become clear that the government had miscalculated. The six states of the EEC were reducing tariffs, harmonizing their external duties, creating central institutions, and achieving substantially faster economic growth than Britain. The then Prime Minister, Harold Macmillan, who had always been more favourably disposed towards Europe than Churchill or Eden, made the first application for membership in 1961. The case was widely expressed in terms of the economic advantages. Western Europe offered the wealthy markets in which Britain could sell her manufactured goods – both EFTA and the Commonwealth were inadequate; her relatively sluggish economy would benefit greatly from the stimulus of EEC membership. However, for Macmillan and pro-Europeans of all parties, the economic arguments were to a large extent a way of popularizing the essentially *political* case for joining. As both the American and the Commonwealth props to Britain's worldwide status were now looking rather precarious, Europe seemed to offer a more realistic vehicle for maintaining her status as a Great Power.

Yet, though Macmillan recognized the force of international trends he could not take too many risks with his own party, which was seriously divided over Europe. Conservatives felt outraged in 1962 when the American former Secretary of State, Dean Acheson, made the perfectly reasonable observation that 'Great Britain has lost an Empire and has not yet found a role'. In view of this sensitivity, Macmillan took care to move pro-Europeans such as Edward Heath, Christopher Soames, and Duncan Sandys into key positions. By September 1961 negotiations were under way under the leadership of Heath. By 1962 considerable agreement had been reached, though largely because Britain simply accepted the common agricultural policy, the commercial policy, and the external tariff. But the process was interrupted in 1963 by a veto on Britain's application delivered by President de Gaulle of France. He was much influenced by Macmillan's recent attempts, following the loss of Skybolt, to re-establish Britain's close defence relationship with the USA; he had seen this as an opportunity for Anglo-French defence cooperation instead. Clearly, there was some validity in de Gaulle's diagnosis. Up to this point Britain had changed her policy out of fear of losing her traditional role, rather than because of her enthusiasm for the European Community.

The *impasse* in British external policy might have been broken by the change of government in 1964, for Harold Wilson, the new Prime Minister, was ostensibly dedicated to modernizing British society. In fact his policies in external affairs, especially towards the USA and defence, proved conservative and traditional. After Macmillan's fresh initiatives there came, for a time, something of a reversion to earlier patterns; Wilson seemed prepared to subject an ailing economy to the burdens of maintaining an increasingly anachronistic world role. Thus Britain's bases and troops 'East of Suez' were supported until, in the late 1960s, economic decline at last forced a reluctant government to withdraw. Wilson also held to the existing nuclear deterrent policy, merely cancelling one out of five Polaris submarines as a political gesture. He was also keen to cultivate the special relationship with America. President Johnson clearly wanted Britain to uphold her role east of Suez, to maintain the value of sterling, and to back the USA in the war in Vietnam. Wilson loyally cooperated in all this for some years, though he gave moral rather than military support over Vietnam.

Once again, Britain had returned to her client relationship. Wilson successfully obscured her demeaning position by deliberate appeals to national pride; 'our frontiers', he once claimed rather absurdly, 'are on the Himalayas.' However, this was not mere posturing on his part. He showed himself willing to devote scarce resources to the fight against subversion. In Malaya, for example, 68,000 troops and a third of the British fleet became engaged against Indonesia. On the other hand, when faced with a more minor challenge by the handful of white Rhodesian settlers who declared illegal independence in 1965, the government backed away from direct intervention.

As the 1960s wore on, however, mounting economic and political pressures forced the Labour government into revisions of its external policy. The Defence Secretary, Denis Healey, attempted to impose tighter control by abolishing the three separate armed-service ministries and amalgamating their interests under a single Ministry of Defence. Several expensive aircraft programmes were scrapped in favour of American substitutes purchased cheaply 'off the shelf'. Healey also began to reduce the navy's reliance on aircraft carriers. The flaw was that commitments had not yet been reduced in line with costs. But in July 1967 the deterioration of the economy led Healey to decide to cut back British forces in Malaya and Singapore by half by 1970–1 as a step towards final withdrawal by 1975. In 1968 a reluctant Cabinet agreed to a complete withdrawal from the east-of-Suez role, with the exception of Hong Kong, by 1971. Coinciding as it did with the painful decision to devalue the pound, this represented something of a turning-point; one minister described it as 'breaking through the status barrier'.

This overdue reassessment had been facilitated by the gradual replacement of elderly Labour ministers and the promotion of the pro-European Roy Jenkins as Chancellor and the modernizing right-winger Tony Benn. Jenkins, like Heath on the Conservative side, had a clear conception that Britain's future lay in embracing the European option and giving up the old imperial role so dear to his colleagues. In 1966 the Prime Minister had announced Britain's intention to make a second application for membership of the EEC provided that the right terms could be negotiated. As in 1961, the fundamental reasoning behind this seemed to be political – there was no other feasible way for Britain to maintain her influence in the world in the long run – though the issue was largely discussed in terms of economics. In 1967 the Cabinet decided to make a second bid for membership, in the belief that Britain's decline would otherwise become precipitate. But, as before, de Gaulle perceived the application more as a matter of expediency than conviction, and he vetoed it in November.

The election of 1970 brought into power for the first time a genuine enthusiast for Europe in the person of Edward Heath. Although the public at large and many members of the two main parties continued to be indifferent or hostile towards the EEC, the determination amongst ministers plus the enthusiasm of some Labour and Liberal leaders gave the European cause fresh momentum. In the mood of disillusion and decline that had set in by the end of the 1960s, Europe seemed to offer the only constructive response to Britain's dilemma. The resignation of de Gaulle in 1969 also helped matters. French fears about the growing economic dominance of Germany began to make Britain's participation as a counterweight more attractive to her.

Britain herself had clearly declined as a manufacturing power when compared with Germany. Heath correctly saw that in American eyes Germany was becoming the more important power, and to that extent the USA would be less interested in the special relationship with Britain. As Britain's negotiator in 1961–2, Heath also had a clear grasp of the difficulties ahead, especially the weakness of Britain's position; in applying to join what was no longer an experiment but a proven success, she could do little to modify the rules of the club or to influence moves to advance from the original customs union towards a more politically coherent community. When negotiations began in January 1970, Heath reassured the French President, Pompidou, about his government's attitude, and by June 1971 the terms for entry had been settled. Once again, the government chose to emphasize the material advantages to the British people, in terms of a higher standard of living, in spite of the higher food prices and large contributions to the EEC budget that membership entailed. In October 1971 the House of Commons approved British entry by 356 votes to 244, and Heath signed the treaty of accession in January 1972. As a detailed bill had still to be passed through Parliament, it was not until January 1973 that Britain actually joined the EEC. It quickly became plain that Heath saw the move as part of a wider change in Britain's role in the world. For example, in negotiations with the USA Britain now joined her eight partners by adopting a common position in place of the old one-to-one talks. A deterioration in Anglo-American relations was underlined by the Arab–Israeli war of 1973, when Britain refused to allow NATO bases in Britain to be used to airlift supplies to Israel. The two countries appeared to be, in Henry Kissinger's words 'at a turning-point in Atlantic relations'.

However, for some years Britain's relations with her new European partners were also strained, partly because the terms for entry into the EEC had been very unfavourable. A high proportion of the EEC budget was devoted to the Common Agricultural Policy, which subsidized small farmers and imposed levies on the cheaper imported food from which Britain had benefited in the past. Moreover, with a smaller and more efficient farming sector, Britain received relatively few benefits from the policy while contributing disproportionately to it. By 1978 she paid 20 per cent of EEC income but received only 8.7 per cent of its spending. The timing of British entry also proved unfortunate. From late 1973 the Arab–Israeli war led to an oil crisis; a 400 per cent increase in the price of oil checked economic growth in Western Europe for the first time in a quarter of a century. Britain experienced recession and inflation, rather than the stimulus she had hoped for after joining Europe.

The economic crisis destroyed the govenment of Edward Heath in 1974 and restored Wilson to office. This placed a question mark over Britain's role in Europe. Since the Labour Party remained badly divided over the EEC, the Prime Minister embarked upon an elaborate process of renegotiating the terms of entry and offering the public a referendum on EEC membership. He achieved no more than marginal modifications, in terms of a rebate on Britain's contributions and easier access for Commonwealth products. But on this basis a national referendum took place in June 1975 which produced a vote of 67.2 per cent to 32.8 per cent in favour of membership. Though the result was very clear, only 64.5 per cent had turned out to vote. The verdict reflected a fear that things might be even worse if Britain left the EEC, rather than a positive conviction about the merits of staying in. Continued inflation, unemployment, and a deteriorating balance of payments during the 1970s did nothing to improve British attitudes towards Europe. On the whole she proved to be an uncooperative member, invariably opposing changes and resisting the application of common policies. Traditionalist leaders like Harold Wilson and his Foreign Secretary, James Callaghan, regretted the loss of the close relationship with the USA and went some way to restoring the Atlantic Alliance. This took the form of nuclear cooperation in which the Polaris submarine system was updated. Thus, in several ways the momentum generated by the Heath government was checked in the later 1970s, and the ground was laid for a further step back into the past under Mrs Thatcher after 1979.

Part V

The Era of Reaction and Decline, 1970–1992

19

The Breakdown of the Postwar Consensus, 1970–1979

Although the 1960s had witnessed a preoccupation with Britain's economic weaknesses, this made little impression upon ordinary people: standards of living continued to improve on the whole. But in the 1970s there was an undeniable deterioration. Inflation, the balance of payments, unemployment, strikes, all became serious matters of concern. Governments scarcely seemed to be in control; both a Labour and a Conservative government broke down over major policies, and within each party an extremist wing began to gather strength. In this brittle and adversarial climate the long-standing political consensus dissolved.

Heath and the Crisis of Conservatism

The surprise election result of 1970 brought Edward Heath to the premiership. Although drawn from more modest origins than most Tory leaders he was, in fact, an insider. He had risen smoothly via Oxford and the army to Parliament, where he spent only eighteen months as a back-bencher before moving on to the whips' office and the Cabinet. Heath stood squarely in the 'One-Nation' tradition of Conservatism; like Macmillan, who promoted him, he believed in maintaining state social services and was detached from the world of business. His most famous remark was an attack on disreputable commercial practices as 'the unacceptable face of capitalism'. Heath was also fairly liberal in his attitudes on such issues as capital punishment, South Africa, and immigration; he demonstrated his feelings about racial prejudice by sacking Enoch Powell from his shadow Cabinet in 1968 for a racially provocative speech. Above all the new Prime Minister was an ardent modernizer, determined to drag a reluctant Tory Party into the twentieth century. Under a previous government he had been responsible for the abolition of resale price maintenance which antagonized small shop-keepers. Health's greatest achievement as Premier was to take Britain into the EEC (see p.282–3), but he also restructured government departments, set up the Central Policy Review Staff or 'Think Tank' under Lord Rothschild, and imposed a sweeping local-government reform which caused much controversy by changing historic boundaries and abolishing some counties. In short, Heath proved to be a twentieth-century Peel. More a civil servant than a politician, he appeared too cold and arrogant ever to be popular with the rank-and-file; ultimately his inability to master the arts of party management was to destroy his career.

In 1970 Heath's arrival in power seemed to represent a clear break with the drift and deviousness of his predecessor. In retrospect some commentators interpreted the new government's strategy as a repudiation of consensus politics and a first instalment of Thatcherism which did not quite come off. Though this is not a valid view, it is easy to see how it arose. Heath did appear to offer a sharp change in style from Wilson; and he was, for a Conservative, unusually concerned with detailed policy planning before entering upon office. In January 1970 he had gathered his shadow Cabinet at the Selsdon Park conference to discuss their future policies in some detail. There was much talk about curtailing state intervention, withdrawing subsidies for industrial 'lame ducks', and reforming the trade unions. To a large extent, however, this was simply the usual rhetoric adopted by Conservatives after a lengthy period in opposition. The sense of change was exaggerated when Harold Wilson attacked the emergence of 'Selsdon Man' as proof of a shift to the right. There were, undoubtedly, some initial indications of a break, though the policies were quite typical of incoming Conservative governments. Expenditure was cut in 1971, income tax was lowered, subsidies on council housing were reduced, free school milk was abolished, and both the Prices and Incomes Board and the Land Commission were wound up.

However, Selsdon Man enjoyed a very short life. By 1971 the government began to reconsider its policies as it was confronted simultaneously by high inflation and unemployment approaching one million. It rapidly became clear that the government could not simply stand aside and allow prices and wage negotiations to go the way market forces dictated. By 1971 employers were surrendering to pressure by raising the wages of manual workers by 15 per cent. The tendency to make concessions was increased by the irresponsible policy of the Chancellor of the Exchequer, Anthony Barber, who tried to thrust the economy into a growth phase by tax cuts and higher expenditure. This only boosted inflation, stoked up a consumer spending spree, and drew imports into the country. In the past Conservative Chancellors had repeatedly done this in order to cultivate a sense of wellbeing among voters before an election. But this time the boom was mistimed: it merely fed the appetite of the trade unions, and the strategy blew up in the government's face. In particular, Heath appeared to reverse his original policy in two striking ways. First, he increased expenditure on education, the NHS, and housing, so much so that the annual average rise exceeded that under Labour during 1964–70. Second, he accepted that major industries could not simply be allowed to succumb to market forces if a national interest was at stake. Thus in 1971 Rolls-Royce was nationalized, the Upper Clyde Shipbuilders received a £35 million subsidy to safeguard 3,000 jobs, and British Steel also obtained extra investment from the state.

But the government's most distinctive initiative was its attempt to control escalating wage demands and check unofficial strikes by means of the Industrial Relations Act. Among other things, this required pre-strike ballots and allowed for the imposition of a 60-day cooling-off period. However, the trade unions determined to defy the law and the TUC threatened to expel any union that registered under the new Act. In the event, 1972 brought the loss of no fewer than 23 million working days in strikes, the highest since 1926. The Act was only rarely invoked, and when it was, as over the railwaymen's dispute in 1972, the workers simply voted in favour of a strike under the terms of the legislation. Several individual leaders of unofficial strikes received prison sentences, but this was subsequently ruled to be illegal and only caused further political difficulty for the government. The most dramatic challenge came when the coal miners rejected an 8 per cent wage offer in the winter of 1972. A well-organized system of 'flying pickets' operated by Arthur Scargill effectively checked the movement of coal, and soon industry found itself restricted to a three-day week. But the real weakness in the government's position lay in the fact that the more moderate miners' leaders, Joe Gormley and Lawrence Daly, commanded considerable public sympathy for their cause. This led to a retreat, in the form of a Commission of Inquiry which offered the miners a massive 21 per cent rise. However, the offer was rejected by the NUM, and the strike did not end until February 1973, when miners won increases worth between 17 and 24 per cent.

The effect of the industrial militancy was extremely serious. It undermined the authority of an elected government, the law, and the police. Moreover, the success of militancy encouraged extremists to rely increasingly upon direct action. The position of moderate union leaders became so insecure that they found it difficult to cooperate with the government over a voluntary pay policy. Conversely, the prestige of militants such as Scargill, who took over as the NUM President in 1973, rose considerably. Within the Labour Party, too, leading figures like Tony Benn were increasingly tempted to support the extraparliamentary forces and their methods. By the winter of 1972–3 the Heath government had been driven to adopt a statutory prices and incomes policy. This involved several stages, the first of which was a complete freeze on wages for three months. By November of 1973 this threatened another confrontation with the miners. In fact the government consciously sought to avoid another clash by offering a 13 per cent rise, but by this time the workers' self-confidence and expectations were too high for them to respond rationally. The NUM refused even to ballot its members on the offer, and imposed an overtime ban.

In spite of the collapse of its economic policies, the Heath government believed it could afford to weather the storm. It still had several years of its term of office, and although it was losing by-elections to the Liberals, it took confidence from the inability of the lacklustre Labour leadership to offer an effective challenge. As the Labour left gained influence, the party appeared divided and unable to recover public support. However, the political cycle was overtaken and disrupted by external factors in the shape of the Arab–Israeli War. This resulted in a huge increase in the price of oil which exacerbated Britain's already pronounced inflation and, in turn, brought industrial relations to their nadir. By December 1973 Heath had declared a state of emergency and a three-day week in view of the shortage of both coal and oil. At the same time the balance-of-payments deficit for the year reached £1.5 billion – the worst so far recorded.

It was understandably tempting to try to break out of the government's political-economic dilemma by holding an early election on the question 'Who Governs Britain?'. There was a presumption that voters would be willing to overrule the unions for their irresponsible behaviour, and that Labour would be caught in an awkward position. As a miners' strike loomed in February, some ministers urged the prompt adoption of this course. But Heath hesitated. To go to the country was an admission of defeat on the government's part. Nor was it clear how far the public blamed the miners rather than the government. In the event Heath delayed a little too long, and eventually gave way to pressure to hold an election which took place on 28 February. The uninspired campaign fought by the Conservatives reflected their lack of confidence in their policies and record. Even so, they won more votes than Labour. But the eccentric British electoral system played another of its tricks upon the politicians. With more votes, the Conservatives won *fewer* seats than Labour – 296 to 301 – and Heath's government was destroyed.

Heath had certainly been unlucky in the circumstances of his premiership, but he had also shown a lack of political skill. His was the last attempt to make the postwar consensus work. The early 1970s ushered in several new elements in British politics. Entry into the EEC was to become a key problem for all successor governments. The spectre of a trade union movement swollen with power posed a serious threat to the economy and to Parliament. Finally the Heath years generated an angry backlash among Conservatives who felt betrayed by their leaders and alienated by the Keynesian-collectivist formula in economic policy, and who were determined to avenge their humiliating defeat at the hands of the miners.

Multi-Party Politics

The heyday of the two-party political system in Britain was between 1945 and 1959. During the 1960s the dominance of the Conservatives and Labour was somewhat diminished, but the 1970s decisively marked the passing of the old pattern. Only the electoral system propped up the two big parties by overrepresenting them in the House of Commons. This was in part the result of repeated failures in economic policy by Tory and Labour governments, and, increasingly, a reflection of the drift towards what was dubbed 'adversarial politics'. The electorate found the debate between the Tory right wing and the Labour left increasingly sterile; and the beneficiaries were the Liberals.

Liberal expansion at the grass roots is measurable in terms of the number of constituencies contested by the party in this period. In the election of February 1974 517 Liberals stood, while in October of that year virtually every seat was contested; and from 1979 Liberals ran everywhere except in Northern Ireland. This marked a reversal of the party's historic decline. Since the 1920s the Liberals had largely lacked the constituency organization and membership needed to capitalize on their parliamentary initiatives. But in the 1970s they successfully employed the methods of 'community politics', which involved links with radical pressure groups outside the party and concentration on involving the electorate in campaigns over local issues. This paid dividends by boosting Liberal representation in local government, and gave the party a strong base in large urban districts like Liverpool, Leeds, and Birmingham that it had not enjoyed since the nineteenth century. It also helped the party to gain several parliamentary seats from both the other parties in the early 1970s.

The Liberals also benefited from able and attractive leaders in Jeremy Thorpe and his successor, David Steel. Thorpe proved to be an asset on television and led his party to a remarkable revival at the February 1974 election, in which 19.8 per cent of the vote was won by the Liberals. This was followed by a 18.8 per cent share in October 1974 and 14.1 per cent in 1979. In view of the unfavourable circumstances of the 1979 election, this was a surprisingly good performance and kept the party at a higher level than it had enjoyed during the 1950s and 1960s. When Heath lost his majority in 1974 he considered a coalition with the Liberals. This was refused largely because Heath was so unpopular that it would have been very damaging to the Liberals' radical credentials to have kept him in office. But the whiff of power was itself flattering to a party so used to the back-benches. From 1976 the new leader, David Steel, concentrated on a parliamentary breakthrough. His opportunity came in March 1977, when the Labour government under James Callaghan lost its majority. A Lib-Lab pact was devised to provide Labour with the vital extra votes it needed. In return the Liberals had regular talks with ministers about government policy, but they got very little in the way of concrete concessions out of the arrangement. In particular, they failed to insist on any firm advance towards proportional representation. In a sense, the real object of the pact was to begin the process of educating the British public about coalition governments which were a normal feature in most European countries but seen only as wartime expedients in Britain. By the summer of 1978 the pact had been abandoned. Although at the time it seemed that the Liberals had been the losers, it was the government that ultimately suffered. They went on to a deal with the Nationalist parties which led them to the electoral defeat of 1979.

It was during the 1970s that the problems of the Irish, Welsh, and Scottish parts of the United Kingdom came to the forefront of politics. This compounded the problem facing the two old parties as a result of the rise of the Liberals. The Irish problem was most troublesome for the Conservatives. Since the 1920s Ulster had been governed by the Unionist-Protestant majority in the Stormont Parliament; the Catholics who lived in the province suffered severe discrimination, particularly in housing and employment. Yet for decades Westminster governments simply turned a blind eye to the situation. In this way they created problems for themselves, for by the 1950s the Catholic population had become sympathetic to the IRA, and in the early 1960s there was enough support for a renewed campaign against British rule in Northern Ireland. The dissatisfaction took the form of a civil-rights movement, inspired partly by a similar campaign in the USA, which organized marches and demonstrations in the province. The resulting clashes between demonstrators and the police led the Home Secretary, James Callaghan, to send in troops in 1969.

Initially British troops were welcomed by the Catholic community as a neutral force. But the central political problem – the Unionists' abuse of power – remained unresolved by the British government. Labour was not particularly interested and the Conservatives were traditionally allies of the Ulster Unionists – hence the absence of a Conservative organization in the province until the 1980s. Thus in time the violence of the campaign began to be turned upon the troops, and in 1971 the first British soldier was killed in Northern Ireland. On 30 January 1972 a demonstration that had been pronounced illegal resulted in the death of thirteen people at the hands of the troops in Londonderry. 'Bloody Sunday', as the incident became known, was a fatal error by the authorities because it completed the alienation of the Catholics.

The response of the Heath government was to abolish the Stormont Parliament in 1973 and impose direct rule upon the province. Though seen as a temporary expedient, this soon became a permanent policy. It proved very costly in terms of resources and the horrendous loss of life; 1,100 people died between 1969 and 1974 alone. Moreover, the move was politically unfortunate in that it moved the British government firmly into the IRA's sights and enabled them to present the Union with Britain as the source of the troubles. The Northern Ireland policy enjoyed bipartisan support in Britain for the next twenty years, but was none the less an unmitigated failure, in that by the 1990s a solution was as far away as ever. The conflict made remarkably little impression upon the British public, even when the violence came to the mainland, and was treated with a stoical disdain as though it took place in a different country rather than within the UK. However the, Ulster problem had one significant domestic political result, for the imposition of direct rule alienated the Unionists from the Conservative Party. When, in 1974, Heath lost his majority in Parliament, his bid to retain office was crippled by his inability to rely on the votes of the Unionist MPs.

While the majority in Northern Ireland wished to stick resolutely to Westminster, the problem in Scotland and Wales was increasingly a disenchantment with control from London. In 1968 the government had thrown the nationalist parties a concession in the shape of the Kilbrandon Commission, which examined the question of a devolution of power from Westminster. Labour had long adopted a conservative attitude towards constitutional reform. Devolution was opposed by most of the party on three grounds. First, socialists traditionally regarded Welsh and Scottish self-government as irrelevant and parochial; it would only detract from effective economic planning from the centre. Second, the experience with a local parliament in Ulster was not very encouraging. Third, there was an element of political expediency. Since the 1920s Labour had largely dominated the Welsh and Scottish constituencies, and believed it could disregard nationalist criticisms. Only the dramatic by-election losses suffered by Labour in industrial Scotland and Wales in the 1960s had overcome this inertia.

Kilbrandon accelerated the pressure by recommending an elected Parliament in Edinburgh and a Welsh assembly. The Scots Nationalists in particular benefited from a new confidence in self-government engendered by the exploitation of rich deposits of oil off their country's coasts. Campaigning under the slogan 'It's Scotland's Oil', they could now argue credibly that an independent economic policy might enable the Scots to capitalize on their material resources and human skills, whereas under London's control they had suffered years of economic decline and unemployment. In the election of October 1974 11 Scottish and 3 Welsh Nationalists were elected – a significant number for a government hanging on by a majority of just 6. Thirty per cent of Scottish electors had voted Nationalist, only 6.4 per cent behind Labour. It was undeniable that the traditional mould of political loyalties had been broken. The Union no longer commanded the approval of the people, and only the undemocratic electoral system had denied the nationalists fair representation in Parliament.

For Labour, political expediency dictated a concession to deprive the nationalists of their momentum. But the Secretary of State for Scotland, William Ross, was a diehard anti-devolutionist. The result was a half-baked compromise that pleased no one. A parliament in Edinburgh would be established, with no revenue-raising powers or real influence over industry and agriculture. An effective veto power was to be left with the Secretary of State. The Welsh proposals were even more paltry. In effect the country would be given a large county council able to allocate a central block grant to the social services. To make matters worse, the government had not really thought out the implications of devolution for Scotland and Wales. Should it lead to a reduction in the number of Welsh and Scots MPs sitting at Westminster? Was it a final solution, or would it pave the way for regional government in the English provinces? There was so much doubt in the Cabinet's mind that when the Bill passed through Parliament in 1976–7 it allowed crucial amendments to be made. One of these provided that, before the new institutions were established, not only must there be a referendum, but at least 40 per cent of all those on the register – not of those who actually voted – must vote in favour. This was extraordinary to the point of absurdity in the context of British electoral practice; but it seemed at the time a neat expedient to ensure the survival of the Bill. In the event it proved fatal to the government.

The campaign over the devolution referendum took place during the winter of 1978–9. By that time the Prime Minister, Callaghan, had lost his pact with the Liberals but relied instead upon the cooperation of the nationalist MPs. In spite of this relationship, the Labour Party in South Wales largely turned its back on the referendum policy, and not surprisingly the Welsh voted 4–1 against devolution. In Scotland the strategy appeared to be working, in that Labour was regaining some ground from the nationalists. But the vote in favour of devolution was very narrow: 51.6–48.4 per cent; and it clearly failed to carry the required 40 per cent of the electorate. This failure left the Scottish Nationalists somewhat discredited, and they had lost the momentum of the mid-1970s. However, the devolution question was in no sense a side-issue cleverly defused by the Westminster Labour Establishment. On the contrary, it led the Nationalists to withdraw their support, which deprived the government of its majority and thus forced it into a disastrous election at the worst possible moment.

The Decline of Labour

From its peak at the general election of 1966, when Labour polled 48 per cent of the vote, the party went into a long-term decline; by the 1970s its share had sunk to the level of 1929–35, and in the 1980s below even that. The seriousness of the decline was obscured by the period in office between 1974 and 1979; but in February 1974 Labour polled half a million *fewer* votes than in 1970. Heath and Wilson fought what was aptly described as an 'unpopularity contest' which Labour had not expected to win.

The Labour leaders of the 1970s, Wilson and Callaghan, faced the same dilemma Gladstone had faced in the Liberal Party a hundred years earlier; they were caught between the general public, which seemed to be moving against radicalism, and the party activists, who were growing increasingly radical. In the event they handled the problem with considerable skill, and the surprising thing is not that they were eventually defeated but that they survived so long on borrowed time.

In retrospect Wilson has been widely blamed for his party's misfortunes; but it was not so much for what he did in the 1970s as for what he failed to do in the 1960s that he was culpable. For by 1970 the party was so disenchanted that membership was collapsing; it continued to dwindle to around 600,000 by 1979, though the official figures almost certainly exaggerate. Not surprisingly, after the economic failures and the right-wing external policies pursued up to 1970, the party's rank-and-file activists began to reassert themselves in the 1970s. Their leader was Tony Benn, now playing Joseph Chamberlain to Wilson's Gladstone. He advocated wider state ownership,

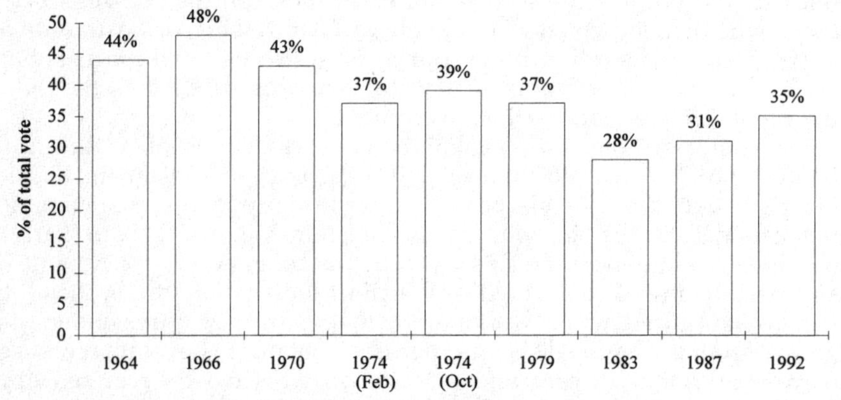

Figure 19.1 Labour Share of the Vote, 1964–1992

unilateral disarmament, withdrawal from the EEC, abolition of the House of Lords, reform of the Official Secrets Act, and greater democracy within the party involving wider participation in the election of the leader and shadow cabinet. It would be a misrepresentation to portray these issues as simply the fads of an extremist minority – several clearly enjoyed wide support beyond the Labour Party. The radicals' position was all the stronger because several major union leaders like Jack Jones and Hugh Scanlon also felt strongly critical of the record of the 1964–70 governments. The National Executive of the party asserted itself and adopted a more socialist policy, including the proposal that the state should acquire a controlling interest in twenty-five major companies. Although Wilson and Callaghan repudiated this and largely emasculated Benn's ideas, they could no longer control the party beyond the immediate parliamentary sphere. By resisting the left's policies they provoked the establishment of the Campaign for Labour Party Democracy and the NEC's decision to lift the long-standing proscription on Communists within the organization. This, combined with the drop in membership, meant that the party was vulnerable, in both the constituencies and local government, to participation by small groups of radical activists of which the best-known was Militant. By the beginning of the 1980s local Labour parties, especially in London and other major towns, had become associated with a number of unpopular causes involving feminism, homosexuality, and republicanism in Northern Ireland, as well as the more conventional left-wing causes, which enabled their opponents to condemn the party as a whole as extremist.

In this situation Harold Wilson scored two successes. The first was simply to consolidate his government during 1974. Labour's strongest card was the 'Social Contract', agreed with the unions in 1973, which offered the prospect that it would be better able than Heath to handle the industrial problem. In 1974 Wilson brought the left-wing rebel, Michael Foot, into office as Employment Secretary. This paved the way for a concession to the miners of the high wage rise they had been demanding, which led to the ending of the state of emergency and a reduction of tension. The promotion of Foot was an inspired move, for as leader Wilson found himself in a much weaker position in managing the party than his predecessors had been because he could no longer rely on the support of the major unions. The alliance with Foot helped to restore relations between government and unions. Wilson capitalized on his success in defusing the sense of crisis by holding another election in October 1974. However, this did not really work, for Labour emerged only slightly stronger, with 319 seats and an overall majority of 3 which was obviously vulnerable to by-elections. In most Western parliamentary systems this situation would have been resolved by a coalition without the need for a second election; but Wilson shared traditional Labour fears about coalitions; and the Queen, who on earlier precedent could quite well have asked her Prime Minister to attempt to gather a clear majority, apparently failed to do so.

Thus, although Wilson returned to office he had no mandate and, in fact, little idea about what he wanted to do. In the event he devoted much time to questions of party management. Herein lay his second major achievement. He largely succeeded in defusing the issue of Britain's position in Europe which seriously threatened Labour unity. Wilson made a point of incorporating into his Cabinet both right-wing pro-Europeans such as Roy Jenkins and Shirley Williams, traditional left-wing critics like Michael Foot and Barbara Castle, and the new generation of radicals represented by Tony Benn, who had charged his position over Europe. It was Benn's idea to hold a referendum over British membership of the EEC, which seemed a likely route to withdrawal, but eventually provided a lifeline for Wilson and party unity. When the House of Commons voted on entry to the EEC in January 1971, 119 Labour MPs opposed the government; moreover, the NEC voted 16–6 against joining. In this situation Wilson made a point of moving a little closer to the anti-Europeans, on the grounds that he could not afford to cut himself off from the left of the party. Although he was basically in favour of joining Europe, Wilson placated the left by agreeing to hold a referendum on the question. This led Shirley Williams to announce that she would leave politics if the referendum went against the EEC. In a subsequent House of Commons division, 69 Labour members defied the whip to support Heath on entry while another 20 abstained. Though this was a substantial number of MPs, it did not conceal the fact that the leading Europeanists, Jenkins and Williams, were becoming rather isolated in the party. The proof of this came in the ballot for the leadership following Wilson's resignation in 1976, when Jenkins received only 54 votes. In the short run the referendum policy provoked resignations from the shadow Cabinet by a number of pro-Europeans, including Jenkins.

In these developments one sees the origins of the subsequent split in the party leading to the formation of the Social Democrats in the 1980s. In this sense Wilson's tactics were ultimately a disaster. But after his return to office in 1974 he appeared to have brought off the difficult trick of maintaining the unity of the party. In the referendum held in June 1975, Wilson posed as a moderate who believed that on balance Britain should remain in the EEC, while other members of his Cabinet freely campaigned in both 'pro' and 'anti' camps. But the 67 per cent vote in favour of membership satisfied the right while also subduing the left. For the time being the Prime Minister could bask in another triumph of management.

On the economic front Wilson had, once again, inherited a serious mess from the Conservatives. During 1974 prices rose by 17 per cent and wages by 22 per cent. The oil shortage and the impact of EEC membership in raising food prices contributed to inflation of 27 per cent in 1975. The Social Contract seemed useless to restrain wage demands. Thus the new Chancellor, Denis Healey, determined to check excessive pay settlements by a combination of higher unemployment, extra taxation, and expenditure cuts. 1975 saw economies of £3 billion in social spending. Wilson and Healey were clearly in retreat from the policies of the 1960s and heading towards monetarism.

The Prime Minister hesitated between a voluntary and a statutory incomes policy. At length he settled for a half-way house. Led by Jack Jones, the TUC agreed to a limit of £6 per week on wage rises, while in his 1976 budget Healey announced a 3 per cent maximum. This incomes policy was a success; inflation fell from 27 to 13 per cent from 1975 to

1976. But unemployment rose to 1.2 million, and between 1975 and 1976 the purchasing power of workers on average wages dropped by 7 per cent. This naturally created friction within the Labour Party, especially as the Cabinet was abandoning the policies for economic planning and the acquisition of shares by the National Enterprise Board championed by Tony Benn. The government did, however, feel obliged to rescue British Leyland from bankruptcy, which in effect meant a state takeover; and Benn himself subsidized a number of workers' cooperatives in collapsing companies. When Wilson interrupted this by moving him from Trade and Industry to Energy, Benn failed to resign, thereby discrediting himself for the remainder of the government's life.

Following Wilson's surprise resignation in 1976, the Labour Party chose Callaghan in preference to Foot as his successor – a significant sign that the parliamentarians remained essentially a right-centre body while the party outside was moving to the left. Callaghan, though experienced and popular with the unions for his opposition to *In Place of Strife*, had been an unsuccessful minister especially at the Treasury. He was essentially a 1950s conservative politician, highly illiberal on social issues, and rather antifeminist in contrast to Wilson, who had promoted more women than any other prime minister. He seemed an inappropriately old-fashioned leader for the kind of party that Labour had become. On the other hand, his avuncular, reassuring style pleased the general public and equipped him to fend off the shrill attacks of the new Conservative leader, Mrs Thatcher. Callaghan was also a more relaxed and secure personality than Wilson; his colleagues appreciated his Cabinet style, involving more open debate and less intrigue. When the government's tiny majority disappeared in 1977 he managed to keep it going first by a pact with David Steel, the new Liberal leader, and then with the Nationalist MPs.

However, by the end of 1976 the economy was very weak. The balance of payments had deteriorated, the pound's value had fallen to $1.70, and Britain's currency reserves came under severe pressure. Healey sought a loan from the International Monetary Fund. This was humiliating in itself, but it also exacerbated the friction in the party because the IMF insisted on expenditure cuts of around $2 billion. The Conservatives condemned this as the result of a succession of economic booms generated by Keynesian finance. Healey substantially agreed with the monetarist diagnosis, and he effectively began the application of what became known as Thatcherite policies two-and-a-half years before the Conservatives returned to power. The result was a fall in living standards and an increase in unemployment to 1.5 million. Though even this was not high by the standards of the 1980s, it seemed a disaster to a generation accustomed to virtually full employment. It led many workers to believe that things could not be worse if the Conservatives ran the government, a belief neatly exploited in 1979 by the slogan 'Labour Isn't Working'.

Yet by the autumn of 1978 there were grounds for thinking that the government's tough policies were effecting some improvement. Income growth was being restrained and the balance of payments improving. By-elections showed that Labour had deprived the Scottish Nationalists of their momentum by its devolution policy. But at that point Callaghan's tactical sense failed him. A maximum wage increase of 5 per cent was announced. This was unnecessarily tight, but the unions would probably have gone along with it in the short term because they assumed it to be the prelude to an autumn general election. However, the Prime Minister was not sure that he could win an election and the moment passed. The control exercised by the unions thereupon rapidly broke down. The rising cost of living seemed to them to justify bigger wage rises, and as a result a mass of strikes occurred, especially among public sector employees. This 'winter of discontent' fatally undermined the Labour government because it destroyed their claim to be competent to handle the unions. But it was the fiasco over the devolution referendum in Scotland that drove the Nationalists out of their pact with Callaghan, thereby exposing him to defeat in Parliament on 28 March 1979.

In the end Labour had failed just as Heath had earlier in dealing with the combined effects of inflation, militant unions, poor investment in industry, and a balance-of-payments deficit. It was tempting to think that Keynesianism no longer worked. The political consequence of this dilemma was to drive much of the Labour Party further to the left while a section of the Conservatives under their new leader moved sharply to the right. Both were repudiating the consensus policies that had dominated the post-1945 era.

The Origins of Thatcherism

In long-term perspective, the ousting of Edward Heath from the Conservative Party leadership by Margaret Thatcher in 1975 appears as a turning-point in ideological terms. This was not, however, at all obvious at the time. The party seemed simply to be ridding itself of an unpopular and unsuccessful leader. In fact a struggle was going on within the Conservative Party at two levels – ideology and class. Although Mrs Thatcher and her followers took socialism and the state as their ostensible targets, in a sense their real enemy lay within the Conservative Party itself.

Ever since the Edwardian period there had been a powerful resentment directed towards the gilded, upper-class leadership by many Conservatives of more modest social standing. This gained force from an ideological attack by the 'radical right' in the aftermath of the Boer War. Balfour became the first victim of this rank-and-file pressure, and from 1911 onwards most Conservative leaders were drawn from the middle class. However, almost all of them attempted to educate their party in a more liberal Conservatism. Influenced by their experience of two world wars and by the enfranchisement of the working class, they shifted the political agenda in the direction of welfare and collectivism. Consequently the party's history, from one perspective, began to seem like a series of retreats dictated by the left which made the party's rank-and-file feel let down by their leaders. Heath's personal remoteness simply exacerbated the feeling; as one Conservative put it: 'Like Macmillan and R. A. Butler, you always had the impression that Heath (and some of his ministers) found some of the activists at Conference distasteful – backward-looking, complacent and reactionary.' In complete contrast, Mrs Thatcher was the first leader who articulated, and apparently believed in, all the prejudices of rank-and-file constituency activists, notably over taxation, welfare 'scroungers', making Britain a great power again, cocking a snook at foreigners, and pandering to racialism. This often involved a simple expression of views rather than a change in policy. In January 1979, for example, Thatcher made a bid for the racist vote when she argued that white Britons had legitimate fears about being 'swamped by people with a different culture'. Heath had sacked Powell for a speech only a little more inflammatory. Herein lay the strength of Thatcher's position as party leader, in spite of the parliamentarians' dislike of her views.

Thus, in spite of the apparent social similarities between them, Heath and Thatcher were very different Conservative leaders. Thatcher was a real outsider, partly because of her sex and partly because she gloried in the lower-middle-class morality and economics of the grocer's shop. The experience of the Second World War also proved to be decisive. Heath emerged from that period prepared to come to terms with the state and recognize it as a force for good. And like other soldiers, he lost his prejudices against foreigners; the key objective after 1945 was the reconstruction of Europe, hence his warmth towards the EEC. By contrast, Thatcher's experience was very narrow. She emerged unaffected by the depression of the 1930s and, having made no contribution to the war effort, lacked either a sense of collective national effort or any understanding of foreign affairs.

However, between 1970 and 1974 Thatcher kept her head down and served as a loyal Minister for Education under Heath. In this period some Tories became steadily more outraged over government policies on immigration, Rhodesia, the EEC, and Northern Ireland. Yet they represented the diehard fringe which most Tory leaders had to suffer. What really undermined Heath with a broad section of the party was his failure to handle the trade unions, and his return to collectivist and Keynesian economic and social policies. In fact, Mrs Thatcher had been as guilty in this respect as most of her colleagues. At Education she had been a big spender, and had abolished more grammar schools than any of her predecessors. Nor did she show any interest publicly in Conservative ideology. She simply became a convert to the new Conservative economic ideas.

Among politicians the real inspiration was Enoch Powell. His belief in classical liberal economics had been carried defiantly through the 1950s and 1960s, in the face of the interventionism practised by Macmillan. By 1974 Powell had become very marginal as a result of his advice to vote Labour in protest against Heath's European policy. But his ideas were taken up by Sir Keith Joseph, one of Heath's leading ministers. Within weeks of the party's defeat in February 1974 Joseph, in a characteristic confession of past error, announced his conversion to Conservatism! What he meant by this was conversion to economic liberalism. He argued that Conservatives had been beguiled by semi-socialism and the mixed economy; in future they must allow market forces to rule and not try to cheat them by state intervention. Above all, Conservatives like Joseph felt dismayed at the effects of Barber's Keynesian policies in terms of inflation and the attempts to impose incomes policies. He was increasingly attracted by monetarism, which had long been the creed of a number of minor economists who believed that governments could limit inflation by controlling the money supply. Joseph began to promote his philosophy through the establishment of the Centre for Policy Studies under the directorship of Alfred Sherman, which effectively acted as a rival to the party's Research Department. Like a number of the men who subsequently became known as Thatcherites, Sherman's origins lay on the left; a former Communist, he brought the extremism and doctrinaire approach of the far left into the far right of Conservatism.

Thus, by the autumn of 1974 Sir Keith Joseph had articulated a sweeping repudiation of Heath's record over inflation, incomes policy, and state subsidies. This did not make a general move to the right inevitable. But in view of the fact that Heath had led his party to three general election defeats, there was bound to be a challenge to his leadership, and that challenge was mounted by the new right. Heath had been the first Tory leader to be elected to his job, and his arrogant refusal to contemplate the need to defend his record was therefore particularly unwise. He did not face an ideological challenge – the majority of the MPs appear to have been content with traditional policies; but by his stubbornness, Heath placed those policies in danger.

Joseph seemed the obvious person to challenge Heath in a ballot of MPs, but he quickly ruined his chances by some tactless remarks and withdrew in November 1974. Mrs Thatcher stepped into his place, since no one else of any stature was prepared to take the risk. She was not remotely expected to win. However, her supporters used shrewd tactics; by emphasizing that she was well behind Heath they made it easier for the large number of MPs who were exasperated with the leader to give vent, safely, to their feelings by voting for the challenger. In the event the vote on 4 February 1975 produced a major shock; Thatcher received 130 votes to only 119 for Heath. This 'peasants' revolt', as it was dubbed, destroyed Heath's leadership. Moreover, although it was not an ideological rejection of his brand of Conservatism, the ballot left the supporters of Heath too demoralized and too far behind to make an effective response. Thus on the second ballot Thatcher comfortably defeated her four opponents and seized the leadership. Her victory was an unintended revolution. She had not been elected because she offered a right-wing policy, but after its defeats the party was ready to welcome a return to basic Conservative principles. By boldly attacking socialism and championing the cause of free enterprise, she began to restore to her party the sense that it occupied the moral high ground. In a speech in 1977 she declared:

> The tide is beginning to turn against collectivism, socialism, statism, dirigisme, whatever you call it . . . It is becoming increasingly obvious to many people who were intellectual socialists that socialism has failed to fulfil its promises, both in the more extreme form in the Communist world, and in its compromise version.

As yet her colleagues did not realize how literally their new leader took these rhetorical flights. For some years she was obliged to surround herself with former Heath supporters, who clearly believed that they would be able to control her once in office. In any case, many felt uncertain whether they would in fact recover power under Mrs Thatcher. No one knew whether the country was yet prepared to accept a woman as Prime Minister, especially one so aggressively middle-class. The avuncular and chauvinistic James Callaghan evidently thought he could afford to patronize her, for he continued to enjoy a big lead in terms of personal popularity. Mrs Thatcher was clearly lucky for a second time. Had Callaghan held an election in the autumn of 1978 he might well have won. Delay, and the industrial chaos of the winter of 1978–9, gave Mrs Thatcher her opportunity. Even so, the Conservatives won only 43 per cent of the vote, not one of their better performances. But it was enough to deliver a parliamentary majority.

20

The Era of Thatcherism

'I have changed everything.' Few would disagree with Mrs Thatcher's own claim that her assumption of the premiership in 1979 marked a turning-point in modern British history, comparable to 1914 or 1931. Her supporters believed it brought an end to the Keynesian era that had been ushered in by the Second World War; her critics saw her as a politician whose ideological clock had stopped somewhere around 1937.

However, even at this stage it is clear that many of these contemporary claims are considerable exaggerations. After twelve years of Conservative rule it became apparent that most of the central objectives of Thatcherism had not, in fact, been achieved (see p.321–3). Moreover, any attempt at assessment is complicated by the fact that 'Thatcherism' developed over time; some of the original elements like monetarism were abandoned after a few years, while themes like privatization that loomed large in the late 1980s had hardly been mentioned in the beginning. Admittedly, in 1979 Mrs Thatcher made it clear that she intended to curtail the role of the state, restore market forces, and create an 'enterprise culture'; but this was the kind of generalized rhetoric often employed by Conservatives on returning to office after a long spell in opposition; it was not clear what it really meant. The 1979 manifesto had not been very specific. It mentioned the reduction of government borrowing, tax cuts, control of the money supply, the sale of council houses, and reform of the law on trade unions; but many of the subsequent controversies – privatization, higher value-added tax, the poll tax, NHS reform – were not discussed; and Mrs Thatcher herself said relatively little during the election campaign.

She had two reasons for caution during the early years of her premiership. First, the Conservative victory represented popular dissatisfaction over the trade unions, inflation, and taxation, but not a positive consensus about the alternatives. At 43.9 per cent the party's vote was quite low, the lowest, in fact, achieved by any party that had won a majority of seats since 1922. This vote subsequently went *down*, so that Mrs Thatcher could never be said to enjoy a genuine mandate from the country; she owed her parliamentary dominance to an electoral system which converted a minority of votes into a majority of seats.

Second, the Prime Minister felt very conscious of leading only a minority within her own party. She was dependent upon the ideas generated by several right-wing think-tanks, notably the Centre for Policy Studies (set up in 1975) and the Adam Smith Institute (1977), because the official party policy-making machinery remained in the hands of the 'One-Nation' Tories. In the new Cabinet only Sir Keith Joseph, Sir Geoffrey Howe, and John Biffen shared her economic philosophy. By contrast, Lord Carrington, William Whitelaw, James Prior, Peter Walker, Ian Gilmour, Norman St John Stevas, Francis Pym, and Mark Carlisle were closer to what Mrs Thatcher was to disparage as the 'wet' school of Conservatism. Her sense of being an outsider was heightened not just by her political ideas but by her sex and by her social origins. She and her followers felt disdainful of the traditional Tory elite, which had betrayed the party's principles in an attempt to imitate socialism. As a result, for much of her premiership Mrs Thatcher sounded more like a leader of the opposition attacking the conventional ideas and institutions of British society. In the course of a decade she alienated many of the key elements in the British political system. The most obvious sign of this was her contempt for nationalist aspirations in Wales and Scotland. Her unpopularity north of the border helped accelerated the headlong decline in the number of Scottish Tory MPs, which had been as high as 36 in 1955 but fell to a mere 10 by 1987. More surprising was the collapse of the relationship between Conservatism and other pillars of the Establishment including the civil service, which the Prime Minister believed to be infected with Keynesianism, the House of Lords, which defeated the government's proposals on several occasions, the Church of England, which was seen as either liberal or Marxist, the BBC, which was considered anti-Conservative, and the universities. The refusal of Oxford University to award Mrs Thatcher an honorary degree, by a vote of 700 to 300, in 1985, was striking proof of the extent to which she had antagonized the academic community. Also, despite her emphasis on enterprise she alienated many businessmen, who felt that she understood only the simplicities of the 1930s grocer's shop in which she had grown up, not the complexities of modern research-based industries which required support from government. Finally, a major gulf opened up between the Prime Minister's supporters and several members of the royal family, who were seen as dangerously liberal in their views; this led to attacks on Prince Charles by the right-wing press, which pronounced him unsuitable to become King. The Prime Minister herself perceptibly aspired to a monarchical role in public life which was underlined by her extraordinary use of the royal 'we', as on the celebrated occasion when she declared to an astonished press: 'We have become a grandmother.'

Monetarism and Deindustrialization

The new government took office determined to purge the economy of inflation; but, partly as a result of national wage settlements agreed earlier, inflation rose to no less than 22 per cent by May 1980. However, Joseph and Howe believed that they had the answer to the problem with what was called 'monetarism' – the idea that price rises could be restrained by restricting the supply of money to the economy. Though professed for many years by marginal figures in the economics profession, monetarism remained an unproved theory; and it was not clear what exactly constituted the money supply. For some years Chancellors of the Exchequer kept changing their definition in an effort to apply the theory in the real world.

However, by 1981 inflation did begin to be tamed because the economy had slipped once again into a depression. In his budget of that year, Chancellor Howe astonished the political world by resorting to a severe dose of orthodox deflation, including big tax increases and a reduction in the public sector borrowing requirement from £13.5 billion to £10.5 billion. This decisive repudiation of Keynesianism was strongly opposed by a group of leading ministers, not least because they feared the large-scale unemployment that would result. Had they taken their opposition to the point of resignation they would probably have compelled Mrs Thatcher to back down. But their failure to do so proved decisive. Thereafter the Prime Minister felt emboldened to begin weeding out the 'wets' – St John Stevas, Gilmour, Carrington, Prior, and Pym – though Whitelaw and Walker survived. They were gradually replaced by more junior figures loyal to Thatcherism, including Norman Tebbit, Cecil Parkinson, Leon Brittan, Nigel Lawson, and Norman Fowler. The Prime Minister now had the measure of her colleagues; only three, Michael Heseltine, Lawson, and Howe, were prepared to resign over a policy difference with her.

The economic consequences of Howe's budget proved to be horrendous. The British economy suffered its worst depression since the 1930s, with unemployment reaching 2.7. million. During 1980–1 the gross national product contracted by 3.2 per cent, and it increased by only 1.7 per cent in 1982. Around 25 per cent of British manufacturing capacity was destroyed in this period. Although the government succeeded in persuading the country that full employment was no longer a practical aim, it could never escape the consequences of this disastrous depression. It led to a major balance-of-payments deficit which, in turn, resulted in high interest rates because of the need to defend the pound; and high unemployment prevented the overall reduction in government spending which the government had hoped for.

However, the government argued that its strategy was to put the economy on a stronger footing in the long run. The improvement in economic growth between 1983 and 1988 was hailed by Howe's successor, Nigel Lawson, as an 'economic miracle'. But beneath the hyperbole, the deep-seated problems of the economy remained. In industrial relations the government could at least point to the taming of the trade unions. Their membership fell from 13 to 10 million in 1980–3, and in 1981 only 4.2 million working days were lost in strikes, compared with an average of 13 million during the 1970s. However, the chief explanation for this was simply the depression and continuously high unemployment which knocked the stuffing out of militancy. At first Mrs Thatcher was careful to avoid provoking the unions; she actually backed down over the miners' demands in 1981, for example. James Prior's trade union legislation in 1980 was moderate, though in 1982 Norman Tebbit went further. He removed the immunity from the financial costs of strikes which the unions had enjoyed since 1906. Unions could be fined for 'unlawful' strikes and pay penalties for secondary picketing, while workers who had been sacked as a result of a closed-shop agreement received compensation. Politically the government's greatest triumph came with the miners' strike, which ran from April 1984 to April 1985 and which allowed them to avenge the defeat suffered by Edward Heath in 1974. This time the Cabinet was well prepared for its confrontation with Arthur Scargill, the NUM leader. Coal stocks were high, and the Cabinet insisted on maintaining full electricity supplies so that the economy could weather the anticipated lengthy dispute. By January 1985 71,000 of the 187,000 miners had returned to work, and the defeat of Scargill decisively broke the political power of the coal industry (see p.311–12).

However, none of these successes made any impact on Britain's long-term problem of wage inflation. Rises in wages and salaries continued to run at a higer level than the increase in output during the 1980s. Although the government renounced any idea of a policy for incomes on ideological grounds, it did in fact have a clear policy. This involved restricting public-sector wages drastically in order to reduce the inflationary effects of increases awarded in the private sector; thus during 1979–85, when private-sector wages rose by 11 per cent, those in the public sector fell by 1 per cent. Within the public sector the government favoured its political supporters – the police and the armed services – with settlements consistently above inflation, while holding down those in teaching, higher education, local government, and nursing; in 1985, for example, the 17.6 per cent rise for the armed forces contrasted starkly with 6 per cent for teachers.

Throughout the 1980s, manufacturing industry proved to be a major weakness in economic strategy. A bewildering succession of ministers occupied the Department of Trade and Industry, each for a short period, all of whom were largely convinced that there was little a government could do to promote industry. This marked out the Thatcher government not only from previous British administrations but from the practice in Germany, France, the USA, and above all Japan, where, regardless of ideological complexion, governments actively supported their major manufacturers with finance where necessary. In Britain, the special opportunity presented by the earnings from North Sea oil and the sales of nationalized industries to finance a state-led investment strategy to re-equip industry were neglected by ministers for ideological reasons. They preferred to allow the extra income generated from these sources to flow into short-term consumption. In addition, they appeared to accept a decline of manufacturing capacity but argued that there would be compensation in the form of an expanding service sector. To some extent this did materialize. However, new jobs in the service sector could not replace those lost in industry because they required different skills, were often for women, not men, and were concentrated in the south-east, where expensive housing continued to be a deterrent to workers. But above all, service-sector employment failed to make inroads into the huge balance-of-payments deficit which resulted from the loss of manufacturing. Indeed, in so far as the growth areas relied upon equipment such as computers they exacerbated the problem by sucking in yet more imports. Thus the weakness of the export sector began to lock Britain into a spiral of decline. Inevitably the trade deficit led to speculation against sterling, which in turn led the government to defend the value of the pound by maintaining high interest rates. But the cost of borrowing only deterred industry from making the investment necessary if their productivity was not to fall further behind that of their rivals.

However, during the mid-1980s the effects of these policies were not fully apparent. Inflation had fallen from around 22 per cent in 1980 to 7 per cent in 1985 and 3 per cent by 1986. The economy grew by approximately 4 per cent a year between 1985 and 1988. Yet unemployment remained at an unprecedentedly high level. In 1985 it stood officially at 3.2 million, though the true figure was half a million above this because the government repeatedly changed the methods of counting the unemployed with a view to excluding some people from the lists. Hitherto governments of all kinds had assumed that the public would not tolerate unemployment of even one million, but the Conservatives succeeded in retaining office at the 1983 and 1987 elections in spite of their record on this issue. The explanation was that, as in 1935, although unemployment remained very high it still affected only a minority of the population concentrated in certain regions; the Conservatives relied for their majority upon the more buoyant parts of the country such as the south-east of England.

The boom of the mid-1980s was essentially the product of another characteristic policy associated with Howe and Lawson: financial deregulation. Its practical effect was to encourage a major increase in lending by building societies in the form of mortgages which often amounted to loans diverted into ordinary expenditure; in so far as they *did* pay for housing alone they helped to generate rapid inflation of property values. In addition, the decade witnessed an considerable expansion of credit cards, bank overdrafts, and hire purchase by consumers. As a result, private household debt rose from £16 billion to £47 billion between 1980 and 1989, and mortgages from £43 billion to £235 billion. Whereas in 1980 debt accounted for 29 per cent of national income, by 1989 it amounted to no less than 62 per cent.

Thus the economic miracle claimed by Lawson rested largely upon an artificially inflated debt which boosted the demand for consumer goods; much of this expenditure disappeared into imports, which resulted in unprecedented trade deficits of around £15 billion by 1988–9. Meanwhile manufacturing industry endured the high interest rates that resulted. It was only a matter of time before the bubble burst. Unfortunately, in his 1988 budget Lawson recklessly exacerbated the problem by cutting taxes, thereby repeating the errors of previous chancellors such as Barber in the 1970s by stoking up a consumer boom. As the economy began to turn downwards, Lawson claimed it was suffering no more than a temporary blip; but it rapidly became clear that a second major depression had begun. In 1990 the economy stagnated, and during 1991 and 1992 it contracted. Monetarism had long since been abandoned as unworkable, so that the government had no weapons at its disposal except for the highly destructive use of interest rates which, at 13 per cent, severely handicapped industry. The appearance of a second depression within the decade undermined all the claims made for Mrs Thatcher's radical economic strategy. Rates of growth during 1979–90 were barely 2 per cent – a poorer record than that of the 1960s or 1970s. After twelve years in power Mrs Thatcher was to leave office with the economy demonstrably weaker than when she first took over.

Delusions of Grandeur

Every twentieth-century British Prime Minister has felt the temptation to boost his standing by capitalizing on the remains of the country's international status. For Mrs Thatcher foreign affairs held more than usual significance. Not only was she relatively ignorant of foreign policy; as a woman, she seemed to some men to be likely to be a liability in this field. As soon as she became leader she consciously tried to compensate for this by adopting an unusually belligerent tone towards the Soviet Union.

In the event Mrs Thatcher's premiership witnessed a resurgence of patriotism within her party, and to some extent in the country, amongst those who regretted Britain's decline as a great power. This was hardly surprising. During the 1950s and 1960s decolonization had been accomplished most expeditiously by Macmillan with virtually none of the sense of trauma experienced by other ex-imperial powers; the bitter determination shown by the French over Algeria or the Portugese in Angloa and Mozambique had no parallel in Britain. Nevertheless, many Conservatives resented the surrender of empire, especially in view of the anti-British views expressed later by some Commonwealth premiers. The appearance of the League of Empire Loyalists and, in the early 1960s, the Monday Club, were signs of the frustration of right-wingers at being driven to the margins even within their own party. Mrs Thatcher proved to be the first and only Conservative leader prepared to articulate these concerns in blunt language. She genuinely despised the diplomacy of the Foreign Office and happily substituted her own methods, which often seemed to involve being rude to foreigners on the assumption that this would appeal to her domestic constituency.

Mrs Thatcher's approach to foreign policy owed something to her autocratic temperament; but it also reflected her early experience of the Foreign Office and its handling of the long-standing problem of the illegal white regime in Southern Rhodesia. The diplomats engineered a conference at which Britain effectively signed away power to an elected – which meant a left-wing, black – majority. This distinctly liberal triumph sat rather uncomfortably with the views of Mrs Thatcher and her supporters. It left her with the feeling that she had been bamboozled by the Foreign Office and Lord Carrington. Thereafter she relied more on her own instincts by adopting an uncompromising line with the Commonwealth. In particular Mrs Thatcher steadfastly resisted the pressure to impose sanctions against the apartheid regime in South Africa, and made plain her pleasure in defying her critics at successive Commonwealth Conferences.

An equally Palmerstonian approach characterized Mrs Thatcher's handling of Britain's relations with the European Economic Community. Not only had she never shared Heath's zeal for Europe, she had become convinced that his policy had damaged the party. Ever since Lord Salisbury's unavailing protest to Macmillan – 'we are an island' – many Conservatives had lacked a major voice to articulate their hostility to Europe. Enoch Powell had gone a little too far in 1974 when he advised Conservatives to vote Labour because of Heath's policy.

Although Mrs Thatcher was powerless to change what her predecessors had done, she had nothing to lose by showing that she shared the misgivings of the rank-and-file about foreigners. But there was a more fundamental consideration at work. The Thatcherites perceived that the EEC was by no means the capitalists' club once depicted by the Labour left. On the contrary, the interventionism entailed in EEC policies was inimical to free market economics, and Mrs Thatcher began to fear the reintroduction of socialism via Britain's membership. In 1979 she took up the issue of Britain's disproportionate contributions to the EEC budget by demanding the repayment of £1,000 million of 'our money'. If megaphone diplomacy had limited results and embarrassed the Foreign Office, it enabled the Prime Minister to glory in her isolation at European summit meetings, fully aware that this would do her no harm with the British public.

Mrs Thatcher consolidated her moral withdrawal from Europe by another step back into an earlier era. Since the American President, Ronald Reagan, shared both her market economics and her hostility towards the Soviet Union, it proved easy to rekindle the embers of the special relationship with the USA. Although the regular meetings between the two leaders were largely public-relations exercises, something of substance emerged. They agreed on a deal whereby Britain allowed the USA to keep cruise missiles on her soil, and in return received the Trident nuclear missile as a substitute for the ageing Polaris submarine system. This kept alive the fiction of Britain as an independent nuclear power, but at a considerable cost. Her conventional weapons suffered from economies which had near-disastrous consequences; when the Falklands War broke out, the government had been on the verge of scrapping Britain's two remaining aircraft carriers, without which the expedition could not have been carried out.

The Falklands War of March–April 1982 was a remarkable throwback to the era of minor Victorian wars; it took place more by accident than by design in an area so remote that most British people had never heard of it, let alone known of its whereabouts, and it was fought at great cost for no substantial reasons, but with a triumphant outcome. This small group of islands, populated by 1,200 people of British origin, had for many years been regarded by British governments as being of no importance, in fact, as rather a nuisance. Their policy was gradually to loosen the ties between Britain and the Falklands and lead the islanders into a closer relationship with Argentina, which claimed the islands as her own territory. Mrs Thatcher's government fully accepted this policy, and to this end they dispatched a Foreign Office minister, Nicholas Ridley, to persuade the islanders of its wisdom. However, as a result of the defence economies in 1982 the one British ship in South Atlantic waters, HMS *Endurance*, was withdrawn. This error on the government's part was interpreted by Argentina as a sign that Britain was backing out. Thereupon General Galtieri foolishly tried to accelerate the process by invading the Falklands in March, in the belief that Britain would either give them up or be unable to intervene.

This misjudgement placed Mrs Thatcher's government in a crisis. She determined to recover the Falklands for essentially political reasons – not to do so could well have destroyed her premiership, taken in conjunction with the economic depression at home. Accordingly, a 28,000 strong task force was mobilized and transported with some difficulty to the South Atlantic. The lack of adequate air cover made the operation very risky. However, by May the islands had been recovered, though with a serious losses of shipping which put Britain's commitment to NATO in jeopardy. Politically the war seemed to have the effect of accelerating the slight improvement in the government's standing that had become apparent during the spring of 1982; it certainly helped to improve the poor morale of Conservatives before the 1983 general election. In the longer term the financial consequences proved to be serious. Since the government could hardly afford to admit that it had fought to retain a territory formerly considered to be worthless, it reversed its previous policy by expending large sums of money on the defence of the Falklands so that the material assets of the region would not be lost to Britain.

Although Mrs Thatcher and her supporters believed that they had restored Britain's standing in the world, the fact was that the 1980s made no substantial difference to Britain's position. The apparently inexorable decline of her economy led to reduced military forces. The special relationship with America could not outlast the Reagan presidency; and in any case the USA became increasingly interested in cultivating its relations with the real power in Europe – Germany. In this context the Falklands proved to be quite peripheral and was not, in fact, typical of Mrs Thatcher's policy. Faced with similar dilemmas created by the remnants of empire in Hong Kong, she simply agreed to surrender the territory to China when Britain's treaties expired in 1997.

But above all Mrs Thatcher's aspirations to restore Britain to her former world role were killed by the European embrace. Although the British government regularly obstructed European legislation, its tactics often proved ineffective and increasingly counterproductive at home. For example, European interventions designed to check the despoliation of the countryside by road-building schemes, or to introduce proposals to bring maternity leave up to European standards, or to raise the minimum wages of very low-paid employees, exposed the extent to which the British government did *not* in fact reflect the views of the British people. On the other hand, the Prime Minister did appear to be in tune with popular attitudes in voicing her dislike of the trend towards economic union, which she expressed in a famous speech at Bruges in September 1988. However, she was essentially voicing a protest against forces that seemed to be beyond her control. Her government committed Britain to the single European market from 1992, and to eventual membership of the European Exchange Rate Mechanism. This latter effectively meant that Britain surrendered her long-cherished control of currency policy. By 1990 the end of British sovereignty as traditionally understood was clearly in sight.

Breaking the Mould of Politics

At first sight the Labour Party's defeat in 1979 appeared an inevitable consequence of the immediate background against which it had been fought: the 'winter of discontent'. But 1979 has to be seen in the context of a protracted decline in the party's support from 48 per cent of the vote in 1966 to 27 per cent in 1983 (see p.294). This suggests more than the impact of passing issues or leaders; it points to long-term social and structural changes that were working to the party's disadvantage. By the 1980s, for example, the generation that had first voted in 1945 – a uniquely pro-Labour group of voters – was disappearing from the electoral register. Moreover, the steady expansion of white-collar occupations at the expense of industrial jobs gradually diminished Labour's traditional sources of support; by 1988 manual workers comprised only 45 per cent of the electorate, and not much more than a third of the workforce still belonged to a trade union. The population of the industrial centres where Labour seats were concentrated dwindled, while the suburbs returned a growing number of MPs who were inevitably Conservatives. Consequently, each new revision of the boundaries of the constituencies produced net gains of around 20 for the Conservatives. Labour also lost touch with the aspirations of some sections of the working class, notably in home owners. By 1985, 62 per cent of the population owned or were buying their own homes, including 58 per cent of all manual workers. By identifying themselves strongly with home-ownership and the sale of council houses, the Conservatives had driven a wedge into traditional Labour support.

These developments need not have been disastrous for the Labour Party – other European socialist parties coped more successfully with changing working-class aspirations – but for the errors of tactics and strategy with which they coincided. Following the resignation of James Callaghan as leader in 1980, the Labour MPs selected Michael Foot as his successor, largely because he seemed most likely to be able to restore the party's unity. But although a traditional left-winger, Foot proved to be a weak leader who was unable to cope with the resurgence of the new left led by Tony Benn, who almost defeated Denis Healey for the deputy leadership in September 1981. In essence the rank-and-file had begun to react against the years of control by Wilson and Callaghan, which now appeared to them to have been a betrayal of the principles of the movement. At the 1981 party conference the grass-roots activists succeeded in imposing a system of regular reselection upon Members of Parliament, and in placing the election of the leader in the hands of an electoral college comprising MPs, trade unions, and constituency parties. Subsequently Labour committed itself to an unusually left-wing programme including widespread nationalization, unilateral disarmament, and opposition to both the EEC and NATO. The resulting manifesto on which Labour fought the disastrous election of 1983 was aptly dubbed 'the longest suicide note in history'.

That setback had been heralded by a major split within the party in February 1981, when three former ministers, Shirley Williams, David Owen, and William Rodgers, joined with Roy Jenkins to issue the 'Limehouse Declaration' which paved the way for the establishment of the Social

Democratic Party. Jenkins, who had taken the first initiatives in breaking away from his old party, seriously considered whether simply to join the Liberal Party rather than set up a new organization. Surprisingly the Liberal leader, David Steel, urged that a separate organization would make a greater impact by attracting support away from Labour. Subsequently it transpired that this was a crucial error. However, during 1981–2 the local cooperation between Liberals and Social Democrats enabled the two parties to promote a single candidate in every constituency. Meanwhile they scored a series of spectacular victories in by-elections; and by November 1981, according to opinion polls, up to 45 per cent of the electorate supported the new 'Alliance'. Twenty-nine Labour MP's joined the Social Democrats, though only one Conservative. Amid the economic failures of doctrinaire Thatcherism and Labour's lurch to the left, it seemed for a time that a major new force was about to break the mould of British politics.

However, the new strategy suffered from several flaws. Although an impressive parliamentary force backed by a sympathetic press, the Social Democratic Party lacked roots in the constituencies. Its leaders were notably unsuccessful in attracting rank-and-file Labour voters, and most of them lost their seats at the next general election. Social Democrats performed better in Conservative territory, but no better than the Liberals already did. Thus the rationale for launching a separate party rather than joining the Liberal Party was never fully justified.

Support for the Alliance reached a peak around March 1982. But Conservative support had already begun to revive prior to the Falklands War, which accelerated the process. This decisively interrupted the progress of the Alliance by depriving it of publicity and initiative. In spite of this the new grouping performed extremely well at the election of 1983, when it won 25.4 per cent of the poll, only slightly behind Labour with 27.6 per cent. However, the effect of the electoral system was to create a complete travesty of democracy: Labour had 209 MPs compared to 23 for the Alliance. Although the Conservatives' share of the vote *dropped* to 42.4 per cent, their seats greatly increased to give them a huge overall majority. These distortions of public opinion were so great that they created widespread support for the introduction of proportional representation during the later 1980s. The Labour Party rank-and-file, accepting that they were most unlikely ever to win a majority again, moved in favour of proportional representation, and the party adopted the reform in connection with the proposed Scottish parliament.

After the fiasco of 1983 Labour replaced Foot by Neil Kinnock as its leader. Though a left-winger, Kinnock recognized that to persist with the policies of the early 1980s would be to relegate Labour to third place behind the Alliance. With some skill and courage, he therefore led the party back towards a more centrist position by abandoning some of its 1983 manifesto, and dramatized this by his public criticism of prominent left-wingers such as Derek Hatton of Liverpool, Ken Livingstone of the London County Council, and Arthur Scargill of the National Union of Miners. The defeat of the miners' strike in 1985 proved helpful in the context of this strategy. Scargill discredited himself initially by refusing to ballot his members on the strike, and by bullying those areas like Nottingham that did hold a ballot.

This had the effect of dividing the miners during the strike and leading to a breakaway union afterwards. In effect, the combination of the strike and the elections dethroned the left in Labour politics. By 1989 the NUM's membership had shrunk to 60,000, and it had lost its automatic right to a seat on Labour's National Executive. An era was coming to an end.

However, the revival engendered by Kinnock proved to be insufficient. At the election of 1987 Labour pushed its vote up modestly to 30.8 per cent but only slightly reduced the Conservative majority; Labour's real achievement lay in widening the crucial gap between itself and the Alliance, whose vote dropped a little to 22.6 per cent. Between 1983 and 1987 the Alliance had clearly missed its opportunity. The responsibility for this failure lay primarily with David Owen. Having succeeded Jenkins as Social Democrat leader, he stubbornly resisted the pressure to merge with the Liberals to form a single party. Yet this offered the only realistic way forward. The constituency members of both parties showed themselves willing; only the pride of a few parliamentary prima donnas denied them. The electorate was reluctant to accept a party without one clear leader at a second election. Owen also proved to be too much influenced by Mrs Thatcher, and failed to grasp that the Alliance would not be able to absorb the non-Conservative vote by contriving to present itself as a moderate version of Thatcherism. Thus in 1987 he tried to move Alliance defence policy to the right, and also made it clear that if no party had a majority he would prefer an arrangement with the Conservatives to one with Labour, a view which was fundamentally inconsistent with the outlook of the Alliance. This was borne out after 1987, when David Steel took the long-awaited initiative to merge the two parties. Owen strongly opposed this, only to be rejected by the majority of his party members. Though Owen rapidly became irrelevant, he had damaged the new Liberal Democrat Party, whose vote dwindled to only 18 per cent in 1992. Although this fourth Conservative victory underlined the unliklihood of Labour being able to win a majority, the party was none the less not about to fold up and be replaced by the Liberal Democrats. The more realistic elements in both opposition parties perceived that their best prospects lay in an electoral pact leading to the introduction of porportional representation.

The Growth of Poverty and Inequality

The 1980s marked a major watershed in British social history. Ever since the innovations of the Edwardian Liberal governments the state had effected a modest but steady redistribution of income, by means of graduated taxation and social welfare, from the richer to the poorer members of the community. After 1979 this was put into reverse. Not only did the decade witness a notable rise in poverty, it also saw a widening gulf between what Disraeli had once described as the 'two nations'. Whereas in 1979 the top 20 per cent of wage earners had enjoyed 37 per cent of all income after tax, by 1988 they had 44 per cent; conversely, the poorest 20 per cent of the population, who had received 9.5 per cent of income in 1979, earned a mere 6.9 per cent by 1988.

The first means of effecting this shift of income towards the wealthy was the adoption of a more regressive system of taxation. By stages the government cut the standard rate of income tax to 30 per cent and then to 25 per cent, and the top rate to 60 per cent and then to 40 per cent. In theory this was supposed to stimulate enterprise and wealth creation. However, no evidence in support of the claim was ever produced, and the most successful economies, such as the Japanese and the German, continued to flourish in spite of higher tax rates than Britain's. These changes did have the effect of further enriching a small number of high earners, but for the majority of people the tax burden *increased* because of additions to other forms of taxation paid by the poor or those on average incomes. For example, National Insurance payments rose so that they yielded three-quarters as much as income tax. In addition the value-added tax, which was paid on consumption, increased from 2.5 per cent to no less than 17.5 per cent.

Although the government criticized welfare expenditure for its undesirable effects on the poor, it did not hesitate to extend the welfare state in so far as it provided subsidies for the middle classes. Much the most costly of these subsidies took the form of income tax relief on mortgage payments, which began to go out of control in this period; from £1,000 million in 1979, mortgage relief cost the exchequer £7,000 million by 1990. The Treasury also sacrificed a further £7,000 million by failing to levy capital gains tax on house sales. In addition the government paid £10 billion a year in tax subsidies to pension funds. They introduced a new subsidy of £800 million for personal pension schemes, and a new tax subsidy for the purchase of shares of £100 million.

All these transfers of resources to the relatively wealthy had to be financed partly by extra taxes on consumption paid by the relatively poor, and by reductions in those parts of the welfare system that benefited the working classes. One of the less obvious ways of achieving this lay in the rules governing unemployment benefits. The level of benefit became linked to *prices*, not to wages, which meant that the growing number of unemployed people fell further behind the rest of the community. A variety of other expedients also helped to restrict the costs of benefits. For example, child benefits, which had replaced the family allowance, were simply frozen for several years so that their value fell sharply. From 1988 the social-security system was reorganized so that those who had depended upon National Assistance were to receive loans from a Social Fund. However, those who were considered unable to repay the loans would be denied this help. Moreover, local Department of Health and Social Security Offices operated within annual expenditure limits, so that the payments had to be made less in accordance with the applicants' need than in the light of the resources available. At this time some 300,000 children lost their right to free school meals. The elderly suffered in two ways; some 700,000 lost their housing benefit, and the link between pensions and national earnings was abandoned. Finally, young people aged sixteen to eighteen lost their claim to unemployment benefit, and students' housing benefit was withdrawn.

The consequences of changes in welfare and taxation policies proved to be dramatic. During the 1980s beggars reappeared on the streets of many British cities, and increasing numbers of people took to sleeping in the open – sights once associated with the Victorian and Edwardian period. In 1989 200,000 people were prosecuted under the Vagrancy Act of 1824. Homeless families had numbered 56,000 in 1979, but as a result of government policies the 1980s saw a fall of nearly 600,000 in the number of council houses available. This, combined with the rise in long-term unemployment and the splitting up of many families, greatly exacerbated the problem; by 1989 no fewer than 128,000 families, including 370,000 people, were officially classed as homeless. According to statistics prepared by the EEC, poverty in Britain had increased from 5 million in 1980 to 6.6 million by 1989.

Further symptoms of poverty and unemployment appeared in the form of higher rates of divorce, suicide, mental illness, and, above all, crime, which increased by 79 per cent under Mrs Thatcher's premiership. The emergence of a large 'underclass' alienated from the mainstream of society was marked by the outbreaks of rioting in many of the depressed urban districts. Beginning with the Bristol riots of 1980, this phenomenon spread to Brixton, Toxteth (Liverpool), and Moss Side (Manchester) in 1981, Handsworth (Birmingham), Tottenham, and Brixton in 1985, and Newcastle in 1991. The government's proposals to require even the poorest to contribute towards the new poll tax resulted not only in massive non-payment, but also in the failure of around a million people to register as parliamentary voters. Thus many of the poorest people lost their political influence just as they had during the Victorian period. By the end of the 1980s British society had clearly begun to break down under the strains of the economic setbacks and the reactionary social policies of the decade.

Rolling Back the State

The Conservative attack on the role of the state concentrated on four main fronts: local government, education, the health services, and nationalized industry. However, most of this work was accomplished during Mrs Thatcher's third term of office, after 1987, largely because by that time much of her original agenda had been exhausted or abandoned and there seemed less reason to fear the unpopularity likely to be aroused.

Conservative mistrust of elective local government went back a long way (see p.202–3). By the 1970s the huge expansion of employment by local authorities, and the publicity given to several left-wing council leaders, including Derek Hatton of Liverpool, Ken Livingstone of the Greater London Council, and Ted Knight of Lambeth, provided Conservatives with an irresistible target. However, the drawback was that attacks upon local councils often antagonized Conservative councillors and Conservative voters who still adhered to the collectivist ideology, at least in this sphere of government. Mrs Thatcher's intervention followed exactly the same lines as those of earlier Conservative governments faced with local authorities whose policies they disliked. In 1986 the GLC and the metropolitan counties, both created by previous Conservative governments, were simply abolished. In addition the government reduced grants and attempted to set limites on the expenditure of all councils. This, however, created conflict, even with Tory-controlled councils, and was dropped in favour of 'rate-capping' which was designed to penalize those authorities which exceeded levels of spending prescribed by Whitehall. The replacement of the rates by the poll tax was originally supposed to make councils fully responsible to their own electorate for their expenditure, but it rapidly transpired that central government would continue to dictate total spending. Several attempts were also made to remove major functions from local authorities. Schools, for example, were given financial incentives to opt out of their control, though very few did so. Councils were forced to sell many of their houses, and were prevented by the government from using the proceeds to build new ones. This created a major crisis in the housing market. By the end of the 1980s central government had thus gone a considerable way towards destroying the effective powers of elected local authorities. This provoked many resignations by Conservative councillors and undermined the party's position at local level. Even in the shire councils – the most pro-Conservative areas – only 11 of 39 still had a Conservative majority by 1993, and the party had only 965 councillors, against 1,425 Labour and 875 Liberal Democrats. In many cities such as Manchester, Liverpool, Newcastle, and Sheffield, Conservative representation had largely been eliminated and the party had been relegated to third place behind the Liberal Democrats.

The Thatcherites regarded the educational establishment, including schoolteachers, lecturers, and the schools inspectorate, with particular distaste, and subjected every part of it to a sustained attack which alienated virtually the whole of the profession. A bewildering succession of new policies was announced almost annually throughout the 1980s, but as most were ill-thought-out and underfunded, ministers were obliged to abandon many aspects of their reforms. The fundamentally doctrinaire motivation behind the educational policies became clear from the otherwise inexplicable decision to abolish the schools inspectorate and leave its function in the hands of non-professionals. Private business was invited to fund City Technology Colleges which were outside local authority control; but since companies generally refused to pay the government's bills, very few of these materialized. By and large the only schools interested in opting out of council control were those with too few students to be viable; this naturally caused chaos and contradicted the government's other declared policy of closing down small schools to make a more cost-effective system. Meanwhile the underlying problems of underfunding were never tackled. Severe shortages of teachers qualified in foreign languages, science, and mathematics became worse as up to 20,000 teachers left the profession each year by the late 1980s. Nursery education continued to be neglected, as it had been by successive governments.

From the beginning of her term of office Mrs Thatcher began to reduce the funding for higher education. The results were a major decline in academic salaries relative to other professions and the loss of thousands of lecturing posts as universities struggled to keep within their resources.

Many senior staff retired early, or emigrated, especially to North America, and, most alarmingly, many young scholars opted not to pursue careers in science because of the poor rewards and opportunities. The difficulties of funding scientific research led to the launching of the 'Save British Science' campaign. Students also suffered financially from the freezing of grants; they were obliged either to take out expensive loans or to jeopardize their studies by earning money from part-time jobs.

In spite of this the government found the universities an intractable opponent, partly because they were largely independent bodies and because middle-class parents were easily antagonized by threats to their children's university education. However, the government adopted a more subtle means of attack in a divide-and-rule policy. They removed the polytechnic sector from local-authority control, granted university status to polytechnics, and relied upon a new breed of managers to put students through degree courses at much lower costs. This had the positive effect of helping to raise the proportion of eighteen-year-olds in higher education from 14 per cent to 28 per cent, though at the cost of a sharp deterioration in the quality of their education. By 1993, however, the government, realizing the impossibility of educating more students without investing more money in the system, simply called a halt to the expansion programme.

Exactly similar strategies were adopted in the National Health Service. The government attempted to show that more patients could be handled at lower unit costs. During the 1980s, cuts in funding resulted in the closure of thousands of hospital beds, the withdrawal of thousands of nurses from the profession every year on account of poor wages, and a major increase in hospital waiting lists; by 1988 the lists had reached 700,000 and many people waited over two years for operations, at grave risk to life in some cases. As in education, the government antagonized all the professionals working in the health service by trying to impose a market system on what was still a service. Hospitals were encouraged to opt out of existing funding arrangements and manage their own finances in order to be more efficient. As in the polytechnics, a new breed of managers was eager to try to push more patients through the system quickly at a lower unit cost; but the effects proved unfortunate, especially for elderly patients who began to be dispatched from hospitals too soon, only to be readmitted later because of incomplete treatment and recovery. Readmissions, however, improved the statistics and were thus taken as proof of efficiency. By 1992 many hospitals had exhausted their annual resources three months before the end of the year and were thus unable to take even seriously ill patients. In this way the health reforms had begun to collapse in chaos. Mrs Thatcher, who was always on the defensive because of the popularity of the NHS, had claimed that the service was 'safe in our hands'. In practice, by the end of the decade it had been at least partly dismantled. For example, growing numbers of dentists now refused to take NHS patients for financial reasons. Prescriptions increased in price from 20p in 1979 to £4.25 by 1993. In 1988 charges were imposed for both dental tests and eye tests, which resulted in a 40 per cent fall in the number of eye tests. Finally, patients using a general practitioner who was not a fundholder often found that they were refused admission to hospital because fundholders enjoyed preference. The irony in all this was that whenever the NHS was investigated by researchers from the USA it was concluded that it provided a much more efficient service than that in America, which relied much more on private health. In Britain the expanding private sector was highly inefficient and charged inflated prices for its services, but more people felt obliged to make use of it because of the underfunding of the NHS.

Although the Thatcher government deeply disliked nationalized industries for purely ideological reasons, it took no steps to return them to the private sector for several years. However, it was eventually realized that sales of such industries could provide the resources required to pay for cuts in income tax. The result was a succession of privatizations of state industries including British Telecom, British Aerospace, Britoil, British Airways, the Trustee Savings Bank, British Gas, Rolls-Royce, Rover, Jaguar, and the Electricity and Water Boards. The most eccentric aspects of privatization included prisons, the use of private security firms by the Ministry of Defence to guard barracks, and proposals by British Heritage to sell off ancient monuments under its care. The greatest scandal was the sale of cemeteries worth £7 million for £1 by the Conservative-controlled Westminster Council.

These decisions were imposed against public opposition; four out of five people opposed the privatization of water, for example. There were a number of reasons for this. First, it became clear that the government obtained very poor deals on the sales because the state industries were disposed of at far below their value for fear that private interests would not wish to undertake the risks of managing them. Second, it was felt that many industries were assets belonging to the nation, paid for over many years by taxpayers, and which, like the state's oil holdings, had earned much revenue for the Exchequer. In a memorable speech the former Conservative prime minister, Harold Macmillan, denounced privatization as 'selling the family silver'. Third, those privatized industries offering services to the public on what was effectively a monopoly basis, such as gas, water, and electricity, subsequently made enormous profits by raising charges to customers and sacking staff. Even before privatization ministers had obliged electricity, for example, to increase prices in order to make the industry appear more profitable to investors. In spite of this, by 1992 only coal and railways remained in the public sector.

The Growth of State Power

Although Mrs Thatcher's government sought to restrict the role of the state in the economic sphere, it adopted a wholly different view of the *political* activities of the state. In many ways it extended the right of the state to interfere in the lives of its citizens and effectively curtailed the civil liberties of at least some of them. For example, trade unionism was banned from GCHQ at Cheltenham, on the grounds that it might be a threat to national security. The miners' strike saw new restrictions imposed by the police and the courts on freedom of movement and protest; and the 1986 Public Order Act extended the powers of the police to restrict public meetings.

The decade also witnessed a huge rise in telephone-tapping – 30,000 cases a year – by the agents of the government.

Of course, the extension of the political work of the British state was not by any means novel, but it was accelerated in the 1980s. The decade saw a major growth in the government's expenditure on advertising – up to £22 million by 1985, and no less than £86 million in 1987–8. Its critics felt the government overstepped the line between providing information about new laws and circulating party propaganda. Similarly, civil servants, especially those employed as information officers, seemed to be used for party-political purposes. This view was corroborated by the Director of the Central Office of Information, who complained in 1989 that the government required civil servants to cost the opposition's programmes – a blatantly partisan activity outside their proper functions. Like all governments, Mrs Thatcher's increasingly began to resent criticism from the mass media. However, since almost all the press supported her, and vociferously in the case of several of those with the largest circulations, this reaction was a sign of paranoia. Conservatives vented their resentment towards the BBC in particular. Norman Tebbit led the way, in his capacity as party chairman, in bullying the BBC governors over such topics as Ulster, the security forces, and the Falklands War.

Yet much the most serious threat to civil and political liberties during the 1980s arose in connection with the secret operations of the security services. A succession of civil servants suffered prosecution for revealing abuses and errors at the heart of the official machine, including Sarah Tisdall, Kathy Massiter, and Clive Ponting. Massiter helped to expose the misconduct of M15, which subjected several trade unionists and officers of the National Council for Civil Liberties to investigations for no other reason than that they seemed to hold left-wing opinions. The Prime Minister became outraged when the publication of *Spycatcher*, the memoirs of a former M15 employee, Peter Wright, appeared to corroborate claims about the abuses of the security services. He claimed to have taken part in burglaries and bugging of the premises of the former Labour Prime Minister, Harold Wilson. Systematic attempts to destabilize the Wilson government were believed to be justified on the assumption that Wilson enjoyed contacts with the Soviet Union and was thus a traitor. This painted an embarrassing picture of a security service that was as naive as it was incompetent. In spite of these revelations there followed a succession of politically motivated burglaries at the premises of Labour and Liberal politicians during the late 1980s and early 1990s, several of which were traced to Conservative sympathizers. However, the government chose to ignore the threat all this posed to democratic politics in Britain and concentrated on attempting to coerce public employees into silence. Wright's book was banned and the press forbidden to quote from it. The government spent large sums of money in a futile and humiliating effort to prevent the circulation of *Spycatcher* in Australia. By 1988 it had failed, a million copies had been purchased abroad, and the Appeal Court eventually ruled that the Thatcher government's action in stopping quotation in the British press had been illegal as well as 'futile and plainly silly'.

One consequence of these events was to stimulate fresh concern about civil liberties in Britain in the form of campaigns for a new Freedom of Information Act and reform of the highly illiberal Official Secrets Act, dating from 1911. The government, however, intervened in order to create a more restrictive Act, because it now appeared that juries were not prepared to convict civil servants such as Ponting who had revealed ministerial lies told during the Falklands War. The resulting Official Secrets Act of 1988 made it illegal for any government employee or newspaper to publish any information which the government itself defined as affecting security. No public-interest defence was allowed, nor were illegal acts such as unauthorized telephone taps to be revealed. In this way Britain moved much closer to East European practice in the area of civil liberties.

The Rise of Environmentalism

For several decades the growing public awareness about the environment had become evident in the rising membership of a variety of organizations including the National Trust, the Royal Society for the Protection of Birds, the Ramblers Association, Friends of the Earth, and Greenpeace. However, attempts to politicize this support in the shape of the Green Party met with only modest success. Not until the 1989 elections to the European Parliament did the Green Party make a major impact, by polling 15 per cent of the votes. Although the existing parties adopted 'Green' policies of varying shades, the movement was fundamentally antithetical to Thatcherite Conservatism because effective remedies for environmental problems required state intervention in the public interest to restrict the free operation of market forces. The dilemma was symbolized by the Prime Minister's pronounced hostility to railways. While desperately needed investment in British Rail was cut, the spread of the motor car continued to be promoted by road-building schemes which frequently destroyed attractive landscapes, archaeological sites, and even officially designated Sites of Special Scientific Importance, largely because such land was the cheapest available. In general the government showed itself reluctant to challenge vested interests, in spite of the evidence of damage to the environment and to health caused by their commercial activities. For example, it refused to place a ban on cigarette advertising, and subsidized hedgerow destruction and prairie farming as well as conifer plantations in upland regions. Faced with a series of scandals over food – listeria in dairy products, salmonella in eggs and chickens, and the disease of cattle known as BSE – the government invariably withheld information for as long as it could, and subsequently took ineffective counter-measures for fear of offending the farming community.

Politically, matters were not helped by the involvement of the EEC in various attempts to force the British government to respect environmental concerns and observe higher standards. For years the British resisted international pressure to protect the ozone layer, the diminution of which had begun to lead to a higher incidence of skin cancer. Britain also broke EEC law by failing to keep her water supplies free of pesticides, and by continuing to dump sewage and industrial pollutants into her rivers and the North Sea. As a result, relatively few British beaches were safe enough to meet European standards, many rivers became more polluted, and water companies were officially permitted to infringe the standards for water purity after privatization.

Perhaps the greatest environmental dilemma, however, was that posed by the nuclear power industry. Mrs Thatcher badly wanted to promote the industry because it would enable the country to free itself from dependence upon coal. However, after the major accident at Windscale in Cumbria in 1957, which was kept secret, nuclear power stations clearly posed a threat to the populations living in their vicinity. Over the years evidence mounted of concentrations of leukaemia and other cancers, especially among children in these areas. On the whole the government argued that these claims were unproven, and that nuclear energy could be produced much more cheaply than alternative sources of power. However, during the 1980s it emerged that the economic case for nuclear power had been based on faulty evidence; the massive costs of disposing of nuclear waste and decommissioning the older power stations made the whole enterprise uneconomic after all. Unfortunately the government had subsidized the industry for many years and at the same time withheld adequate investment for less damaging wind- and wave-power projects and research. Embarrassingly for the government, the truth was eventually flushed out by attempts to privatize the electricity industry, since the costs involved proved too daunting for commercial companies to take on. The decay of the industry thus stood as a monument to the blighted hopes of the decade.

21

The Downfall of Thatcherism

After three successive election victories in 1979, 1983, and 1987, the dominance of the Thatcherite brand of politics both in the Conservative Party and in the country might have been considered to be complete. In reality, however, it had always been the creed of a minority within a minority. Each victory had been won on the basis of only 42–3 per cent of the popular vote – historically a rather low share for the Conservatives. Moreover, their vote failed to improve during the years of Mrs Thatcher's leadership; indeed it *diminished*, albeit slightly. Thus her parliamentary position was very much the product of the British electoral system, which converted a minority of votes into substantial majorities in terms of seats. Exactly the same pattern had occurred during the 1920s, when the even division of the non-Conservative vote between Labour and the Liberals put the Conservatives in office for most of the period.

The Limited Revolution

In view of the precarious electoral base of the Conservative governments, it is not surprising to discover that the impact of Mrs Thatcher's ideas on British society proved to be limited. Studies of the public's views conducted by the British Social Attitudes Survey, for example, consistently showed that the philosophy associated with the Prime Minister had made little mark. Throughout the 1980s the public's faith in collectivist policies – state welfare, the National Health Service, public spending, state education – remained as strong as ever; and by the beginning of the 1990s the appetite for interventionism in social and economic affairs appeared to be growing, encouraged no doubt by the prolonged economic depression. Nor was there any evidence that an enterprise culture had been fostered in British society. Between 1980 and 1989 the self-employed section of the labour force rose from 8 to 12 per cent. However, as there was no increase in the number of people interested in becoming self-employed, this shift seems to have been a by-product of high unemployment; many of those who had lost their jobs used their redundancy money to start small business, which perished subsequently during the depression of 1990–3. Another possible measurement of attitudes towards enterprise lay in share-ownership. Although more individuals became owners of shares as a result of the sales of nationalized industries during the 1980s, the number was dwindling by the later years of the decade, and the proportion of all shares owned by individuals fell from 28 to 21 per cent. The explanation is that when buying shares the public had not been taking a risk on backing British entrerprise. On the contrary, they had purchased shares in privatized companies precisely because there was no risk involved; the initial share prices had been set deliberately low so that a rise, and thus an immediate profit for the purchaser, was guaranteed.

On the other hand, if Mrs Thatcher's rule failed to change public attitudes, her policy innovations may well have left a more enduring mark. This seems likely to be true in two respects. First, the trade unions lost much of their membership and their political influence; moreover, their capacity to lead strikes was severely curtailed by legal reforms. However, the prime cause of this decline was to be found in the sharp contraction of manufacturing industry, which simply destroyed much of the unions' traditional recruiting grounds, and in the high and prolonged unemployment which inhibited the will to strike. Secondly, most of the nationalized industries passed into private ownership. In political terms this proved to be important as the most tangible expression of the new Conservatism. On the other hand, the *economic* failure of the privatized industries (see p.331) raised doubts about the value of this strategy and about the wisdom of extending privatization to such industries as the railways and coal.

Beyond these two examples, however, the chief objectives identified by Mrs Thatcher herself in the 1970s proved to be almost wholly beyond the reach of her governments during over a decade in power. She had repeatedly insisted that Britain must pay its way in the world; but the modest balance-of-payments deficit she inherited increased massively, such that by 1989 it stood at a daunting £13.6 billion. This would not have been so serious had it been a short-term problem. But the decimation of the British manufacturing base made a big deficit an established, not to say immovable, feature of the economy. Mrs Thatcher had also placed great emphasis on curtailing government expenditure, which she saw as essentially a burden on the productive economy. In fact, state spending absorbed a *higher* proportion of gross national product during the 1980s. This reflected the costs of the two depressions and protracted unemployment, the inability of the British government to control such items as subsidies for agriculture, and the political obstacles to dismantling the welfare state. Eventually public finance got so far out of control that by 1992 public-sector spending exceeded current revenue for the first time since 1964, when the data first began to be published; thus a government that had embarked upon a policy of careful budgeting ended up with its account embarrassingly in the red.

The drift away from financial responsibility reflected a contradiction at the heart of Thatcherism which was reflected in society generally by the growth of debt during the 1980s. Whereas in 1981 16.5 per cent of national income had been saved, by 1987 the total had fallen to a mere 2.4 per cent. The excessive use of credit, encouraged by banks and building societies, resulted in widespread social problems by the end of the decade. A similar loss of control became apparent in the steady rise in crime, the reduction of which had been seen as a major Conservative objective. During Mrs Thatcher's premiership, crime increased by no less than 79 per cent. Both Home Office research and the police argued that much of this increase was the product of unemployment, especially among the young, thus contradicting the Conservative view which held it to be a reflection of moral decline.

Another prime objective throughout the decade was the reduction, indeed the elimination, of inflation. In the event, inflation rose sharply in the early years, then fell back as a result of the economic depression in the early 1980s, but rose again before dwindling in the second depression around 1990. This was no more than the familiar cyclical pattern. Periodic ministerial claims that inflation had been conquered turned out to be premature. Indeed, the underlying problem remained in the sense that, in spite of the diminished role of the unions, wage rates continued to rise faster than output and at two to three times the rate of their European competitors.

Finally, Mrs Thatcher had devoted great attention to the need to stimulate enterprise in Britain by reducing the level of taxation. Indeed, this was widely claimed by Conservatives as one of their achievements. Yet the government's own figures showed the claim to be quite false. Mrs Thatcher inherited a situation in which just over 35 per cent of gross domestic product was consumed by taxes of all kinds. By 1984 this *risen* to almost 38 per cent, though it fell slightly to just under 37 per cent by 1990. All that the government had done was to shift taxes from one to another; but throughout the 1980s it took a consistently higher share of GNP in taxes than it had during the 1970s. This was, perhaps, the most signal failure of all. What was remarkable, however, was that the government largely escaped the political penality of its high-tax policies because the opposition failed to challenge effectively the claims made by the Conservatives.

All Power Corrupts

As a result of its divisions and repeated defeats during the 1980s, the Labour Party suffered serious demoralization and largely failed to offer effective opposition in Parliament. In these circumstances the government gradually ceased to be responsive to criticism, and the whole quality of parliamentary government deteriorated. One symptom of this was the marked reluctance of the Prime Minister to speak in the House of Commons. This avoidance was obscured by the publicity given to the regular Prime Minister's Question Times; these were merely brief, stage-managed affairs, however, at which Mrs Thatcher delivered well-rehearsed statements for television. In fact she made only two full speeches in the four years up to 1990, and overall recorded about one-fifth as many contributions as her three predecessors. The longer she remained Prime Minister the less frequently she spoke in the House – a sign that she had begun to retreat into a political laager where she could be surrounded by loyal colleagues and advisers, insulated from her critics both inside the Conservative Party and beyond.

However, Mrs Thatcher took care to bolster her regime by the restoration of political honours. This was indeed another sense in which her practice closely resembled that of Lloyd George. Like him, she lavishly rewarded the editors and proprietors of major newspapers who supported her. She also awarded ten times as many knighthoods to Conservative back-bench MPs as Edward Heath had done. Those who had donated large sums to the party's funds were, as in the past, recognized; for example, peerages were given to 17 industrialists whose companies had contributed over £5 million to the Conservative Party or to one of its front organizations. Two-thirds of the newly created knights belonged to companies whose political donations totalled £11 million during Mrs Thatcher's premiership. Although this practice was by no means new, the scale of the awards to supporters seemed to represent a throwback to an earlier era.

Similarly, Mrs Thatcher's ex-ministers proved to be more than usually successful in gathering lucrative company directorships. In particular, there was a conspicuous connection with the newly privatized companies, in that several former ministers obtained paid positions with the very companies which took over the industries for whose privatization they had been responsible. Although movement between official employment and private industry was not new, nor did it contravene any rule, there was some surprise and comment on the extent to which politicians enriched themselves as a consequence of their government service.

In course of time the absence of effective checks upon government began to show in terms of a deterioration in the quality of decision-making. Though originally in a minority in her 1979 Cabinet, Mrs Thatcher gradually changed its composition by introducing her own followers. Moreover, like Lloyd George and Neville Chamberlain, she proved to be disloyal to her ministers. They were undermined by leaks to the press from Number 10 Downing Street. This practice gradually destroyed normal Cabinet government by largely eliminating those ministers who were prepared to challenge the Premier. Increasingly, policy decisions emanated from Mrs Thatcher's small body of personal advisers.

Much the most disastrous example of this trend was the Community Charge, better known as the poll tax. It represented the fulfilment of a long-standing pledge by Mrs Thatcher to abolish the system of rates. But by imposing the same charge upon poor householders as on the very wealthy the poll tax offended the sense of fairness which was still widespread in British society. Although Mrs Thatcher's ministers felt the new policy to be a mistake, none except for Nigel Lawson were prepared to express their opposition. In the event the poll tax was introduced first in Scotland, where its immediate unpopularity contributed to the Conservatives' heavy losses at the 1987 general election. During 1988 and 1989 it began to emerge that local councils in Scotland were unable to collect the new tax. By 1989, 1.2 million Scots poll-tax-payers had been sent sheriffs' warrants for failure to pay. By 1990 the extension of the scheme to England had produced a similar resistance which, crucially, affected many middle-class, Conservative areas of the country. For example, non-payment rates in 1990 were 51 per cent in Liverpool but also 39 per cent in Bath. As a result, the system of local government was in danger of breaking down altogether, for each year's non-payment forced councils to raise the poll tax subsequently for those who did pay; collection thus became more difficult. This fiasco played a key part in undermining Conservative support in the country and thereby in destroying Mrs Thatcher's hold over the MPs. By 1990 she had become a serious political liability.

The Economic Depression

Important as the poll tax was, the central explanation for Mrs Thatcher's downfall lay in a massive failure of economic policy. The origins of the second depression under her premiership were twofold. First, the eclipse of one-fifth of British manufacturing capacity in 1980–1, combined with low levels of investment and an overconcentration on the service sector, left the country with a severe and unsustainable balance-of-payments problem. The deficit, which reached £13.6 billion in 1989, rose to £16 billion in 1990. Second, financial deregulation and the reckless lending to individuals by banks and building societies during the 1980s had got out of hand. This itself resulted in increases in imports, huge rises in house prices, and general inflation. By 1990 inflation had reached 10.6 per cent, which was higher than the level inherited by Mrs Thatcher in 1979.

The problems thus created were exacerbated by the government's insistence on responding with a single strategy – the raising of interest rates. They considered this a necessary means of protecting the value of the pound, which had been rising in anticipation of a British decision to join the European Exchange Rate Mechanism. But higher interest rates inevitably had the effect of reducing investment and profits, thereby helping to precipitate a recession. In addition, the government restricted public spending by cutting the funds for industrial training, by capping local authorities' expenditure, and by refusing to fund major national projects such as the high-speed rail network. This latter was a necessary scheme if Britain was to derive economic advantages from the construction of the Channel Tunnel. However, the government's philosophy dictated that such investment should be financed by the private sector. Unfortunately, this was the kind of long-term investment which the private sector was never keen to undertake. This threatened to leave the British rail system unable to match the superior Continental one, which was state-financed, and thus to curtail the benefits to the economy as a whole. It was a perfect example of the damage done to the national interest by strict adherence to a political doctrine.

As a result of the government's unwillingness to use countercyclical measures, unemployment began to rise again during the spring of 1990. Mortgage rates rose to 15.5 per cent, with the result that the housing market and the building industry experienced a sharp decline; after massive borrowing many people now found themselves facing large debts and higher repayment charges at a time of rising unemployment. Consequently, by the summer of 1990 it became clear that Mrs Thatcher's policies had not engineered an economic miracle so much as a second major economic recession. This largely destroyed her political credibility and the belief in her economic doctrines. Official figures now showed that the performance of the economy during the Conservatives' tenure of office had been far worse than had been realized. Between 1979 and 1991 the average annual rate of economic growth was only 1.75 per cent. This compared with the 2.4 per cent average for the previous decade which, though derided by Conservative propaganda, now appeared modestly impressive.

During the second half of 1990 it became clear that the government had few weapons with which to tackle the recession beyond the higher interest rates which were exacerbating the problem. As investment fell, unemployment continued to mount, the homes of defaulting mortgage-payers began to be repossessed, and some 29,000 businesses went bankrupt in the course of the year. The former Chancellor, Mr Lawson, blamed Mrs Thatcher for her refusal to join the Exchange Rate Mechanism. This was, in fact, the one policy initiative which was seized upon in October 1990 by Mr John Major, the new Chancellor. He determined that Britain should join the ERM with the pound valued at 2.95 German marks. This was intended, over a period of some years, to have the effect of forcing British companies to reduce their costs to maintain competitiveness because they could no longer expect to be rescued by a devaluation of the currency. Since British wage and salary increases continued to be much higher than those in the EEC, this was a powerful threat to living standards, at least in the short term. In fact, Britain had not only entered the ERM at a bad time, in view of her mounting recession, but at too high a rate. Major, it soon transpired, had committed a fatal error in overvaluing the pound. His decision was comparable to the return to the gold standard under Churchill in 1925, and it was to bring in its wake a similar economic and political crisis.

The Turning-Point

It was Europe that eventually provided the trigger for Mrs Thatcher's overthrow. Since 1988 Sir Geoffrey Howe, the Foreign Secretary, had been publicly supporting the idea that Britain should join the ERM, and in June 1989 he and the then Chancellor, Lawson, manœuvred a reluctant Prime Minister into accepting this policy in principle. However, the struggle within the government continued. In July 1989 Howe was sacked as Foreign Secretary, and at the party conference that year he attacked Mrs Thatcher's views on Europe. At the European summit conference in Rome she antagonized the pro-Europeans in her party by denouncing the whole idea of economic and monetary union, despite the fact that she had committed her government to it already. Subsequently she made Howe, now Deputy Prime Minister, the target of leaks and rumours inspired by her close advisers. As in the case of other ministers who had crossed Mrs Thatcher, Howe was being publicly undermined with a view to his eventual removal.

However, in November 1990 Howe seized the initiative by resigning, and in a devastating speech in the Commons he made it clear that the Prime Minister's determination to sabotage the European policy had made life intolerable. This proved to be one of the most damaging speeches ever

The downfall of Thatcher

delivered in Parliament. It finally forced Michael Heseltine, who had been quietly preparing to replace Mrs Thatcher after the next election, to launch an open challenge for the party leadership. In the event Heseltine won a surprising total of 152 votes, and although Mrs Thatcher comfortably outpolled him, she narrowly failed to obtain sufficient votes to retain the leadership on the first ballot.

What was the explanation for this remarkable result? The strength of Mrs Thatcher's position lay partly, as for all Prime Ministers, in her control of patronage, and partly in the immense loyalty she commanded among the rank-and-file Conservative members in the constituencies, many of whom kept their MPs under pressure to support her. Against this, however, only a minority of the MPs had ever become fully converted to Mrs Thatcher's brand of Conservatism. They had accepted her as long as she delivered victory at the elections; but by 1990 they judged that the public disenchantment over the poll tax and the economy was so deep that she would lead the party to defeat next time. This feeling spread rapidly once the Prime Minister had failed to win outright on the first ballot. Moreover, she now began to pay the price for her disloyalty towards her Cabinet colleagues, the majority of whom told her that she should abandon the leadership.

As a result of her collapsing support amongst the MPs, Mrs Thatcher withdrew from the second ballot and was replaced by John Major. This was by no means the first occasion in the twentieth century when a Tory Prime Minister found himself driven from office by his own followers. But it was the most remarkable coup, because Mrs Thatcher had delivered three victories to her party and not lost an election. However, the party's instinct for survival proved to be very sound: as in the past, the abandonment of a leader made it possible also to avoid the unpopularity caused by that leader's policies. It was not long before the party's standing in the polls effected a sharp recovery.

The new Prime Minister enjoyed the advantage of being an unknown quantity because of his short experience in office. On the one hand the public warmed to him because he seemed such a welcome change from Mrs Thatcher, while on the other hand, the Thatcher loyalists in Parliament believed him to be one of them, particularly as Mrs Thatcher herself had given him her support. However, by February 1991 it had become clear that they had been misled, or perhaps had misled themselves. Mr Major's decision to appoint a distinctly liberal Tory, Chris Patten, as the new party chairman indicated that he did not intend to fight a general election on a Thatcherite programme. A clear indication of this came with the abandonment of the poll tax in March. With no less than £1.6 billion now unpaid, the tax was becoming virtually unworkable, and it seemed likely to be fatal to Conservative election prospects. Beyond that Mr Major indicated that, while he accepted much of his predecessor's economic thinking, he would break with her social policies; in particular he believed in the need to improve public-sector services such as health and education. This meant that he had little use for the right-wing think-tanks such as the Adam Smith Institute and the No Turning Back Group, which wished to preserve and extend the Thatcher revolution. Consequently, a new body, the Conservative Way Forward, was established by such former ministers as Lord Tebbit, Cecil Parkinson, and John Moore, to uphold the ex-Prime Minister's ideas. This portended a serious struggle for the party's conscience, especially when Mr Major began to emerge as far more pro-European than Mrs Thatcher.

However, for the time being these divisions were suppressed by the urgent need to rally support at the general election, which eventually took place in April 1992. Since it was fought against a background of serious economic depression and a small but sustained Labour lead in the opinion polls, the Conservatives were expected to lose. Their victory, albeit on a much reduced majority, proved to be one of the greatest surprises of modern electoral history. The Conservative share of the poll was virtually unchanged at 42 per cent to Labour's 35 per cent and the Liberal Democrats' 18 per cent. Yet although the Conservative vote had held up, the party lost seats heavily – a typically perverse product of the electoral system. With a net loss of 34, their majority fell to 21 seats. The explanation lay in the widespread adoption of tactical voting by Liberal Democrat and Labour supporters; this was an indication of the success that the opposition parties might have obtained had they put together a formal electoral pact.

The Conservatives' success in retaining their previous share of the vote was attributable to three main causes. First, they derived a key advantage from structural changes in the electorate. By 1991 no fewer than a million voters had dropped off the electoral register out of a desire to escape the poll tax. Thus the proportion of eligible adults registered in England and Wales fell from 97.4 per cent in 1987 to 95.5 per cent in 1991. To a lesser extent the Conservatives also gained from a 1989 reform which allowed British expatriates living abroad to vote if they had resided in Britain during the previous twenty years.

The second reason for the outcome of the election was that most voters blamed Mrs Thatcher, not Mr Major, for the depression, in spite of the latter's share of the responsibility as Chancellor of the Exchequer. A third factor lay in the lack of confidence in the Labour Party's economic and taxation policies. Both Labour and the Liberal Democrats handicapped themselves by making detailed promises involving increases in taxation. Moreover, the fact that a Labour victory was widely anticipated led some disillusioned Conservatives who had moved to the Liberal Democrats to return to their old party out of fear of higher taxation. Although these shifts came very late in the campaign and were therefore difficult to detect, it was also clear that the opinion polls had failed to register the full extent of Conservative support over a long period of time; to that extent, the polls, by persuading voters that Labour would win, actually had an influence on the outcome of the election.

The Crisis over Europe

The election victory of 1992 rapidly proved to be a poisoned chalice. It obliged the Conservative government to face up to the legacy of an economic depression and to become impaled upon the divisive issue of

Britain's role in Europe. Since 1990 Mrs Thatcher had begun to lose control of British policy towards the EEC. This was partly due to the logic of events. The collapse of the Communist regimes in Eastern Europe and the reunification of Germany accelerated the existing pressure both to expand the EEC and to advance towards a federal system. This involved accepting a single currency, a central bank, and ultimately a common foreign and defence policy. As a result of her rooted dislike of such changes, the Prime Minister had lost several of her leading colleagues, notably Lawson and Howe. But in party terms the most damaging resignation was that of Nicholas Ridley, one of Mrs Thatcher's strongest supporters, in June 1990. In a bitter and emotional outburst, Ridley denounced economic union as 'a German racket to take over the whole of Europe'. In this he undoubtedly reflected fundamental fears held amongst Tory supporters that Britain would simply lose her sovereignty. Moreover, the closer the EEC approached to economic union the less feasible would it be to pursue a *laissez-faire* economic policy in Britain. In this sense the Thatcherites correctly diagnosed Europe as being ultimately inimical to their whole approach to government; it soon began to seem that they were attempting to hold out against a tide that was too powerful for a country that suffered from a weak economy. Ridley also spoke for many people in the country; the generation that had *not* fought in the Second World War felt impotent and outraged that Germany had gone from strength to strength since 1945, while Britain had steadily declined in spite of her victory.

Although Mrs Thatcher had sacrificed Ridley in 1990, she shared his views. Thus, within months of losing the premiership she joined with other former ministers, including Tebbit and Parkinson, to lead what was effectively a campaign against Major's European policy in order to save the free-market philosophy. A new Conservative organization, the Bruges Group, urged the Prime Minister to use Britain's veto to check progress towards economic union. However, during 1991 Major declared himself by pledging that Britain would work 'at the heart of Europe'. In December of that year he negotiated an agreement at Maastricht which allowed Britain to opt out of some aspects of the common economic policy; but this was not enough to satisfy the Thatcherites, who demanded a popular referendum on whether Britain should accept the treaty. Seventy Tory MPs signed a motion hostile to Masstricht, and opinion polls suggested that less than a quarter of the electorate favoured economic union.

For Major the dilemma was deepened by two factors. Although he himself wished to move away from Thatcherism, his party was now becoming increasingly Thatcherite. This was because the newly elected Conservative MPs, who had been influenced by the politics of the 1980s, were much more supportive of Mrs Thatcher than their predecessors, who were Tories of the Macmillan–Heath era. Major was also undermined by the deepening economic recession. During 1991 Britain's gross domestic product diminished by 2.5 per cent and continued to fall in 1992. Some 48,000 businesses had gone bankrupt in 1991 alone, and the rate of failure increased in the following year. Unemployment had reached 2.85 million by the autumn of 1992, over 10 per cent of the labour force. However, the official figures grossly underestimated unemployment, because the method of counting had been repeatedly changed and because half a million people had been diverted on to official schemes and thus kept off the unemployment registers. The real total was at least 3.5 million.

As a result of the chronic weakness of the economy, the pound lost value within the Exchange Rate Mechanism. In this situation the government followed the traditional tactics used in the 1920s, and with equally fatal results. It attempted to sustain an unrealistically high exchange rate by raising interest rates. Economic considerations dictated that Britain should either withdraw from the ERM or seek a realignment of currencies within it. However, pride and reputation was now at stake, for Mr Major had been the Chancellor who had originally valued the pound at DM 2.95. In the event, though the government pledged itself to raise interest rates as far as was necessary to defend the pound, by September 1992 the dealers were selling the currency so fast that sterling fell inexorably. The Bank of England spent £60 billion in an attempt to save the pound, and the Chancellor pushed interest rates as high as 15 per cent; but all in vain. The inevitable outcome was a humiliating climb-down. Sterling was withdrawn from the ERM, allowed to float, and thus in effect devalued by around 15 per cent.

So comprehensive a reversal of the government's economic strategy exposed Mr Major to attack at his party's conference in October 1992. But the attack took the form of a renewed challenge to the European policy led by Lord Tebbit, though effectively supported by Mrs Thatcher, who made it clear that she believed her successor to be betraying the revolution she had begun. The Conservatives had become as fundamentally divided over Europe as they had been over India and appeasement during the 1930s. Major handled the party just as Baldwin had done then. By repeatedly referring to 'Britain' and using such phrases as 'a thousand years of history', he sought to deflect criticism that he was putting an end to the sovereignty of Britain by joining an economic union in which Westminster would no longer have effective control of policy. The only compensation for this bitter dispute was that it diverted some attention from economic problems.

By the winter of 1992 the government had begun to readjust to the more positive consequences of its policy reversals: more competitive exports and lower interest rates. However, the continued rise in unemployment prevented any recovery from the depression, now of three years' duration. Thousands of jobs continued to be lost in banking, British Rail, Ford, Rolls-Royce, and British Aerospace, to name only a few. It now became clear that those privatized industries that engaged in real competition – Rover, Rolls-Royce, Jaguar, British Aerospace, British Steel, the Trustee Savings Bank – had done poorly since leaving the state sector. The contraction of British Aerospace in particular threatened to bring an end to Britain's proud record in civil aviation. As an *economic* expedient, privatization had failed. However, the political rationale continued to drive it forward, for it represented one of the very few aspects of Thatcherism still within the government's control. The government pushed ahead with plans to privatize the coal and railway industries. This, however, brought fresh disaster when it was announced that in preparation for coal privatization no fewer than 31 of the surviving 50 pits were to be closed, with the loss of 30,000 jobs. The explanation for this was that the privatized power companies had locked themselves into agreements to buy gas and nuclear power, the latter being quite uneconomic but heavily subsidized by the government. As a state industry, coal had dramatically improved its productivity, so much so that British coal was produced at £42 per ton compared to £90 per ton in Germany. However, the government's political ideology dictated that state ownership was wrong, however efficient the industry; it was therefore prepared to allow coal to shrink even though this would expose the country to the prospect of having to import coal whose price would inevitably rise in the future. The dilemma underlined the futility of applying privatization to strategic industries like energy, where the national interest required a long-term, coordinated policy rather than one determined by short-term profit motives. Coal in fact served to crystallize public misgivings about the whole trend of economic policy, not least because it threatened another increase in unemployment which would prolong the depression still further. It produced a remarkable display of support for maintaining the existing mines even in the middle-class parts of the country. Sensing the political danger, the Cabinet retreated by offering to reconsider its energy policy.

Thus by the end of 1992 the accelerating decline of the British economy had cast the whole experiment of the 1980s into perspective. Although Mrs Thatcher had appeared to be an immensely powerful Prime Minister during the 1980s, she had wholly failed to prevent Britain being drawn inexorably into Europe; indeed the country had, by 1990, lost many of the attributes traditionally associated with a sovereign state. Her economic strategy had been a doomed attempt to recapture a half-remembered, half-imagined era in which the state had played a minimal role in economic and social affairs. By the early 1990s a reaction in favour of Keynesianism had set in. Thatcherism in this context seemed increasingly a Quixotic attempt to tilt at such windmills as Europe and socialism; essentially an aberration in the course of British history.

Epilogue

The Management of Britain's Decline

In 1870 Britain had been the proud supplier of manufactured goods to the rest of the world. By 1992 she had ceased to produce a wide range of sophisticated items of the sort that had once been the basis of her wealth: typewriters, computers, washing machines, television sets, sewing machines, and motor cycles. Her companies had been driven out of motor cars and shipbuilding except for small remnants; and aerospace, railways, and coal were fast going the same way. Whereas in the late 1950s 8 million people had worked in British manufacturing industry, by 1991 fewer than 5 million still did so. Not only had rivals such as France, Germany, Japan, and the USA achieved a higher rate of economic growth for several decades; during the 1980s Britain's economy was overtaken by that of Italy, and was in danger of being overhauled by that of Spain, not to mention the rapidly expanding economies of the 'Pacific rim' – Taiwan and Korea. No longer among the first-rank powers in the world, Britain had begun to decline precipitately within the middle-ranking states.

In these circumstances the task for British statesmen was how to manage the process of Britain's decline – since none had been able to arrest the trend. Much of the population continued to be shielded from the consequences of decline by means of pay settlements well above productivity and by huge imports of goods that were not earned. Even so, the country had suffered a sharp relative decline in its wealth. In 1950 standards of living in Britain had been second only to those of the USA and Switzerland. By 1985 real income per head in Britain was between a half and two-thirds that in Germany and France; she had fallen behind most of the countries of Western Europe and Japan, and was on a level with other Asian states such as Taiwan.

There was no prospect of a check to this long-term decline in the 1990s, for Britain laboured under a huge balance-of-payments deficit which was the unavoidable result of the loss of a large proportion of her manufacturing industry during the 1980s. The underlying reason for poor performance lay in much lower – and declining – levels of investment and innovation in Britain. The opportunity presented by the oil revenues of the 1980s to catch up on Britain's rivals had been dissipated in consumer spending. Poor investment was a reflection of both political and commercial attitudes. Governments largely failed to promote primary research to the extent that other countries, notably the Japanese, had done. The neglect of university science research had fundamental and long-term consequences in terms of a loss of leading personnel, diminished recruitment, and a loss of British dominance in several areas of science. Nor did the financial system in Britain promote industry and innovation as it did in other countries. The City of London rewarded companies that concentrated on short-term profits and takeovers, but undervalued the shares of companies that invested in research and development. What was good for shareholders was not necessarily in the national interest. The only qualification to this picture was provided by fresh Japanese investment in Britain in the 1980s. However, whereas the Japanese companies offered the skill, technology, and capital, Britain supplied comparatively cheap labour and access to markets; this reversed the nineteenth-century pattern by casting Britain in the role of a colonial dependency.

By far the most powerful single method of breaking out of the cycle of decline ought to have been Europe. But during the two decades of Britain's membership of the EEC the economic rationale failed to work, largely because the other European countries penetrated the British market with their exports while Britain failed to take advantage of the opportunities on the Continent. The result was a major growth of the trade deficit with with EEC. The combination of a single European market and the neglect of Britain's transport infrastructure appeared likely to exacerbate the problem during the 1990s. The growing danger was that, like Ireland, Britain would increasingly export her educated people, attracted by the higher salaries available in neighbouring countries. This did not mean that there were no compensations. But they were increasingly in the form of grants and subsidies available for depressed and economically backwards regions such as Portugal and southern Italy. In this way Europe seemed able to assist Britain in adjusting to her diminished position in the world by cushioning the rate of decline.

Guide to Further Reading

Chapter 1

D. H. Aldcroft and H. W. Richardson, *The British Economy, 1870–1939* (1969)
D. C. Coleman, 'Gentlemen and Players', *Economic History Review*, 26 (1973)
M. J. Daunton, 'Gentlemanly Capitalism and British Industry, 1820–1914', *Past and Present*, 122 (1989)
D. S. Landes, 'Technological Change and Development in Western Europe, 1750–1914', in *Cambridge Economic History of Europe*, vol. vi, pt. 1 (1966)
P. L. Payne, *British Entrepreneurship in the Nineteenth Century* (1974)
S. Pollard, *Britain's Prime and Britain's Decline: The British Economy, 1870–1914* (1989)
J. Raven, 'British History and the Enterprise Culture', *Past and Present*, 123 (1989)
S. B. Saul, *The Myth of the Great Depression, 1873–1896* (1969)
A. J. Taylor, *Laissez-Faire and State Intervention in Nineteenth-Century Britain* (1972)
J. Tomlinson, *Problems of British Economic Policy, 1870–1945* (1981)
M. J. Weiner, *English Culture and the Decline of the Industrial Spirit, 1850–1980* (1981)

Chapter 2

D. W. Bebbington, *The Nonconformist Conscience* (1982)
R. Blake, *Disraeli* (1966)
——, *The Conservative Party from Peel to Churchill* (1970)
H. A. Clegg, A. Fox, and A. Thompson, *A History of Trade Unions since 1889*, i (1964)
B. Coleman, *Conservatism and the Conservative Party in Nineteenth-Century Britain* (1988)
J. P. Cornford, 'The Transformation of Conservatism in the Late Nineteenth Century', *Victorian Studies*, 7 (1963)
E. J. Feuchtwanger, *Gladstone* (1975)
——, *Democracy and Empire, 1865–1914* (1985)
D. A. Hamer, *The Politics of Electoral Pressure* (1977)
H. J. Hanham, *Elections and Party Management: Politics in the Age of Disraeli and Gladstone* (1959)
R. Harrison, *Before the Socialists: Studies in Labour and Politics, 1861–1881* (1965)
J. Hinton, *Labour and Socialism* (1983)
D. Howell, *British Workers and the Independent Labour Party, 1888–1906* (1983)
E. H. Hunt, *British Labour History, 1815–1914* (1981)
R. Jay, *Joseph Chamberlain* (1981)
G. I. T. Machin, *Politics and the Churches in Great Britain, 1869–1921* (1987)
R. McKibbin, 'Why Was There No Marxism in Great Britain?', *English Historical Review*, 99 (1984)
P. Marsh, *The Discipline of Popular Government: Lord Salisbury's Domestic Statecraft, 1881–1902* (1978)
C. Matthew, *Gladstone, 1809–1874* (1986)

K. O. Morgan, *Keir Hardie: Radical and Socialist* (1975)
R. Moore, *The Emergence of the Labour Party, 1880–1924* (1978)
J. P. Parry, *Democracy and Religion: Gladstone and Liberalism, 1867–1875* (1986)
H. Pelling, *A History of British Trade Unionism* (1963)
M. Pugh, *The Making of Modern British Politics, 1867–1939* (1982; repr. 1993)
——, *The Tories and the People, 1880–1935* (1985)
G. R. Searle, *The Liberal Party: Triumph and Disunity, 1886–1929* (1992)
R. Shannon, *The Age of Disraeli, 1868–1881* (1992)
F. B. Smith, *The Making of the Second Reform Act* (1967)
P. Smith, *Disraelian Conservatism and Social Reform* (1967)
J. R. Vincent, *The Formation of the British Liberal Party, 1857–1868* (1966)

Chapter 3

C. Booth, *Life and Labour of the People in London* (1902–4)
J. Burnett, *Plenty and Want* (1966)
M. A. Crowther, *The Workhouse System, 1834–1929* (1981)
A. Davin, 'Imperialism and Motherhood', *History Workshop Journal*, 5 (1978)
H. V. Emy, *Liberals, Radicals and Social Politics, 1892–1914* (1973)
H. Fraser, *The Coming of the Mass Market* (1973)
B. B. Gilbert, *The Evolution of National Insurance in Great Britain* (1966)
J. Harris, *Unemployment and Politics* (1972)
P. Hollis, *Ladies Elect* (1987)
J. Lewis, *Women and Social Action in Victorian and Edwardian England* (1991)
M. Llewelyn Davies, *Life As We Have Known It* (1931)
A. McBriar, *Fabian Socialism and British Politics, 1884–1918* (1962)
G. C. Peden, *British Economic and Social Policy: Lloyd George to Margaret Thatcher* (1985)
H. Pelling, 'The Working Class and the Origins of the Welfare State', in *Popular Politics and Society in Late Victorian Britain* (1979)
M. Pember Reeves, *Round About a Pound a Week* (1913)
R. Roberts, *The Classic Slum* (1971)
M. Rose, *The Relief of Poverty, 1834–1914* (1972)
B. S. Rowntree, *Poverty: A Study of Town Life* (1901)
G. Stedman Jones, *Outcast London* (1971)
P. Thane, 'Women and the Poor Law in Victorian and Edwardian England', *History Workshop Journal*, 6 (1978)
——, 'The Working Class and State Welfare in Britain, 1880–1914', *Historical Journal*, 27/4 (1984)
R. Tressell, *The Ragged Trousered Philanthropists* (1955)

Chapter 4

P. Bailey, *Leisure and Class in Victorian England* (1978)
J. A. Banks and O. Banks, *Prosperity and Parenthood* (1954)
J. M. Golby and A. W. Purdue, *The Civilisation of the Crowd: Popular Culture in England, 1750–1900* (1984)
W. H. Greenleaf, *The British Political Tradition*, i (1983)
B. Harrison, *Drink and the Victorians* (1971)
R. Holt, *Sport and the British* (1989)
P. Levine, *Victorian Feminism, 1850–1900* (1987)
J. Lewis, *Women in England, 1870–1950* (1984)
R. McKibbin, 'Working-Class Gambling in Britain 1880–1939', *Past and Present*, 82 (1979)

G. Marsden (ed.), *Victorian Values: Personalities and Perspectives in Nineteenth-Century Society* (1990)
J. Perkin, *Women and Marriage in Nineteenth-Century England* (1989)
E. Roberts, *Women and Work, 1840–1940* (1988)
R. A. Soloway, *Birth Control and the Population Question in England, 1872–1930* (1982)
F. M. L. Thompson, *The Rise of Respectable Society* (1988)
L. Tilly and J. Scott, *Women, Work and Family* (1978)
P. J. Waller, *Town, City and Nation: England, 1850–1914* (1983)
J. Walton, *The English Seaside Resort* (1983)
J. Walvin, *Victorian Values* (1987)
J. Weeks, *Sex, Politics and Society* (1981)

Chapter 5

G. Anderson, *Victorian Clerks* (1976)
D. Cannadine, *The Decline of the Aristocracy* (1991)
G. Crossick, *An Artisan Elite in Victorian Society* (1978)
—— (ed.), *The Lower Middle Class in Britain* (1977)
R. Faber, *High Road to England* (1985)
J. A. Garrard, *The English and Immigration, 1880–1910* (1971)
J. M. Golby and A. W. Purdue, *The Monarchy and the British People* (1988)
R. Gray, *The Aristocracy of Labour in Nineteenth Century Britain* (1981)
B. Harrison, 'Traditions of Respectability in British Labour History', in *Peaceable Kingdom* (1982)
C. Harvie, *Scotland and Nationalism: Scottish Society and Politics, 1707–1977* (1977)
E. Hobsbawm and T. Ranger (eds.), *The Invention of Tradition* (1983)
C. Holmes, *John Bull's Island: Immigration and British Society, 1871–1971* (1988)
R. Holt, *Sport and the British* (1989)
R. McKibbin, 'Why Was There No Marxism In Great Britain?', *English Historical Review*, 99 (1984)
G. E. Mingay, *The Gentry: The Rise and Fall of a Ruling Class* (1976)
K. O. Morgan, *Rebirth of a Nation: Wales 1880–1980* (1982)
——, *Wales in British Politics, 1868–1922* (1963)
H. Perkin, *The Origins of Modern English Society, 1780–1880* (1969)
——, *The Rise of Professional Society* (1989)
K. Robbins, *Nineteenth Century Britain: Integration and Diversity* (1988)
W. D. Rubinstein, *Men of Property* (1981)
R. Samuel (ed.), *Patriotism: The Making and Unmaking of British National Identity* (1989)
G. Stedman Jones, *Languages of Class* (1983)
L. Stone, *An Open Elite? England 1540–1880* (1986)
R. Swift, *The Irish in Britain, 1815–1914* (1990)
——, and S. Gilley, *The Irish in Britain, 1815–1939* (1989)
J. Walvin, *Passage to Britain: Immigration in British History and Politics* (1984)
J. N. Wolfe (ed.), *Government and Nationalism in Scotland* (1969)

Chapter 6

I. F. Clarke, *Voices Prophesying War, 1763–1984* (1970)
F. Coetzee, *For Party or Country: Nationalism and the Dilemmas of Popular Conservatism in Edwardian England* (1990)
P. A. Duane, '"Boys" Literature and the Idea of Empire, 1870–1914', *Victorian Studies*, 24 (1980)

C. C. Eldrige, *Victorian Imperialism* (1978)
D. K. Fieldhouse, *Economics and Empire, 1830–1914* (1973)
D. French, *The British Way in Warfare, 1688–2000* (1990)
P. Kennedy, *The Rise of the Anglo-German Antagonism, 1860–1914* (1980)
——, *The Realities Behind Diplomacy* (1981)
—— (ed.), *The War Plans of the Great Powers, 1880–1914* (1979)
H. W. Koch, 'The Anglo-German Alliance Negotiations', *History*, 54 (1969)
T. Lloyd, 'Africa and Hobson's Imperialism', *Past and Present*, 55 (1972)
J. M. Mackenzie, *Propaganda and Empire: The Manipulation of British Public Opinion, 1880–1960* (1984)
H. Pelling, 'British Labour and British Imperialism', in *Popular Politics and Society in Late Victorian Britain* (1979)
D. C. M. Platt, 'Economic Factors in the New Imperialism', *Past and Present*, 39 (1968)
B. Porter, *The Lion's Share: A Short History of British Imperialism, 1850 1970* (1975)
R. Price, *An Imperial War and the British Working Class* (1972)
R. Robinson and J. Gallagher, *Africa and the Victorians* (1961)
R. Samuel (ed.), *Patriotism: The Making and Unmaking of British National Identity*, 3 vols. (1989)
G. R. Searle, *The Quest for National Efficiency* (1971)
L. Senelick, 'Politics as Entertainment: Victorian Music Hall Songs', *Victorian Studies*, 19 (1975)
E. Spiers, *Army and Society, 1815–1914* (1980)
J. Springhall, *Youth, Empire and Society: British Youth Movements, 1883–1940* (1977)
Z. Steiner, *Britain and the Origins of the First World War* (1977)
A. Summers, 'Militarism in Britain before the Great War', *History Workshop Journal*, 2 (1976)

Chapter 7

K. D. Brown, *Labour and Unemployment, 1900–1914* (1971)
P. Cain, 'Political Economy in Edwardian England: The Tariff Reform Controversy', in A. O'Day (ed.), *The Edwardian Age* (1979)
D. Collins, 'The Introduction of Old Age Pensions in Great Britain', *Historical Journal*, 8 (1965)
H. V. Emy, *Liberals, Radicals and Social Politics, 1892–1914* (1973)
B. B. Gilbert, *The Evolution of National Insurance* (1966)
J. Harris, *Unemployment and Politics* (1972)
J. R. Hay, *The Origins of the Liberal Welfare Reforms, 1906–1914* (1975)
——, 'Employers and Social Policy in Britain: The Evolution of Welfare Legislation, 1905–1914' *Social History*, 2 (1977)
B. K. Murray, *The People's Budget, 1909–10* (1986)
G. C. Peden, *British Economic and Social Policy* (1985)
M. Pugh, 'Lloyd George, the Working Class and Social Reform', *History Review*, 10 (1991)
M. E. Rose, *The Relief of Poverty, 1834–1914* (1972)
P. Thane (ed.), *The Origins of British Social Policy* (1978)
P. Thane, 'The Working Class and State Welfare in Britain 1880–1914', *Historical Journal*, 27 (1984)
D. Vincent, *Poor Citizens: The State and the Poor in the Twentieth Century* (1991)

Chapter 8

F. Bealey and H. Pelling, *Labour and Politics, 1900–1906* (1958)

N. Blewett, *The Peers, the Parties and the People* (1972)
P. F. Clarke, *Lancashire and the New Liberalism* (1971)
M. Freeden, *The New Liberalism* (1978)
R. Gregory, *The Miners and British Politics, 1906–1914* (1968)
R. McKibbin, *The Evolution of the Labour Party, 1910–1924* (1974)
H. Matthew, R. McKibbin, and J. Kay, 'The Franchise Factor in the Rise of the Labour Party', *English Historical Review*, 91 (1976)
A. J. A. Morris (ed.), *Edwardian Radicalism* (1974)
H. Pelling, *Social Geography of British Elections, 1885–1910* (1967)
M. Pugh, *The Making of Modern British Politics, 1867–1939* (1982)
G. R. Searle, *The Liberal Party: Triumph and Disintegration, 1886–1929* (1992)
D. Tanner, *Political Change and the Labour Party, 1900–1918* (1990)

Chapter 9

I. F. Clarke, *Voices Prophesying War, 1763–1984* (1966)
H. Clegg, A. Fox, and A. Thompson, *A History of British Trade Unions since 1889*, i (1964)
G. Dangerfield, *The Strange Death of Liberal England* (1935)
J. Grigg, *Lloyd George: The People's Champion, 1902–1911* (1978)
——, *Lloyd George: From Peace to War, 1912–1916* (1985)
R. Jenkins, *Mr Balfour's Poodle* (1954)
——, *Asquith* (1964)
S. Koss, *Asquith* (1976)
S. Meacham, *A Life Apart: The English Working Class, 1880–1914* (1977)
A. O'Day (ed.), *The Edwardian Age* (1979)
H. Pelling, *Popular Politics and Society in Late Victorian Britain* (1979)
M. Pugh, *Lloyd George* (1988)
D. Read, *Edwardian England* (1972)
K. Robbins, *Sir Edward Grey* (1974)
A. Rosen, *Rise Up Women!* (1974)
G. Searle, *Corruption in British Politics, 1895–1930* (1987)
R. Strachey, *The Cause* (1928)
A. Sykes, *Tariff Reform in British Politics, 1903–1913* (1979)
R. Roberts, *The Classic Slum* (1971)

Chapter 10

R. Blake, *The Unknown Prime Minister* (1955)
J. Bourne, *Britain and the Great War, 1914–1918* (1989)
J. E. Cronin, *The Politics of State Expansion* (1991)
D. French, *The British Way in Warfare, 1688–2000* (1990)
Lord Hankey, *The Supreme Command, 1914–1918* (1961)
M. Hart, 'The Liberals, the War and the Franchise', *English Historical Review*, 97 (1982)
J. Keegan, *The Face of Battle* (1976)
R. I. McKibbin, *The Evolution of the Labour Party, 1910–1914* (1974)
A. Marwick, *The Deluge* (1965)
M. Middlebrook, *The First Day on the Somme* (1971)
S. Pollard, *The Development of the British Economy, 1914–1967* (1967)
M. Pugh, *Electoral Reform in War and Peace, 1906–1914* (1978)
——, *The Making of Modern British Politics, 1867–1939* (1982)
G. R. Searle, *The Liberal Party: Triumph and Distintegration, 1886–1929* (1992)
J. Stevenson, *British Society, 1914–1945* (1984)

J. Turner, *Lloyd George's Secretariat* (1980)
——, *British Politics and the Great War, 1915–1918* (1992)
T. Wilson, *The Downfall of the Liberal Party, 1914–1935* (1966)
——, *The Myriad Faces of War* (1986)
J. M. Winter, *Socialism and the Challenge of War* (1974)
——, *The Great War and the British People* (1985)

Chapter 11

D. Aldcroft and H. Richardson, *The British Economy, 1870–1939* (1969)
B. W. E. Alford, *Depression and Recovery? British Economic Growth, 1918–1939* (1972)
P. F. Clarke, *The Keynesian Revolution in the Making, 1924–1936* (1988)
H. J. Hancock, 'The Reduction of Unemployment as a Problem of Public Policy, 1920–1929', *Economic History Review*, 15 (1962–1963)
R. Lowe, *Adjusting to Democracy: The Role of the Ministry of Labour in British Politics, 1916–1939* (1986)
R. I. McKibbin, 'The Economic Policy of the Second Labour Government', *Past and Present*, 36 (1975)
A. Marwick, 'Middle Opinion in the Thirties: Planning, Progress and Political Agreement', *English Historical Review*, 79 (1965)
A. S. Milward, *The Economic Effects of the Two World Wars on Britain* (1970)
G. C. Peden, *British Economic and Social Policy* (1985)
——, *Keynes, the Treasury, and British Economic Policy* (1988)
S. Pollard, *The Development of the British Economy, 1914–1967* (1969)
M. Stewart, *Keynes and After* (1967)
D. Winch, *Economics and Policy* (1969)

Chapter 12

S. Ball, *Baldwin and the Conservative Party: The Crisis of 1929–31* (1988)
J. Campbell, *Lloyd George: The Goat in the Wilderness* (1977)
C. Cook, *The Age of Alignment* (1975)
R. I. McKibbin, *The Evolution of the Labour Party, 1910–1924* (1974)
D. Marquand, *Ramsay MacDonald* (1977)
K. O. Morgan, *Consensus and Disunity: The Lloyd George Coalition Government, 1918–1922* (1979)
B. Pimlott, *Labour and the Left in the 1930s* (1977)
M. Pugh, *Women and the Women's Movement in Britain, 1914–1959* (1992)
——, *The Making of Modern British Politics, 1867–1939* (1982)
J. Ramsden, *The Age of Balfour and Baldwin, 1902–1940* (1978)
P. Renshaw, *The General Strike* (1975)
G. R. Searle, *The Liberal Party: Triumph and Disintegration, 1886–1929* (1992)
R. Skidelsky, *Politicians and the Slump* (1967)
——, *Oswald Mosley* (1975)
T. Stannage, *Baldwin Thwarts the Opposition: The British General Election of 1935* (1980)
J. Stevenson and C. Cook, *The Slump* (1977)
A. Thorpe, *The British General Election of 1931* (1991)
P. Williamson, *National Crisis and National Government* (1992)
C. Wrigley, *Arthur Henderson* (1990)

Chapter 13

D. Beddoe, *Back to Home and Duty: Women Between the Wars* (1991)

J. Burnett, *A Social History of Housing, 1815–1970* (1978)
S. Constantine, *Unemployment in Britain Between the Wars* (1980)
A. Crowther, *British Social Policy, 1914–1939* (1988)
M. Daunton, *A Property Owning Democracy? Housing in Britain* (1987)
C. Dyehouse, *Feminism and Family Planning in England, 1880–1939* (1989)
B. B. Gilbert, *British Social Policy, 1914–1939* (1970)
D. Gittins, *Fair Sex: Family Size and Structure, 1900–1939* (1982)
W. Greenwood, *Love on the Dole* (1933)
J. Lewis, *The Politics of Motherhood* (1980)
G. Orwell, *The Road to Wigan Pier* (1937)
G. C. Peden, *British Economic and Social Policy* (1985)
M. Pugh, *Women and the Women's Movement in Britain, 1914–1959* (1992)
J. Stevenson, *British Society, 1914–1945* (1984)
——, and C. Cook, *The Slump* (1977)
M. Stopes, *Married Love* (1918)
C. Webster, 'Health, Welfare and Unemployment During the Depression', *Past and Present*, 109 (1985)
J. Weeks, *Sex, Politics and Society* (1981)
D. Vincent, *Poor Citizens: The State and the Poor in Twentieth Century Britain* (1991)

Chapter 14

A. Adamthwaite, *The Making of the Second World War* (1977)
C. Bartlett, *British Foreign Policy in the Twentieth Century* (1989)
B. Bond, *British Military Policy Between Two World Wars* (1980)
J. Brown, *Modern India* (1985)
M. Ceadel, *Pacifism in Britain, 1914–1945* (1980)
J. Charmley, *Chamberlain and the Lost Peace* (1989)
D. French, *The British Way in War, 1688–2000* (1990)
K. Middlemas, *Diplomacy of Illusion: The British Government and Germany, 1937–1939* (1972)
R. J. Moore, *The Crisis of Indian Unity, 1917–1940* (1974)
B. Porter, *The Lion's Share: A Short History of British Imperialism, 1850–1983* (1984)
——, *Britain, Europe and the World, 1850–1986* (1987)
D. Reynolds, *Britannia Overruled* (1991)
G. Schmidt, *The Politics and Economics of Appeasement: British Foreign Policy in the 1930s* (1986)
R. P. Shay, *British Rearmament in the Thirties* (1977)
M. Smith, *British Air Strategy Between the Wars* (1984)
B. R. Tomlinson, *The Political Economy of the Raj, 1914–1947* (1979)
D. C. Watt, 'Appeasement: The Rise of a Revisionist School', *Political Quarterly*, 36 (1965)

Chapter 15

P. Addison, *The Road to 1945* (1975)
C. Barnett, *The Audit of War* (1986)
S. Brooke, *Labour's War* (1992)
A. Calder, *The People's War* (1971)
M. Gilbert, *Winston S. Churchill*, vii (1986)
J. Harris, *William Beveridge* (1977)
T. Harrison, *Living Through the Blitz* (1976)
K. Jeffreys, *The Churchill Coalition and Wartime Politics, 1940–1945* (1991)
R. B. MacCallum and A. Readman, *The British General Election of 1945* (1947)

H. Pelling, *Britain and the Second World War* (1970)
H. Smith (ed.), *War and Social Change: British Society and the Second World War* (1987)
J. Stevenson, *British Society, 1914–1945* (1984)

Chapter 16

B. W. E. Alford, *British Economic Performance, 1945–1975* (1988)
T. Burridge, *Clement Attlee* (1985)
D. Butler, *British General Elections Since 1945* (1989)
D. Dutton, *British Politics since 1945* (1992)
T. Gourvish and A. O'Day (eds.), *Britain Since 1945* (1991)
J. Harris, *William Beveridge* (1977)
K. Harris, *Attlee* (1982)
P. Hennessy and A. Seldon (eds.), *Ruling Performance: British Governments from Attlee to Thatcher* (1987)
A. Horne, *Macmillan*, iii (1990)
D. Kavanagh and P. Morris, *Consensus Politics* (1989)
B. Lenman, *The Eclipse of Parliament* (1992)
K. O. Morgan, *Labour in Power, 1945–1951* (1984)
——, *The People's Peace, 1945–1990* (1992)
G. C. Peden, *British Economic and Social Policy: Lloyd George to Thatcher* (1985)
B. Pimlott, *Harold Wilson* (1992)
S. Pollard, *The Development of the British Economy, 1914–1967* (1969)
C. Ponting, *Breach of Promise: Labour in Power, 1964–1970* (1989)
A. Seldon, *Churchill's Indian Summer: The Conservative Government, 1951–1955* (1981)
M. Shanks, *The Stagnant Society* (1963)
A. Sked, *Britain's Decline: Problems and Perspectives* (1987)
R. Taylor, *Against the Bomb: The British Peace Movement, 1958–1965* (1988)

Chapter 17

D. Bouchier, *The Feminist Challenge* (1983)
A. H. Halsey (ed.), *British Social Trends Since 1900* (1988)
C. Holmes, *John Bull's Island: Immigration and British Society, 1871–1971* (1988)
C. Jones, *Immigration and Social Policy in Britain* (1977)
R. Jowell, S. Witherspoon, and L. Brook (eds.), *British Social Attitudes: The 1987 Report* (1987)
A. Marwick, *British Society Since 1945* (1982)
K. O. Morgan, *The People's Peace, 1945–1990* (1992)
M. Pugh, *Women and the Women's Movement in Britain, 1914–1959* (1992)
M. Sanderson, *Educational Opportunity and Social Change in England* (1987)
A. Sked, *Britain's Decline: Problems and Perspectives* (1987)
J. Weeks, *Sex, Politics and Society* (1981)

Chapter 18

C. J. Bartlett, *British Foreign Policy in the Twentieth Century* (1989)
A. Bullock, *Ernest Bevin: Foreign Secretary* (1983)
D. Carlton, *Britain and the Suez Crisis* (1988)
J. Darwin, *Britain and Decolonisation* (1988)
——, *The End of the British Empire* (1991)
D. French, *The British Way in Warfare, 1688–2000* (1990)
J. D. Hargreaves, *Decolonisation in Africa* (1988)

R. Holland, *The Pursuit of Greatness: Britain and the World Role, 1900–1970* (1991)
A. Horne, *Macmillan*, ii (1990)
B. Porter, *The Lion's Share: A Short History of British Imperialism, 1850–1970* (1975)
D. Reynolds, *Britannia Overruled: British Policy and World Power in the Twentieth Century* (1991)
D. Sanders, *Losing an Empire, Finding a Role: British Foreign Policy since 1945* (1990)

Chapter 19

V. Bogdanor (ed.), *Liberal Party Politics* (1983)
H. Drucker (ed.), *Multi-Party Britain* (1978)
S. Finer (ed.), *Adversary Politics and Electoral Reform* (1975)
M. Franklin, *The Decline of Class Voting in Britain* (1985)
C. Harvie, *No Gods and Precious Few Heroes* (1981)
D. Healey, *The Time of My Life* (1989)
P. Hennessy and A. Seldon (eds.), *Ruling Performance: British Governments from Attlee to Thatcher* (1987)
K. O. Morgan, *Rebirth of a Nation: Wales, 1880–1980* (1981)
——, *The People's Peace, 1945–1990* (1992)
A. B. Philip, *The Welsh Question* (1975)
B. Pimlott, *Harold Wilson* (1992)
A. Roth, *Heath and the Heathmen* (1972)
B. Sarlvik and I. Crewe, *Decade of Dealignment* (1983)

Chapters 20 and 21

M. Adeney and J. Lloyd, *The Miners' Strike, 1984–1985* (1987)
D. Butler and D. Kavanagh, *The British General Election of 1983* (1984)
——, *The British General Election of 1987* (1988)
——, *The British General Election of 1992* (1993)
P. Dunleavy and C. Husbands, *British Democracy at the Crossroads* (1985)
L. Freedman, *Britain and the Falklands War* (1988)
M. Holmes, *The Thatcher Government, 1979–1983* (1985)
R. Jowell, S. Witherspoon, and L. Brook (eds.), *British Social Attitudes: The 1987 Report* (1987)
D. Kavanagh, *Thatcherism and British Politics* (1990)
—— and P. Morris, *Consensus Politics* (1983)
W. Keegan, *Mrs Thatcher's Economic Experiment* (1984)
P. Riddell, *The Thatcher Government* (1987)
N. Wapshot and G. Brock, *Thatcher* (1983)
H. Young, *One of Us* (1989)

Index